AF781685

MOLECULAR ORBITAL THEORY

FRONTIERS IN CHEMISTRY

Ronald Breslow and Martin Karplus, Editors
Columbia University

CONTRIBUTIONS TO THE THEORY OF CHEMICAL KINETICS
T. A. Bak

MOLECULAR ORBITAL THEORY
C. J. Ballhausen — *Københavns Universitet*
H. B. Gray — *Columbia University*

THERMODYNAMICS OF SMALL SYSTEMS: *Parts I and II*
T. L. Hill — *University of Oregon*

LECTURES ON QUANTUM THEORY OF MOLECULAR ELECTRONIC STRUCTURE
R. G. Parr — *The Johns Hopkins University*

THE BIOSYNTHESIS OF STEROIDS, TERPENES, AND ACETOGENINS
J. H. Richards — *California Institute of Technology*
J. B. Hendrickson — *Brandeis University*

OXIDATION MECHANISMS: Applications to Organic Chemistry
R. Stewart — *University of British Columbia*

COMPUTER PROGRAMMING FOR ORGANIC CHEMISTS
K. J. Wiberg — *Yale University*

Don Fisher

MOLECULAR ORBITAL THEORY

An Introductory Lecture Note and Reprint Volume

C. J. BALLHAUSEN

Københavns Universitet

and

HARRY B. GRAY

Columbia University

W. A. BENJAMIN, INC. 1965

New York Amsterdam

MOLECULAR ORBITAL THEORY: An Introductory Lecture Note and Reprint Volume

Copyright © 1964 by W. A. Benjamin, Inc.
All rights reserved

Library of Congress Catalog Number 65–12062
Manufactured in the United States of America

This manuscript was put into production on May 7, 1964;
this volume was published on December 7, 1964;
second printing with corrections September 20, 1965

W. A. BENJAMIN, INC.
New York, New York 10016

Preface

These notes are based on lectures on molecular orbital theory that we have presented at the University of Copenhagen and Columbia University. They were designed primarily for advanced-undergraduate and first-year graduate students as an introduction to molecular orbital theory.

It is apparent that the molecular orbital theory is a very useful method of classifying the ground and excited states of small molecules. The transition metal complexes occupy a special place here, and the last chapter is devoted entirely to this subject. We believe that modern inorganic chemists should be acquainted with the methods of the theory, and that they will find approximate one-electron calculations as helpful as the organic chemists have found simple Hückel calculations. For this reason, we have included a calculation of the permanganate ion in Chapter 8. On the other hand, we have not considered conjugated pi systems because they are excellently discussed in a number of books.

Our intuitive approach in the use of symmetry methods is admittedly nonrigorous and therefore will be unsatisfactory to purists, but we believe this is the best way to introduce symmetry ideas to the majority of students. Once the student has learned how to use symmetry methods, it will be easier for him to appreciate more formal and rigorous treatments.

Several reprints of papers on molecular orbital theory are included in the back of the book. The papers treat a substantial number of the important molecular geometries. The reader should be able to follow the discussions after reading through the lecture notes.

We thank our colleagues in New York and Copenhagen for help with the manuscript. We gratefully acknowledge the help of Dr. Arlen Viste and Mr. Harold Basch in preparing Appendix 8-B. Finally, it is a pleasure to acknowledge the expert assistance of Mrs. Diane Celeste in preparing the final manuscript.

C. J. BALLHAUSEN, *København*
HARRY B. GRAY, *New York*

October 1964

Contents

Reprint Collection

General Papers

The publisher wishes to thank the American Institute of Physics, the American Chemical Society, Taylor and Francis, and the Chemical Society (London) for permission to reprint material from their journals.

MOLECULAR ORBITAL THEORY

1 / Atomic Orbitals

One of the most important questions confronting the chemist is the problem of describing how atoms are held together in molecules. Such a description can only be obtained by using quantum mechanical methods. We are in general interested in accounting for the properties of excited states, as well as the ground state. A simple and useful method of describing the electronic structures of molecules starts with wave functions which are localized on individual atoms in the molecule, and then proceeds to combine these functions in various trial combinations. The combinations are called molecular orbitals and are used as approximate solutions to the molecular Schrödinger equation. We test their "goodness" as wave functions for a molecule by calculating observable quantities with them and comparing the results with experiment. In these notes we shall treat the fundamental principles governing the construction and properties of such molecular orbitals.

We must realize, however, that such a description of a molecule involves drastic approximations; thus only approximate numerical results can be obtained. It is possible by performing elaborate numerical calculations to obtain better and better approximations for the molecular wave functions. Here we shall be interested only in semiquantitative approximate schemes which allow us to place the low-lying electronic states of molecules.

1-1 THE SCHRÖDINGER EQUATION

The Schrödinger equation must be understood in the same way as Newton's laws, and, just like these, it cannot be proved by reasoning.

The presence of Planck's constant, h, shows that the equation cannot be derived from classical mechanics. We know that all observable quantities which contain this constant are discontinuous and, consequently, not classical. The proof of its validity lies, therefore, in its applicability. Our calculations yield results that can be tested experimentally. If we do our calculations as accurately as possible, it turns out that our calculated results agree with experience. It is, however, obvious that the more we apply approximate solutions to the Schrödinger equation, the less we can expect to obtain results that "fit perfectly."

When we, therefore, in many cases are happy just to "understand" what happens, it is natural that we must rely on the validity of the Schrödinger equation. There is nothing in our experience which indicates that the equation should not be true, provided the particles under consideration do not have such large velocities that relativistic effects become important.

Let us first review the different atomic orbitals. These orbitals are obtained by solving the Schrödinger equation for a particle of mass m, the electron, in a central field produced by the nucleus:

$$\nabla^2\psi + \frac{2m}{\hbar^2}(W - V)\psi = 0 \tag{1-1}$$

where W is the total energy (a constant), V is the potential energy as a function of the position (x,y,z) of m, and $\hbar$ is Planck's constant h divided by 2π. The solutions are the wave functions $\psi(x,y,z)$ which satisfy Eq. (1-1). Since $\nabla^2 = (\partial^2/\partial x^2) + (\partial^2/\partial y^2) + (\partial^2/\partial z^2)$, (1-1) is a second-order differential equation.

The wave functions, ψ, must fulfill certain conditions:

1. They must be single-valued, continuous, and differentiable at every point in space.
2. They must be finite for all values of x,y, and z.

For most purposes it is convenient if the wave functions are normalized. This means that $\int|\psi|^2\, d\tau = 1$; that is, the numerical quadrate of the wave function integrated over all space must equal one.

The physical meaning of the wave functions is as follows: The probability for finding the particle that the wave function describes in a small volume element $d\tau$ is given by $|\psi|^2\, d\tau$. It is clear, then, that the probability of finding the particle in the complete region of configuration equals one (the condition of normalization).

Furthermore, it can be shown that if we want to find the expectation value for some observable quantity $\hat{P}(x,y,z)$, we have the expression

$$P = \int \psi^* \hat{P} \psi \, d\tau \qquad (1\text{-}2)$$

if ψ is normalized. $\hat{P}$ is the operator expression for the quantity; $\hat{P}$ operates on ψ. ψ^* stands for the complex conjugate of the wave function in the case in which the wave function contains complex variables.

From the Schrödinger equation we obtain for the energy W of the system

$$W = \int \psi^* \left(- \frac{\hbar^2}{2m} \nabla^2 + V \right) \psi \, d\tau \qquad (1\text{-}3)$$

It is obvious that $-(\hbar^2/2m)\nabla^2 + V$, which is called the Hamiltonian operator ($\mathcal{H}$), is the operator that gives us the energy of the system:

$$W = \int \psi^* \mathcal{H} \psi \, d\tau \qquad (1\text{-}4)$$

A very important relationship is that different wave functions of the *same* Hamiltonian operator are orthogonal. This can be proved mathematically. Orthogonality means that the integral

$$\int \psi_n^* \psi_m \, d\tau = \delta_{n,m} \qquad (1\text{-}5)$$

is equal to one for $n = m$, but equals zero for $n \neq m$. This is expressed by the so-called Kronecker delta:

$$\delta_{n,m} = \begin{cases} 1 & \text{for } n = m \\ 0 & \text{for } n \neq m \end{cases} \qquad (1\text{-}6)$$

The orthogonality of the wave functions plays an extremely important role in practice.

1-2 "HYDROGEN-LIKE" ORBITALS

If we consider a spherically symmetric field as occurs around an atomic nucleus, it is convenient to use a polar-coordinate system (Figure 1-1).

The following conversions are easily obtained for the coordinates.

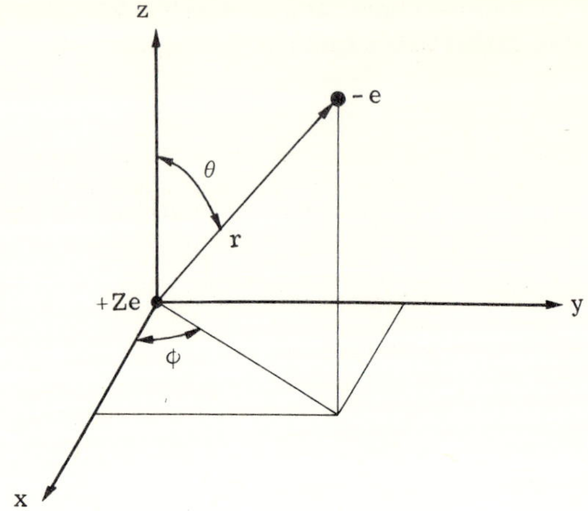

Figure 1-1 Polar and rectangular coordinates.

$$x = r \sin\theta \cos\phi \tag{1-7}$$

$$y = r \sin\theta \sin\phi \tag{1-8}$$

$$z = r \cos\theta \tag{1-9}$$

A little more difficulty is encountered with the conversion of the volume element $d\tau = dx\,dy\,dz$:

$$d\tau = r^2\,dr \sin\theta\,d\theta\,d\phi \tag{1-10}$$

If the nucleus (which is placed in the center) has a positive charge of Ze, where e is the numerical value of the charge of the electron, we obtain the Hamiltonian operator for an electron in this central field,

$$\mathcal{H} = -\frac{\hbar^2}{2m}\nabla^2 - \frac{Ze^2}{r} \tag{1-11}$$

With ∇^2 written out in polar coordinates, the Schrödinger equation for this system is

$$\left[\frac{1}{r^2}\frac{\partial}{\partial r}\left(r^2\frac{\partial}{\partial r}\right) + \frac{1}{r^2\sin\theta}\frac{\partial}{\partial\theta}\left(\sin\theta\frac{\partial}{\partial\theta}\right) + \frac{1}{r^2\sin^2\theta}\frac{\partial^2}{\partial\phi^2}\right]\Psi + \frac{2m}{\hbar^2}\left[W + \frac{Ze^2}{r}\right]\Psi = 0 \tag{1-12}$$

As expected, the equation is a differential equation of second order in three variables. Let us try to guess a solution.

$$\psi(r,\theta,\phi) = R(r)\,Y(\theta,\phi) \tag{1-13}$$

This is substituted into the equation, and the r dependency is collected on the left side of the equation, the θ and ϕ dependency on the right:

$$\frac{1}{R(r)}\frac{\partial}{\partial r}\left(r^2\frac{\partial R(r)}{\partial r}\right) + \frac{2m}{h^2}\left(W + \frac{Ze^2}{r}\right)r^2$$
$$= -\frac{1}{Y(\theta,\phi)\sin\theta}\frac{\partial}{\partial\theta}\left(\sin\theta\,\frac{\partial Y(\theta,\phi)}{\partial\theta}\right)$$
$$-\frac{1}{Y(\theta,\phi)\sin^2\theta}\frac{\partial^2 Y(\theta,\phi)}{\partial\phi^2} \tag{1-14}$$

Because r, θ, and ϕ are independent coordinates which can vary freely, each side can be equated to a constant, λ. Let Y stand for $Y(\theta,\phi)$ and R for $R(r)$. We obtain then

$$\frac{1}{\sin\theta}\frac{\partial}{\partial\theta}\left(\sin\theta\,\frac{\partial Y}{\partial\theta}\right) + \frac{1}{\sin^2\theta}\frac{\partial^2 Y}{\partial\phi^2} + \lambda Y = 0 \tag{1-15}$$

$$\frac{1}{r^2}\frac{d}{dr}\left(r^2\frac{dR}{dr}\right) + \left[\frac{2m}{\hbar^2}\left(W + \frac{Ze^2}{r}\right) - \frac{\lambda}{r^2}\right]R = 0 \tag{1-16}$$

We have split the Schrödinger equation into two independent second-order differential equations, one which depends only on the angles θ and ϕ and one which depends only on r. There are a number of solutions which satisfy our conditions 1 and 2 (continuity and "good behavior" for $r \to \infty$), but a condition for such solutions is that $\lambda = \ell(\ell + 1)$, where ℓ is zero or an integer: $\ell = 0, 1, 2, 3, \ldots$.

For $\ell = 0$, Y = constant is easily seen to be a solution of the angular equation. Normalizing this solution to one over the solid angle $\sin\theta\, d\theta\, d\phi$, we get $Y = \sqrt{1/4\pi}$. Looking now at the radial equation for large values of r, (1-16) reduces to the asymptotic equation

$$\frac{d^2R}{dr^2} + \frac{2m}{\hbar^2}WR = 0$$

Calling $(2m/\hbar^2)\,W = -\alpha^2$, we see that $R = e^{\pm\alpha r}$ is a solution. We must discard the plus sign here, since for $r \to \infty$ the wave function would be ill-behaved.

We now try by substitution to see if the asymptotic solution also is a solution to the complete radial differential equation with $\ell = 0$, and find this to be so provided that $\alpha = Zme^2/\hbar^2 = Z/a_0$, where $a_0 = \hbar^2/me^2$, the Bohr radius.

The first normalized solution to the radial equation with $\ell = 0$ is then found from the condition

$$\int_0^\infty N^2 e^{-2Zr/a_0} r^2\, dr = 1$$

and is

$$R_{1,0}(r) = 2\left(\frac{Z}{a_0}\right)^{3/2} e^{-(Z/a_0)r} \tag{1-17}$$

All orbitals with $\ell = 0$ are called s orbitals, and the subscripts 1,0 associated with $R(r)$ refer to this function as the first radial solution to the complete set of s orbitals. The next one (similarly with $\ell = 0$) will be $R_{2,0}(r)$, and so on. The total wave function is the product of the radial and the angular function and we have

$$\psi_{1s} = \sqrt{\frac{1}{\pi}}\left(\frac{Z}{a_0}\right)^{3/2} e^{-(Z/a_0)r} \tag{1-18}$$

The energy W_1 of an electron in a 1s orbital is found by substituting the value of α into the expression containing W:

$$-\alpha^2 = \frac{2W_1}{a_0 e^2} = \frac{-Z^2}{a_0^{\,2}} \tag{1-19}$$

We obtain $W_1 = -Z^2e^2/2a_0$.

This is the most stable orbital of a hydrogen-like atom—that is, the orbital with the lowest energy. Since a 1s orbital has no angular dependency, the probability density $|\psi|^2$ is spherically symmetrical. Furthermore, this is true for all s orbitals. We depict the boundary surface† for an electron in an s orbital as a sphere (Figure 1-2). The radial function ensures that the probability for finding the particle goes to zero for $r \rightarrow \infty$.

The probability of finding a 1s electron within a spherical shell with radii r and r + dr is found from $R(r)$ to be equal to

$$4\left(\frac{Z}{a_0}\right)^3 e^{-(2Z/a_0)r}\, r^2\, dr \tag{1-20}$$

By differentiation the above function is found to have a maximum

†Boundary surface pictures outline a large fraction of $|\Psi|^2$, and also give the signs on the lobes given by Ψ.

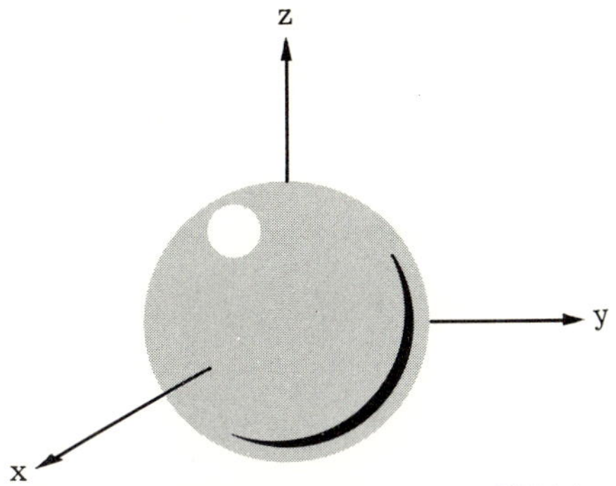

Figure 1-2 The boundary surface of an s orbital; an s orbital has no angular dependency.

at $r = a_0/Z$. At this distance (which for a hydrogen atom equals a_0) the electron density is largest and the chance of finding an electron is, therefore, greatest.

$R(r) = 2(Z/a_0)^{3/2}\, e^{-(Z/a_0)r}$ is, however, not the only function which satisfies the Schrödinger equation for $\ell = 0$. A second radial function which goes with the solution $Y = \sqrt{1/4\pi}$ can be found by substitution of a trial function into the radial differential equation and normalizing. It is

$$R_{2,0}(r) = \frac{1}{2\sqrt{2}}\left(\frac{Z}{a_0}\right)^{3/2}\left(2 - \frac{Z}{a_0}r\right) e^{-(Z/2a_0)r} \tag{1-21}$$

and it is orthogonal on $R_{1,0}(r)$.

The total wave function is again obtained by multiplication of $R_{2,0}(r)$ and the "angular part" $\sqrt{1/4\pi}$. This orbital is called a 2s orbital,

$$\psi_{2s} = \frac{1}{4\sqrt{2\pi}}\left(\frac{Z}{a_0}\right)^{3/2}\left(2 - \frac{Z}{a_0}r\right) e^{-(Z/2a_0)r} \tag{1-22}$$

The 2s orbital has the same angular dependency as 1s, but the probability distribution has two maxima, of which the larger lies farther out than the one which was found for 1s. The energy of 2s is also considerably higher than 1s. The energy of a 2s orbital is $W_2 = -Z^2e^2/8a_0$. The general expression for the energy is given by the Bohr formula

$$W_n = -\frac{Z^2e^2}{2n^2a_0} \tag{1-23}$$

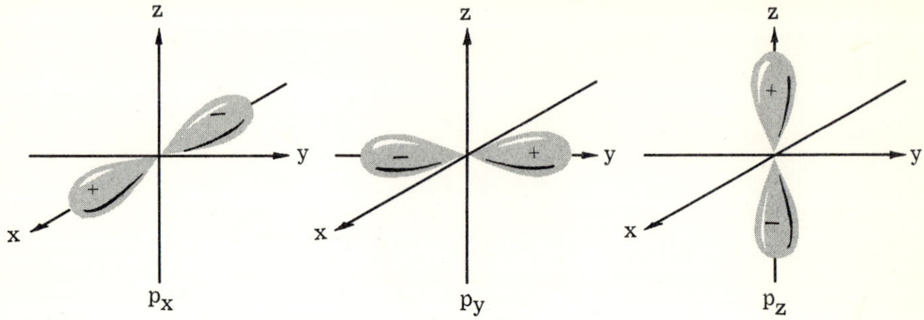

Figure 1-3 Boundary surfaces of the p orbitals.

We call n the principal quantum number, and n takes integer values 1,2,3, ...

The energy difference between the 1s and 2s orbitals is found to be

$$\Delta W = W_{2s} - W_{1s} = \frac{Z^2e^2}{2a_0}\left(-\tfrac{1}{4} + 1\right) = \frac{3Z^2e^2}{8a_0} \tag{1-24}$$

which for $Z = 1$ gives $\Delta W = 3e^2/8a_0 \approx 10$ electron volts, abbreviated 10 eV.

We now turn to the possible solutions of the Schrödinger equation for $\ell \neq 0$. For $\ell = 1$ the angular equation gives three solutions which are orthogonal to each other. These are the so-called p orbitals:

$$p_z = \cos\theta = \frac{z}{r} \tag{1-25}$$

$$p_y = \sin\theta\,\sin\varphi = \frac{y}{r} \tag{1-26}$$

$$p_x = \sin\theta\,\cos\varphi = \frac{x}{r} \tag{1-27}$$

The subscripts x, y, and z indicate the angular dependencies. As already mentioned, the three p orbitals are orthogonal to each other, and it is obvious that they are not spherically symmetrical about the nucleus. A boundary surface for each of the three p orbitals is given in Figure 1-3. The radial function is of course the same for all three p orbitals, and the first radial function is

$$R_{2,1} = \frac{1}{2\sqrt{6}}\left(\frac{Z}{a_0}\right)^{5/2} re^{-(Z/2a_0)r} \tag{1-28}$$

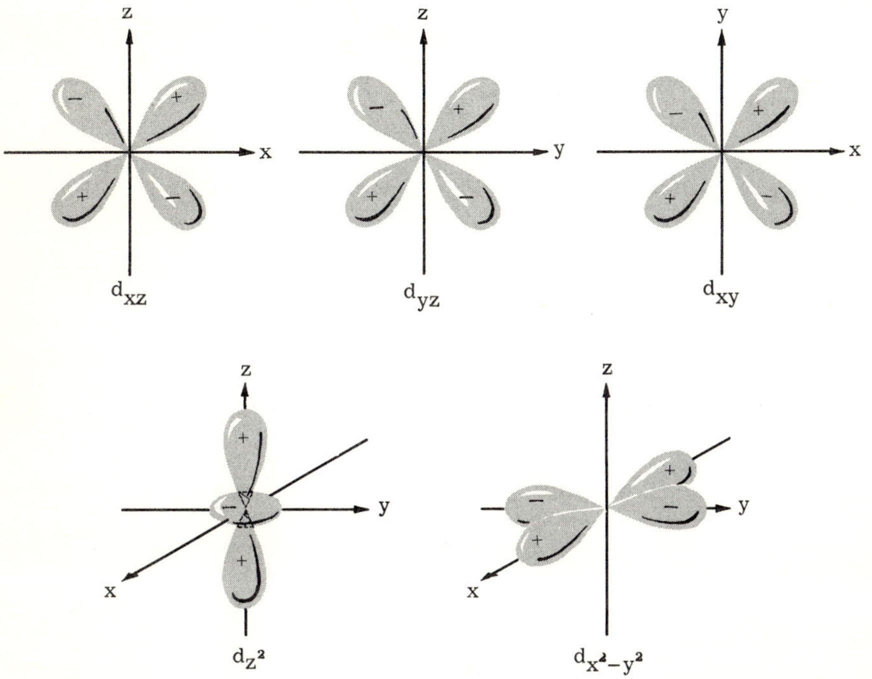

Figure 1-4 Boundary surfaces of the d orbitals.

The energy of the 2p wave function is found to be the same as 2s; that is, $W_2 = -(Z^2e^2/8a_0)$.

In an atom with more than one electron, the energy of the 2p hydrogen-like orbitals is in fact a little higher than the energy of the 2s orbital, because Z in an atom that contains both 2s and 2p electrons is effectively larger for 2s than for 2p. This is partly due to the fact that the 2s electrons "shield" some of the positive charge of the nucleus from the 2p electrons.

For $\ell = 2$ we obtain the following five linearly independent solutions to the angular equation; they are called the d orbitals:

$$d_{x^2-y^2}\text{:}\ \frac{\sqrt{3}}{2}\sin^2\theta\,(\cos^2\varphi - \sin^2\varphi) = \frac{\sqrt{3}}{2}\,\frac{x^2-y^2}{r^2} \qquad (1\text{-}29)$$

$$d_{z^2}\text{:}\ \frac{1}{2}(3\cos^2\theta - 1) = \frac{1}{2}\,\frac{3z^2-r^2}{r^2} \qquad (1\text{-}30)$$

$$d_{xy}\text{:}\ \sqrt{3}\sin^2\theta\cos\varphi\sin\varphi = \sqrt{3}\,\frac{xy}{r^2} \qquad (1\text{-}31)$$

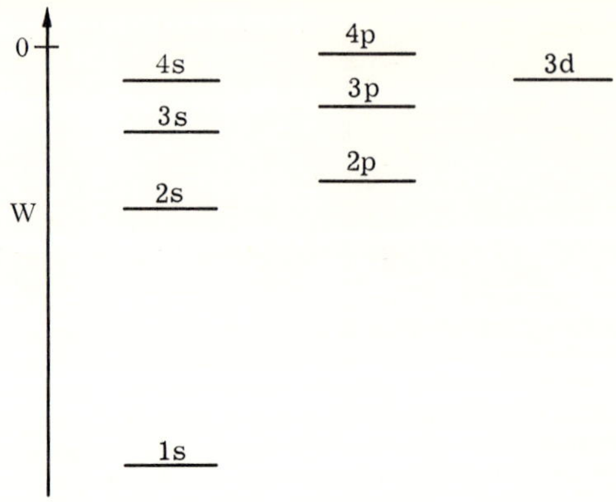

Figure 1-5 The single-electron energies for a many-electron atom. The drawing is only very approximate. The single s, p, and d levels are separated for the sake of clarity.

$$d_{xz}: \sqrt{3}\sin\theta\cos\theta\cos\varphi = \sqrt{3}\,\frac{xz}{r^2} \tag{1-32}$$

$$d_{yz}: \sqrt{3}\sin\theta\cos\theta\sin\varphi = \sqrt{3}\,\frac{yz}{r^2} \tag{1-33}$$

The boundary surfaces for the different d orbitals are given in Figure 1-4.

The radial function for a 3d orbital is

$$R_{3,2} = \frac{4}{81\sqrt{30}}\left(\frac{Z}{a_0}\right)^{7/2} r^2 e^{-(Z/3a_0)r} \tag{1-34}$$

with energy $W_3 = -Z^2e^2/18a_0$.

It is possible in any single case to calculate each of the orbital energies in an atom with many electrons. The results may be presented as an energy-level scheme such as is given in Figure 1-5.

1-3 THE PAULI EXCLUSION PRINCIPLE

The Pauli principle states that a maximum of two electrons can reside in each orbital, and we distinguish them by means of a spin quantum number. Two electrons in one orbital must then have different spin quantum numbers. The spin quantum number can have one of two values, designated α and β, or + and −. The physical basis for the introduction of the spin is that the electron turns out to have an intrinsic magnetic moment.

The total wave function for an electron is therefore $\psi(\text{orbit}) \cdot \psi(\text{spin})$. Further, the two spin functions corresponding to α and β spin are normalized and orthogonal to each other; that is,

$$\int |\psi(\alpha)|^2 \, d\tau = \int |\psi(\beta)|^2 \, d\tau = 1$$

and

$$\int \psi(\alpha)\psi(\beta) \, d\tau = 0$$

In the preceding section we found the single-electron energies of the different atomic orbitals. Many-electron atomic configurations are now constructed by placing the electrons one after the other in the single orbitals, observing the Pauli principle. In the ground state, for self-evident energetic reasons, the electrons first occupy the lowest energy orbitals. There can be two electrons in each orbital, and the ground states for the first few atoms are given in Table 1-1.

The ground state of carbon presents a new problem: Which orbital does the sixth electron go into? This question can be answered by doing either an experiment or a calculation. The answers are summarized in Hund's first rule:

When electrons go into orbitals that have the same energy (degenerate orbitals), the state that has the highest number of equal spin quantum numbers will have the lowest energy. Thus electrons prefer to occupy different orbitals, if possible, since in many cases it is energetically favorable to avoid *spin pairing*.

The lowest energy for two electrons placed in a doubly degenerate set of orbitals, ψ_1 and ψ_2, is then, e.g., $\overset{+}{\psi}_1(1)\overset{+}{\psi}_2(2)$, while $\overset{+}{\psi}_1(1)\bar{\psi}_1(2)$ and $\bar{\psi}_1(1)\overset{+}{\psi}_2(2)$ have higher energy.

The ground state for carbon is, therefore, $(1s)^2(2s)^2(2p_x)(2p_y)$,

Table 1-1 Ground States for Some Atoms

Atom	Ground-state electronic configuration
H	$(1s)$
He	$(1s)^2$
Li	$(1s)^2(2s)$
Be	$(1s)^2(2s)^2$
B	$(1s)^2(2s)^2(2p)$

or $(1s)^2(2s)^2(2p_x)(2p_z)$, or $(1s)^2(2s)^2(2p_y)(2p_z)$, all having two unpaired spins. We see that there are three *orbital* configurations for the two electrons which have the same energy. In other words, the ground state of carbon is orbitally threefold degenerate.

For nitrogen the ground state is given by the configuration $(1s)^2(2s)^2(2p_x)(2p_y)(2p_z)$, having three unpaired spins. The ground state of nitrogen is not orbitally degenerate. For oxygen the ground-state configuration is $(1s)^2(2s)^2(2p_x)^2(2p_y)(2p_z)$, *or* one of the two other combinations. As for carbon, the ground state of oxygen is threefold degenerate. The $(2p_x)^2$ electrons are called a "lone pair," as are the $(2s)^2$ electrons. The name "lone pair" is commonly used for all pairs of electrons in an outer shell that are not involved in bonding to other atoms.

If the difference in energy between orbitals that are to receive electrons is small, as for example between the 4s and the 3d orbitals, it sometimes is more energetically favorable (refer to Hund's rule) to distribute the electrons in both orbitals. As an example, we find the lowest configurations for V, Cr, and Mn to be ([Ar] stands for a closed 18-electron shell)

V: $[Ar](3d)^3(4s)^2$

Cr: $[Ar](3d)^5(4s)$

Mn: $[Ar](3d)^5(4s)^2$

Configurations such as $(3d)^5(4s)$ are, however, the exception rather than the rule. They occur for the first time in the transition elements.

Hund's rule can be traced back to the fact that electrons repel each other. A complete Hamiltonian function for an atom with several electrons must contain a term of the form $\frac{1}{2}\Sigma_{i\neq j}(e^2/r_{ij})$, to account for the repulsion between any two electrons. Unfortunately, the presence of such a term prevents the separation of the variables in the wave equation. As a result, all calculations can only be done by approximate methods. However, in principle these can be done with all the accuracy we wish.

In atomic problems involving several electrons it is common to construct the trial wave functions for the whole system, using the "hydrogen-like" wave functions as basis wave functions for the single electrons. Because the single orbitals are found for only *one* electron, terms of the type e^2/r_{12} are not included. After constructing the total wave function on the basis of the single orbitals and observing the Pauli principle, we consider the e^2/r_{12} terms by means of a "perturbation" calculation. In this method one considers the total energy of the system to be given by the sum of the energies of the single "hydrogen-like" orbitals, and then corrects for the energy contributions that originate from terms of the $1/r_{12}$ type.

After choosing an approximate set of atomic orbitals, the first problem is to combine these in such a way as to obtain a good wave function for an atom with more than one electron. We choose a product function, the reason being that the energy of the system then is equal to the sum of the energies of the single orbitals, if the e^2/r_{12} term is disregarded. This can be verified simply by substituting the product wave function into the Schrödinger equation for the system. One product function for the ground state of He is: $(1\overset{+}{s})(1)\cdot(1\overset{-}{s})(2)$, where 1 and 2 stand for electrons 1 and 2. But we could just as well have chosen $(1\overset{-}{s})(1)\cdot(1\overset{+}{s})(2)$. When we exchange the numbering of the two electrons, the electron density function must not change. Since this is determined by the quadrate of the wave function, this means that an exchange of electrons can only change the sign of the total wave function.

Since we cannot distinguish between the electrons, a wave function for the system could be one of two possibilities:

$$\psi = \frac{1}{\sqrt{2}}[(1\overset{+}{s})(1)(1\overset{-}{s})(2) \pm (1\overset{-}{s})(1)(1\overset{+}{s})(2)] \tag{1-35}$$

where $1/\sqrt{2}$ is included to normalize the wave function to 1. Mathematically, we cannot decide whether the + or − sign (or maybe both) will give solutions to the Schrödinger equation which correspond to states in nature. Empirically, it turns out that only the minus sign corresponds to a reasonably accurate description. Consequently,

$$\psi_{He} = \frac{1}{\sqrt{2}}[(1\overset{+}{s})(1)(1\overset{-}{s})(2) - (1\overset{-}{s})(1)(1\overset{+}{s})(2)] \tag{1-36}$$

By exchange of electrons number (1) and (2), we obtain

$$\begin{aligned}\psi'_{He} &= \frac{1}{\sqrt{2}}[(1\overset{+}{s})(2)(1\overset{-}{s})(1) - (1\overset{-}{s})(2)(1\overset{+}{s})(1)]\\ &= -\frac{1}{\sqrt{2}}[(1\overset{+}{s})(1)(1\overset{-}{s})(2) - (1\overset{-}{s})(1)(1\overset{+}{s})(2)]\\ &= -\psi_{He}\end{aligned} \tag{1-37}$$

We say that the wave function is antisymmetric. As it turns out, nature demands many-electron wave functions to be antisymmetric. Such an antisymmetric wave function can be written as a determinant:

$$\psi_{\mathrm{He}} = \frac{1}{\sqrt{2}} \begin{vmatrix} (1\overset{+}{s})(1) & (1\overset{+}{s})(2) \\ (1\bar{s})(1) & (1\bar{s})(2) \end{vmatrix} \tag{1-38}$$

For three electrons, such as are present in the ground state of Li, the total wave function becomes

$$\psi_{\mathrm{Li}} = \frac{1}{\sqrt{6}} \begin{vmatrix} (1\overset{+}{s})(1) & (1\overset{+}{s})(2) & (1\overset{+}{s})(3) \\ (1\bar{s})(1) & (1\bar{s})(2) & (1\bar{s})(3) \\ (2\overset{+}{s})(1) & (2\overset{+}{s})(2) & (2\overset{+}{s})(3) \end{vmatrix} \tag{1-39}$$

If we exchange two electrons in this wave function, the wave function changes sign (a property of a determinant). Thus, determinantal wave functions are always *antisymmetric*. But in the same way, if two orbitals are equal, the determinant has two identical rows, and such a determinant equals zero. In other words, we note that the Pauli principle is satisfied for determinantal wave functions.

We usually do not write a determinantal wave function in its entirety, but only give the diagonal term. For example, for the ground state of lithium

$$\psi_{\mathrm{Li}} = |(1\overset{+}{s})(1\bar{s})(2\overset{+}{s})| \tag{1-40}$$

When writing it in this way, we assume the presence of the normalizing factor $1/\sqrt{6}$ and also the numbering of the electrons. But it is important to remember that the latter is present. For example, we have that

$$\psi_{\mathrm{Li}} = |(1\overset{+}{s})(1\bar{s})(2\overset{+}{s})| = -|(1\bar{s})(1\overset{+}{s})(2\overset{+}{s})| = |(1\bar{s})(2\overset{+}{s})(1\overset{+}{s})|, \text{ etc.} \tag{1-41}$$

In principle, a change in sign in ψ has no significance regarding the calculation of measurable quantities, but it has significance in other connections (see, e.g., page 46).

Nature requires—as we have seen—that the complete wave function (space and spin) be antisymmetric. Consequently, for two electrons in one orbital ψ_a,

$$\psi_a(1)\psi_a(2)[\alpha(1)\beta(2) - \alpha(2)\beta(1)] = |(\overset{+}{\psi}_a)(\bar{\psi}_a)| \tag{1-42}$$

If two electrons are in different orbitals ψ_a and ψ_b, the total wave function must either be antisymmetric in space coordinates and symmetric in spin coordinates or opposite, so that the product is antisymmetric.

The orbital function

$$\psi_a(1)\psi_b(2) + \psi_a(2)\psi_b(1)$$

is symmetric, but

$$\psi_a(1)\psi_b(2) - \psi_a(2)\psi_b(1)$$

is antisymmetric.

For the spin wave functions, the products

$$\begin{array}{l} \alpha(1)\alpha(2) \\ \alpha(1)\beta(2) + \alpha(2)\beta(1) \\ \beta(1)\beta(2) \end{array}$$

are symmetric, but

$$\alpha(1)\beta(2) - \alpha(2)\beta(1)$$

is antisymmetric.

The complete antisymmetric wave functions, including both space and spin coordinates, are now obtained by multiplication, and after application of our shorthand notation for determinantal functions we have the four possibilities

$$\text{(sym. orb.)(antisym. spin)} = |(\overset{+}{\psi}_a)(\bar{\psi}_b)| - |(\bar{\psi}_a)(\overset{+}{\psi}_b)| \qquad (1\text{-}43)$$

or

$$\text{(antisym. orb.)(sym. spin)} = \begin{cases} |(\overset{+}{\psi}_a)(\overset{+}{\psi}_b)| \\ |(\overset{+}{\psi}_a)(\bar{\psi}_b)| + |(\bar{\psi}_a)(\overset{+}{\psi}_b)| \\ |(\bar{\psi}_a)(\bar{\psi}_b)| \end{cases} \qquad (1\text{-}44)$$

Let us find the energies of these states. We notice that since the Hamiltonian operator is not a function of the spin, integration over the spin coordinates can immediately be carried out. Owing to the spin orthogonality, we see that the three wave functions that occur as (antisym. orb.)·(sym. spin) all have the same energy. Such a state, which evidently is threefold-degenerate in the spin, is called a spin triplet. Since the spin degeneracy is equal to $2S + 1$, we get $2S + 1 = 3$ or $S = 1$. We note that the state contains two unpaired electrons. Therefore, an atom in such a state has a paramagnetic moment.

On the other hand, the product (sym. orb.)·(antisym. spin) is nondegenerate. A state that is nondegenerate in spin is called a spin singlet; we have $2S + 1 = 1$, and thus $S = 0$. A singlet state does not have a paramagnetic moment.

2 / Diatomic Molecules

The problems that are connected with the solution of the electronic structures of molecules are in principle the same as those which occur in the treatment of atomic structures. The single-electron orbitals for molecules are called molecular orbitals, and systems with more than one electron are built up by filling the molecular orbitals with electrons, paying proper attention to the Pauli principle. Thus, we always require that the total wave function be antisymmetric.

2-1 MOLECULAR ORBITALS FOR DIATOMIC MOLECULES

The first step is to find a set of molecular orbitals. Let us consider two nuclei, A and B, each with a positive charge and separated by a distance R (Figure 2-1). In this skeleton of nuclei we allow an electron to move.

If the electron is in the neighborhood of A, we expect that the atomic orbital ψ_A gives a good description of the electron. In the same way, the atomic orbital ψ_B should give a good description of

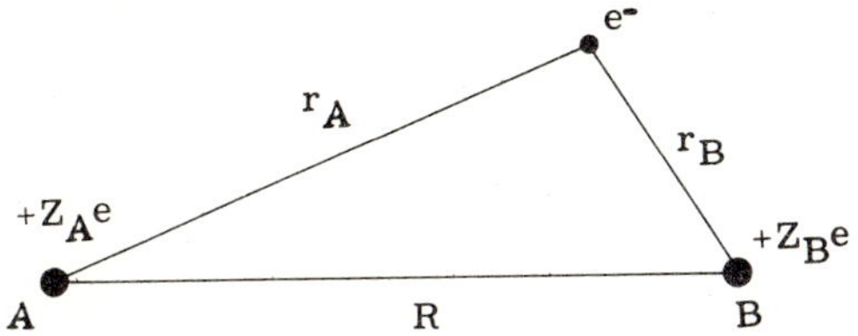

Figure 2-1 An electron in the skeleton of nuclei of a diatomic molecule.

the behavior of the electron in the neighborhood of B. This means that in a molecular-orbital description our molecular wave function must be approximately ψ_A in the neighborhood of A and ψ_B in the neighborhood of B. We then *guess* that a linear combination of ψ_A and ψ_B will be a good wave function for the *molecule* AB. Thus, we have, assuming that ψ_A and ψ_B are real functions,

$$\psi = c_1\psi_A + c_2\psi_B \tag{2-1}$$

Such a wave function is called an LCAO wave function ("linear combination of atomic orbitals").

The Hamiltonian operator, which determines the energy of the system, is given by the sum of the *kinetic* and *potential* energies of the electron (see Figure 2-1). To this we must add—as the last term—the potential energy of the nuclei:

$$\mathcal{H} = -\frac{\hbar^2}{2m}\nabla^2 - \frac{Z_Ae^2}{r_A} - \frac{Z_Be^2}{r_B} + \frac{Z_AZ_Be^2}{R} \tag{2-2}$$

For convenience we shall normalize our wave function to 1 before we make any calculation of the energy. Then

$$N^2\int (c_1\psi_A + c_2\psi_B)(c_1\psi_A + c_2\psi_B)\, d\tau = 1 \tag{2-3}$$

or

$$(c_1)^2\int |\psi_A|^2\, d\tau + (c_2)^2\int |\psi_B|^2\, d\tau + 2c_1c_2\int \psi_A\psi_B\, d\tau = \frac{1}{N^2} \tag{2-4}$$

Assuming that the atomic orbitals ψ_A and ψ_B are normalized, we have immediately

$$(c_1)^2 + (c_2)^2 + 2c_1c_2\int \psi_A\psi_B\, d\tau = \frac{1}{N^2} \tag{2-5}$$

The definite integral $\int \psi_A \psi_B \, d\tau$ is called the *overlap integral,* S, between the two wave functions ψ_A and ψ_B. The reason for this name is clear. For example, if both ψ_A and ψ_B are s orbitals, we obtain the picture shown in Figure 2-2. We see that the two wave functions "overlap" each other in the region between the two nuclei.

By substituting S for the overlap integral in Eq. (2-5) we obtain

$$(c_1)^2 + (c_2)^2 + 2c_1 c_2 S = \frac{1}{N^2} \tag{2-6}$$

The normalized wave function is then

$$\psi = \frac{1}{\sqrt{(c_1)^2 + (c_2)^2 + 2c_1 c_2 S}} \left(c_1 \psi_A + c_2 \psi_B\right) \tag{2-7}$$

The energy of this wave function is given by the expression

$$\begin{aligned} W &= N^2 \int \left(c_1 \psi_A + c_2 \psi_B\right) \mathcal{H} \left(c_1 \psi_A + c_2 \psi_B\right) d\tau \\ &= \frac{c_1^2 \int \psi_A \mathcal{H} \psi_A \, d\tau + c_2^2 \int \psi_B \mathcal{H} \psi_B \, d\tau + 2c_1 c_2 \int \psi_A \mathcal{H} \psi_B \, d\tau}{c_1^2 + c_2^2 + 2c_1 c_2 S} \end{aligned} \tag{2-8}$$

We now abbreviate the definite integrals as follows:

$$H_{AA} = \int \psi_A \mathcal{H} \psi_A \, d\tau \tag{2-9}$$

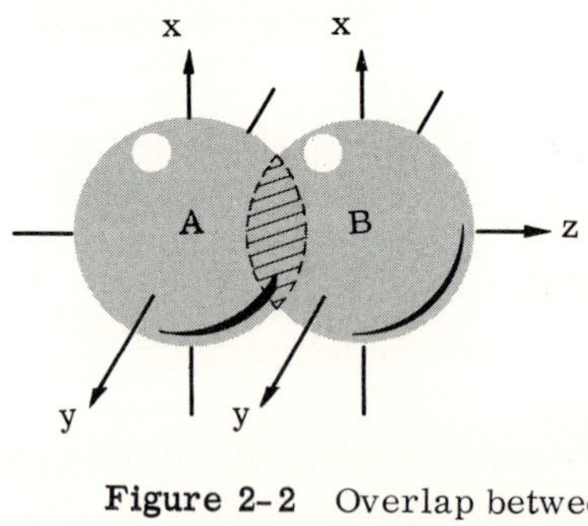

Figure 2-2 Overlap between two s functions. The cross-hatched area has electron density which originates from both ψ_A and ψ_B.

$$H_{BB} = \int \psi_B \mathcal{H} \psi_B \, d\tau \tag{2-10}$$

$$H_{AB} = H_{BA} = \int \psi_A \mathcal{H} \psi_B \, d\tau \tag{2-11}$$

Since a definite integral is a number, we obtain W as a function of the parameters c_1 and c_2 (thus indirectly a function of Z_A, Z_B, and R):

$$W = \frac{(c_1)^2 H_{AA} + (c_2)^2 H_{BB} + 2c_1 c_2 H_{AB}}{(c_1)^2 + (c_2)^2 + 2c_1 c_2 S} \tag{2-12}$$

We are interested in finding the values of c_1 and c_2 which, for given atomic wave functions and constant R (that is, for constant H_{AA}, H_{BB}, and H_{AB}), will minimize the orbital energy W. This is of interest because it can be shown that the minimized solution W always is larger than or equal to W_0, where W_0 is the "correct" solution to the problem. Thus, the smaller W is the better. To determine c_1 and c_2 we differentiate W with respect to these parameters.

Recall that, from Eq. (2-12),

$$W[(c_1)^2 + (c_2)^2 + 2c_1 c_2 S] = (c_1)^2 H_{AA} + (c_2)^2 H_{BB} + 2c_1 c_2 H_{AB} \tag{2-13}$$

and by differentiation

$$\frac{\partial W}{\partial c_1}[(c_1)^2 + (c_2)^2 + 2c_1 c_2 S] + W(2c_1 + 2c_2 S) = 2c_1 H_{AA} + 2c_2 H_{AB} \tag{2-14}$$

Since the extremum value is found for $\partial W/\partial c_1 = 0$, we immediately obtain

$$c_1(H_{AA} - W) + c_2(H_{AB} - WS) = 0 \tag{2-15}$$

The equation obtained from $\partial W/\partial c_2$ can be written by inspection:

$$c_1(H_{AB} - WS) + c_2(H_{BB} - W) = 0 \tag{2-16}$$

These two simultaneous, homogeneous equations have the solutions (not considering the trivial solution, $c_1 = c_2 = 0$), given by the determinant

$$\begin{vmatrix} H_{AA} - W & H_{AB} - WS \\ H_{BA} - WS & H_{BB} - W \end{vmatrix} = 0 \tag{2-17}$$

This determinantal equation gives the extremum energy W of the "best" linear combination of ψ_A and ψ_B for the system. Expanding the equation gives

$$(H_{AA} - W)(H_{BB} - W) - (H_{AB} - WS)^2 = 0 \quad (2\text{-}18)$$

A second-order equation has two roots, and we find the two values of W by solution of Eq. (2-18).

Let us discuss this equation. First we assume that $H_{AA} = H_{BB}$. In other words, we have a homonuclear diatomic molecule, where "both ends are equal." Examples are H_2, O_2, and Cl_2. Then

$$H_{AA} - W = \pm (H_{AB} - WS) \quad (2\text{-}19)$$

or

$$W = \frac{H_{AA} \pm H_{AB}}{1 \pm S} \quad (2\text{-}20)$$

Since S is less than 1 (if the A and B nuclei are together, S = 1; if they are infinitely separated, S = 0), we get one energy which is *smaller* than $H_{AA} = H_{BB}$ and one which is *larger*. These are called, respectively, the *bonding* and the *antibonding* solutions. This is depicted in Figure 2-3. It is customary to designate the bonding solution as W^b and the antibonding solution as W^*. We see that the terms bonding and antibonding refer to the energy loss or gain which is obtained when two atomic nuclei are brought together.

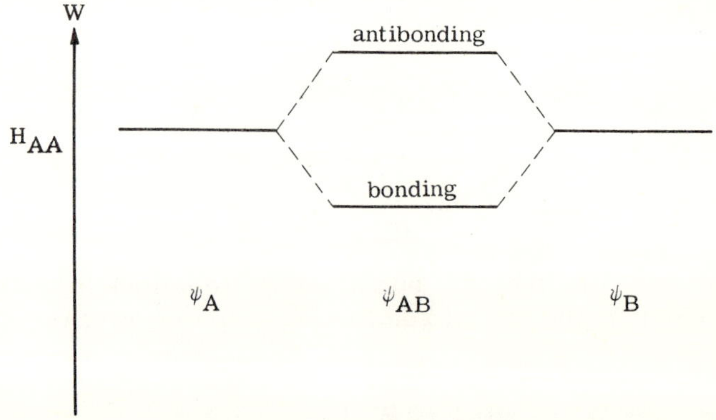

Figure 2-3 Bonding and antibonding orbitals in the AB molecule.

The sign of H_{AA} is negative, as is required for an electron bound to a positive nucleus. Calculations show that H_{AB} is also negative provided S is positive. Calling $H_{AB}/H_{AA} = \gamma$, we have $W = H_{AA}[(1 \pm \gamma)/(1 \pm S)]$. $W < H_{AA}(<0)$ for $(1 \pm \gamma)/(1 \pm S) > 1$ or $\pm\gamma > \pm S$. In other words if $\gamma > S$ the combination $\psi_A + \psi_B$ is bonding. This is always the case for diatomic molecules, but is not necessarily true for polyatomic molecules. Owing to the denominator $1 - S$ in the antibonding combination, we see that the antibonding orbital is destabilized more than the bonding orbital is stabilized.

The energy of the bonding (and antibonding) orbital is a function of R. We can minimize W^b with respect to this parameter and obtain a theoretical value for the energy and for the equilibrium internuclear distance in the H_2^+ molecule. This procedure yields for the equilibrium distance $r_0 = 1.3$ A and the dissociation energy $D_e = 1.76$ eV (the experimental values are 1.06 A and 2.79 eV). In Figure 2-4 are shown the lowest orbital energies of the H_2^+ molecule. Better results are obtained by also minimizing W^b with respect to Z_{eff}.

Let us determine the coefficients c_1 and c_2 in our LCAO wave function, for $H_{AA} = H_{BB}$. This is done by substituting the values found for W in the original set of equations, (2-15) and (2-16):

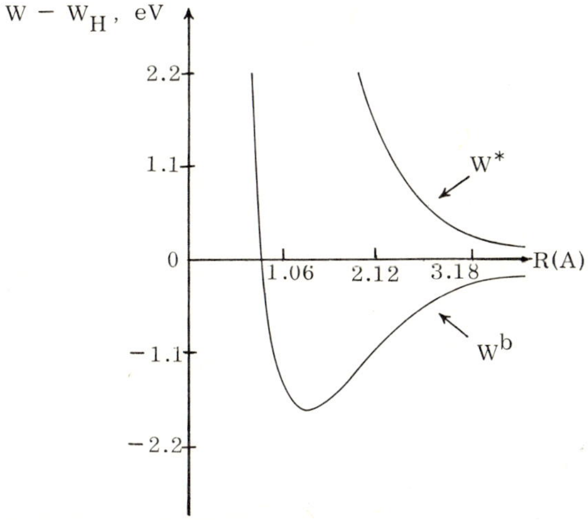

Figure 2-4 Lowest calculated energy levels for the H_2^+ molecule as a function of R. The energy minimum in the W^b curve occurs at $R = 1.3$ A with $D_e = 1.76$ eV.

$$c_1 = -c_2 \left(\frac{H_{AB} - WS}{H_{AA} - W} \right) \tag{2-21}$$

$$c_1 = -c_2 \left(\frac{H_{AA} - W}{H_{AB} - WS} \right) \tag{2-22}$$

Since

$$(H_{AB} - WS)^2 = (H_{AA} - W)^2 \tag{2-23}$$

or

$$(H_{AB} - WS) = \pm (H_{AA} - W) \tag{2-24}$$

we get by substitution,

$$c_2 = -c_1 \quad \text{and} \quad c_2 = c_1 \tag{2-25}$$

The two normalized wave functions are therefore

$$\frac{1}{\sqrt{2 + 2S}} \left(\psi_A + \psi_B \right) \qquad \text{(low energy)} \tag{2-26}$$

$$\frac{1}{\sqrt{2 - 2S}} \left(\psi_A - \psi_B \right) \qquad \text{(high energy)} \tag{2-27}$$

The coefficients c_1 and c_2 are in this case determined by the condition $H_{AA} = H_{BB}$. This is actually a *symmetry condition*—we take advantage of the fact that the molecule contains equivalent nuclei.

Let us now examine the general case given by the following equation:

$$(H_{AA} - W)(H_{BB} - W) - (H_{AB} - WS)^2 = 0 \tag{2-28}$$

H_{AA} and H_{BB} are the energies of the atomic orbitals ψ_A and ψ_B, respectively, in the molecular skeleton. For convenience, let us as-assume that $S = 0$. With H_{AB} numerically smaller than $|H_{AA} - H_{BB}|$, we have on expansion of (2-28),

$$(H_{AA} - W)(H_{BB} - W) - H_{AB}^2 = 0 \tag{2-29}$$

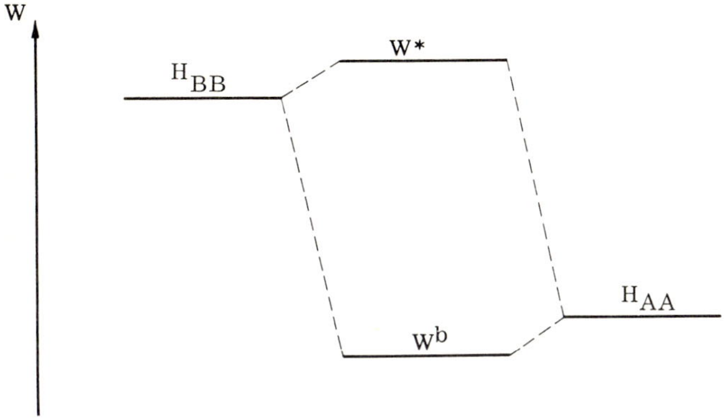

Figure 2-5 Bonding and antibonding energy levels for $H_{BB} \gg H_{AA}$.

or

$$W \approx \begin{cases} H_{AA} - \dfrac{H_{AB}^2}{H_{BB} - H_{AA}} \\ \\ H_{BB} + \dfrac{H_{AB}^2}{H_{BB} - H_{AA}} \end{cases} \tag{2-30}$$

One solution will give an orbital slightly more stable than ψ_A, while the other one will be a little less stable than ψ_B (Figure 2-5).

From (2-28), we see that if H_{AB} and S are zero, no covalent bonding occurs—that is, the energy of the ground state will not be lowered. Also, if S and H_{AB} are small, the covalent bonding is proportionately small. The overlap is zero if the atomic nuclei are far from each other. It becomes larger the shorter the distance is between the two nuclei (for two s orbitals see Figure 2-6). A large

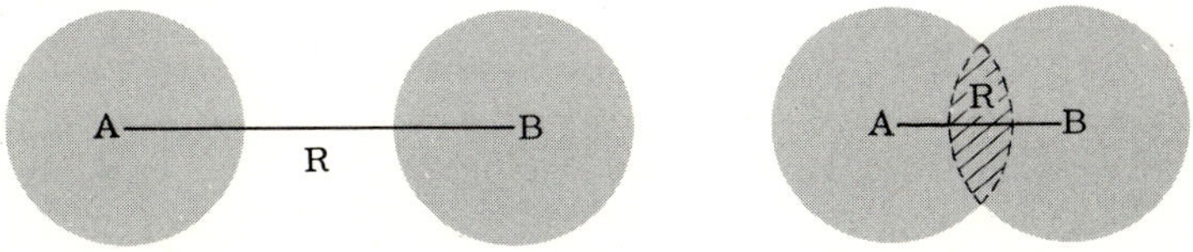

Figure 2-6 The values of S and H_{AB} depend strongly on R.

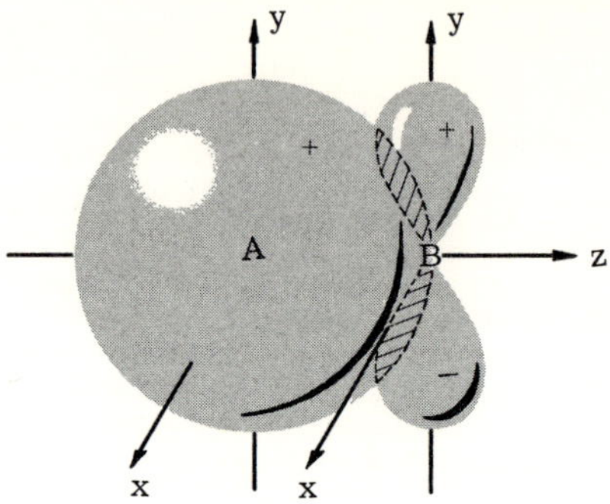

Figure 2-7 Overlap between an s and a p_y orbital.

overlap between valence orbitals usually means a strong bond, and for rough calculations one often assumes that the bond strength is proportional to the numerical value of the overlap integral.

With S given by the integral $\int \psi_A \psi_B \, d\tau$ we can see that the *atomic* orbitals used for forming molecular orbitals must have the *same symmetry* around the line between A and B—otherwise, S will be zero. Consider, for example, the overlap between an s and a p_y orbital. Figure 2-7 shows that each small volume element "in the top" of the p_y orbital is positive, while the corresponding volume element "below" is negative. Therefore, a summation (the integration!) over all volume elements will give zero:

$$\int (s_A)(p_{y_B}) \, d\tau = 0 \tag{2-31}$$

The overlaps of an s orbital on A with various orbitals on B are shown in Figure 2-8. We have $\int (s_A)(p_{z_B}) \, d\tau \neq 0$, $\int (s_A)(d_{z^2}B) \, d\tau \neq 0$, but $\int (s_A)(d_{x^2-y^2}B) \, d\tau = 0$. Thus the general rule emerges that the overlap integral is zero if the two orbitals involved have different symmetries around the connecting axis.

We shall now show that the symmetry conditions which cause an overlap integral to disappear also require that $H_{AB} = \int \psi_A \mathcal{H} \psi_B \, d\tau$ be zero.

Since

$$\mathcal{H} = -\frac{\hbar^2}{2m}\nabla^2 - \frac{Z_A e^2}{r_A} - \frac{Z_B e^2}{r_B} + \frac{Z_A Z_B e^2}{R} \tag{2-32}$$

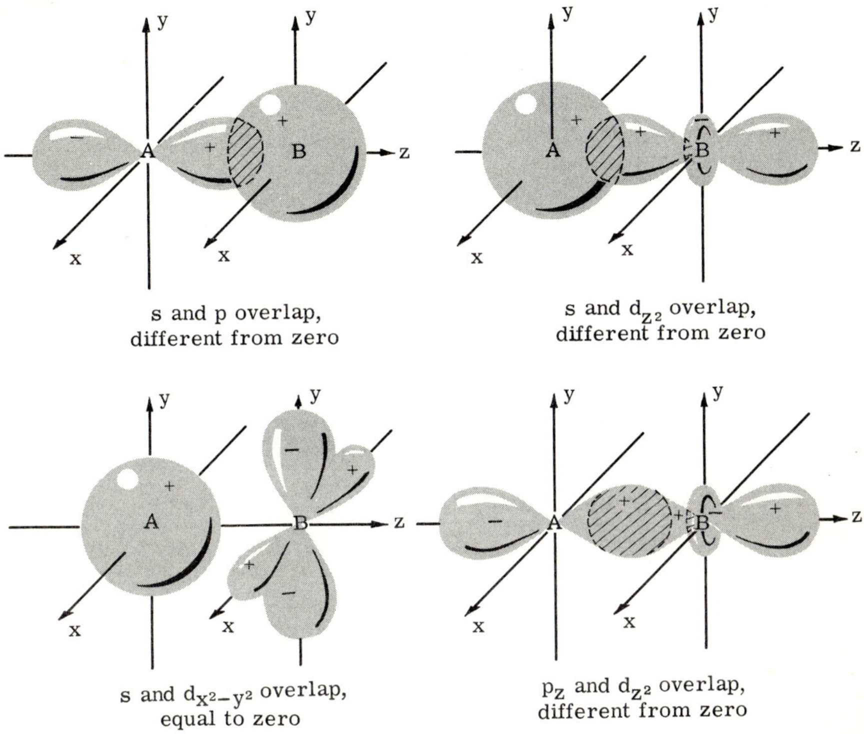

Figure 2-8 Some different overlaps between atomic orbitals on A and B.

we have, recalling that ψ_A and ψ_B are solutions to the atomic problem,

$$\mathcal{H}\psi_B = \left(-\frac{\hbar}{2m}\nabla^2 - \frac{Z_B e^2}{r_B}\right)\psi_B - \frac{Z_A e^2}{r_A}\psi_B + \frac{Z_A Z_B e^2}{R}\psi_B$$

$$= W_B \psi_B - \frac{Z_A e^2}{r_A}\psi_B + \frac{Z_A Z_B e^2}{R}\psi_B \qquad (2\text{-}33)$$

Thus,

$$\int \psi_A \mathcal{H} \psi_B \, d\tau = \left(W_B + \frac{Z_A Z_B e^2}{R}\right) S - Z_A e^2 \int \frac{\psi_A \psi_B}{r_A} \, d\tau \qquad (2\text{-}34)$$

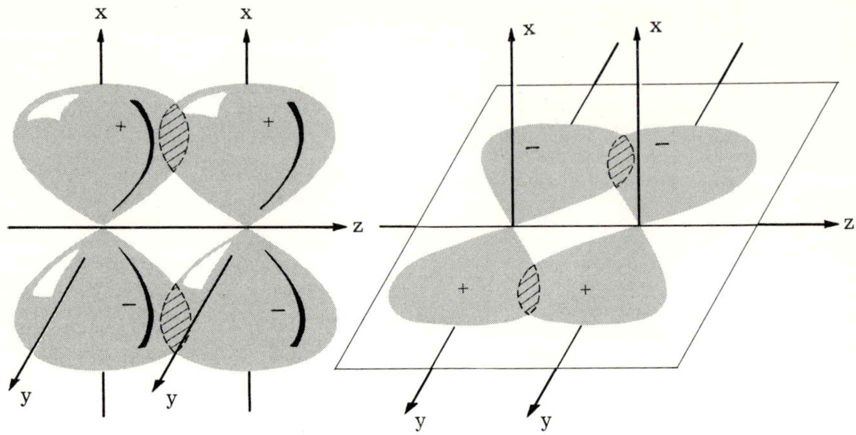

Figure 2-9 π-type overlap.

If ψ_A and ψ_B have different symmetries about A-B, S is zero. The integral $\int(\psi_A\psi_B/r_A)\,d\tau$ is also zero, since every volume element, divided by the distance from A, has a corresponding volume element of opposite sign, also divided by the same distance to A. H_{AB} follows S; both are zero if the two wave functions, centered respectively on A and B, do not have the same symmetry around the line joining the nuclei.

We distinguish among various types of molecular orbitals by means of the "symmetry" of the overlap. If the overlap is symmetric for rotation (as for example, s-s, s-p_z, p_z-p_z, p_z-d_{z^2}, etc.), the resulting molecular orbitals are called sigma (σ) orbitals. If the overlap gives a nodal plane along the connection line p_{xA}-p_{xB}, p_{yA}-p_{yB}), the resulting molecular orbitals are called pi (π) orbitals. If the overlap has two nodal planes which intersect along the connection line (d_{xy}-d_{xy}, $d_{x^2-y^2}$-$d_{x^2-y^2}$), the molecular orbitals are called delta (δ) orbitals.

Note that three p orbitals on each of two atoms give one p_σ (which uses the p_z orbital on each atomic nucleus and two p_π overlaps (which use p_{xA}, p_{xB} and p_{yA}, p_{yB}). The resulting π_x and π_y molecular orbitals have the same energy, since p_x and p_y are equivalent in the molecule. Furthermore, since the S_π overlap is *smaller* than the S_σ overlap, we expect π bonding to be weaker than σ bonding. It also follows that the π antibonding orbitals have lower energy than the σ antibonding orbitals (Figure 2-11).

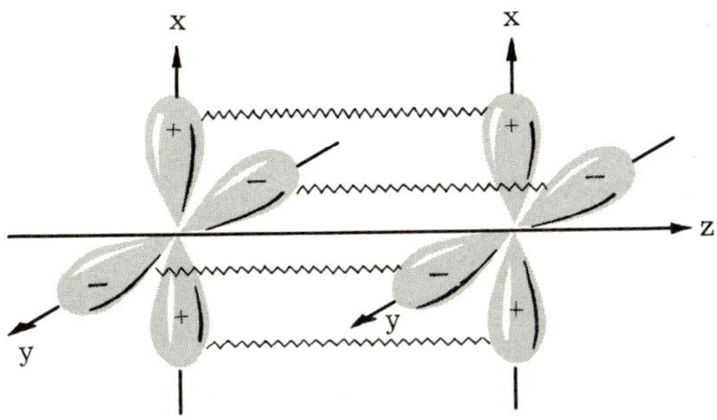

Figure 2-10 δ-type overlap.

It is now possible to draw a complete energy diagram for a diatomic molecule in which the atomic nuclei are alike. Since $W(1s) < W(2s) < W(2p)$, etc., we obtain a schematic energy diagram, which is given in Figure 2-12.

In a way quite analogous to the building up of the electronic structures of atoms, we now build up the electronic configurations of diatomic molecules. The electronic configurations are obtained by placing electron after electron into the empty energy levels, filling up first the lowest energy levels. For example, the ground states of the very simple diatomic molecules and ions are

H_2^+: $(1s\sigma^b)^1$

H_2: $(1s\sigma^b)^2$

He_2^+: $(1s\sigma^b)^2(1s\sigma^*)^1$

He_2: $(1s\sigma^b)^2(1s\sigma^*)^2$ (no net bonding)

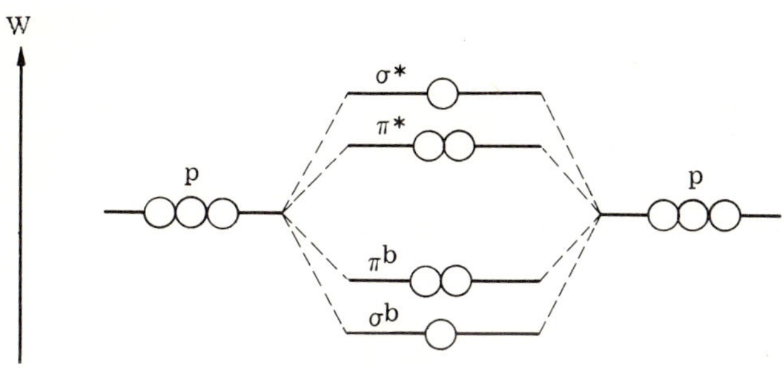

Figure 2-11 Bonding and antibonding molecular orbitals composed of the p orbitals in a homonuclear diatomic molecule.

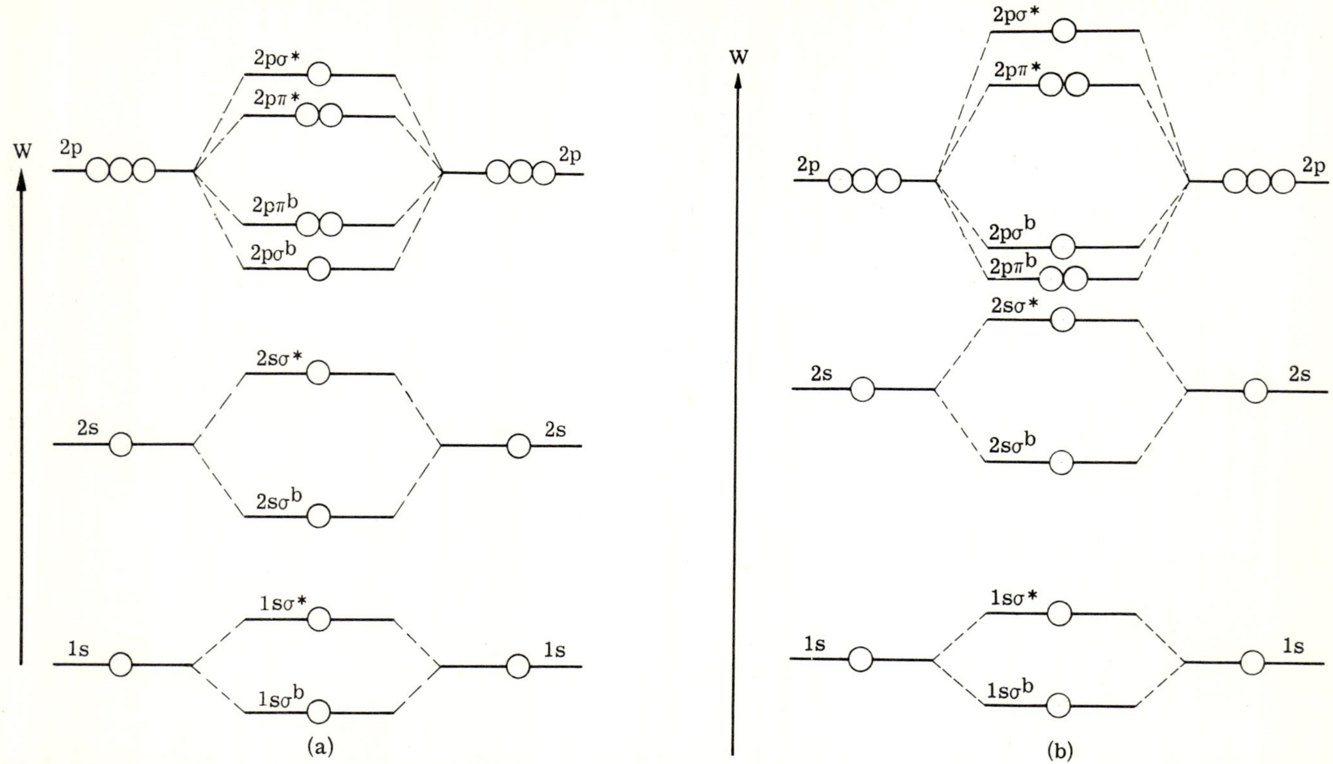

Figure 2-12 Energy diagrams for a homonuclear diatomic molecule. Note that the differences in the energy levels of the atoms are larger than the energy differences between the molecular orbitals. Diagram (a) is appropriate for no interaction between 2s and 2p levels, and diagram (b) is appropriate for substantial interaction between 2s and 2p levels. Refer to pp. 36-38.

We know that $1s\sigma^*$ is placed higher above the atomic 1s level than $1s\sigma^b$ is placed below. Thus, no energy is released—on the contrary, energy must be expended—in forming the He_2 molecule. Therefore, two isolated He atoms have lower energy than a hypothetical helium molecule; this is in agreement with the fact that He_2 does not exist in nature.

Other representative diatomic molecules have the following ground states:

$$Li_2: (1s\sigma^b)^2(1s\sigma^*)^2(2s\sigma^b)^2$$

$$N_2: (1s\sigma^b)^2(1s\sigma^*)^2(2s\sigma^b)^2(2s\sigma^*)^2(2p\pi^b)^4(2p\sigma^b)^2$$

$$O_2: (1s\sigma^b)^2(1s\sigma^*)^2(2s\sigma^b)^2(2s\sigma^*)^2(2p\sigma^b)^2(2p\pi^b)^4(2p\pi^*)^2$$

Since $(2p\pi^*)$ can accommodate a total of four electrons, the two electrons of O_2 that go in $(2p\pi^*)$ will go into different π orbitals, with spins parallel (Hund's rule)! The ground state of O_2 has two unpaired electrons. Thus, the oxygen molecule has a permanent magnetic moment, and we say that it is paramagnetic.

Let us again consider the He_2 configuration $(1s\sigma^b)^2(1s\sigma^*)^2$. With $\psi(1s\sigma^b) = (1/\sqrt{2+2S})(\psi_A + \psi_B)$ and $\psi(1s\sigma^*) = (1/\sqrt{2-2S}) \times (\psi_A - \psi_B)$, where $\psi_A = \psi_B = \psi(1s)$, the charge distribution for this configuration is obtained by summing the contributions of all four electrons:

$$2\left(\frac{1}{2+2S}\right)(\psi_A + \psi_B)^2 + 2\left(\frac{1}{2-2S}\right)(\psi_A - \psi_B)^2$$

$$= \frac{1}{1+S}\left(\psi_A^2 + \psi_B^2 + 2\psi_A\psi_B\right) + \frac{1}{1-S}\left(\psi_A^2 + \psi_B^2 - 2\psi_A\psi_B\right)$$

$$= \frac{2}{1-S^2}\left(\psi_A^2 + \psi_B^2 - 2S\psi_A\psi_B\right) \qquad (2\text{-}35)$$

For small S values the charge distribution is approximately $2\psi_A^2 + 2\psi_B^2$. This is just the value for the case in which there is no "bonding" between the atoms. In general, the overlap between the atomic 1s orbitals in molecules containing large atoms is very small (high nuclear charge draws the 1s electrons close to the nucleus). Consequently, we can neglect the bonding between 1s orbitals in such molecules.

2-2 SYMMETRY CONSIDERATIONS

In general, it is advantageous to use the symmetry elements of a molecule in dealing with the molecular orbitals. For example, consider the symmetry properties of a homonuclear diatomic molecule

(Figure 2-13). If we rotate the molecule around the x or the y axis through an angle of 180°, then A will go into B and B into A. But if A and B are the same, the rotated molecule cannot be distinguished from the starting molecule. In addition, we can rotate the molecule any arbitrary angle around the z axis without changing it. A reflection in the plane which contains A-B also makes the molecule "go into itself." An inversion of A and B through the center of the connection line takes A into B and B into A, and again (since A = B) there would not in a physical sense be any difference in the molecule after the symmetry operation.

The symmetry operation of rotation by 180° around the x axis is called a rotation around a twofold axis and is written $\hat{C}_2(=\hat{C}_{360/180})$. A rotation of $\varphi°$ around the z axis is written $\hat{C}_\varphi$, where φ is an arbitrary angle. If we, for example, rotate around the twofold axis two times, we return to the original position. This is the same operation as leaving the molecule alone. This latter symmetry operation is called $\hat{E}$; we have, then, $\hat{E} = \hat{C}_2\hat{C}_2 = (\hat{C}_2)^2 = \hat{C}_2^2$. The symmetry operation that reflects the molecule in a plane which contains the principal axis is called $\hat{\sigma}_V$. Further, an inversion in the center is called $\hat{i}$, with $\hat{i}\cdot\hat{i} = (\hat{i})^2 = \hat{i}^2 = \hat{E}$. In general we see that symmetry operations change the numbering of atoms in a molecule. But such a change of coordinates cannot alter the value of any physical quantity. Therefore, only the molecular integrals that are invariant under all symmetry operations of the molecule are different from zero.

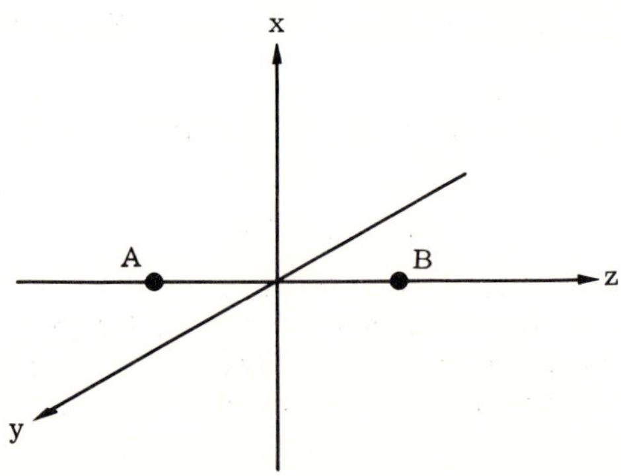

Figure 2-13 Diatomic molecule. The origin of the coordinate system is halfway between A and B.

Let us now examine how the molecular wave functions $\psi_1 = (1/\sqrt{2+2S})(s_A + s_B)$ and $\psi_2 = (1/\sqrt{2-2S})(s_A - s_B)$ behave under the symmetry operations of a homonuclear diatomic molecule. Since the normalization factor is a number (N), we have

$$
\begin{aligned}
\hat{E}\psi_1 &= \psi_1 = 1\cdot\psi_1 & \hat{E}\psi_2 &= \psi_2 = 1\cdot\psi_2 \\
\hat{\sigma}_v\psi_1 &= N\left(s_A + s_B\right) = 1\cdot\psi_1 & \hat{\sigma}_v\psi_2 &= N\left(s_A - s_B\right) = 1\cdot\psi_2 \\
\hat{C}_\varphi\psi_1 &= \psi_1 = 1\cdot\psi_1 & \hat{C}_\varphi\psi_2 &= \psi_2 = 1\cdot\psi_2 \\
\hat{i}\psi_1 &= N\left(s_B + s_A\right) = 1\cdot\psi_1 & \hat{i}\psi_2 &= N\left(s_B - s_A\right) = -1\cdot\psi_2 \\
\hat{C}_2\psi_1 &= N\left(s_B + s_A\right) = 1\cdot\psi_1 & \hat{C}_2\psi_2 &= N\left(s_B - s_A\right) = -1\cdot\psi_2
\end{aligned}
\tag{2-36}
$$

The two wave functions change phase during certain symmetry operations. We call an arbitrary symmetry operation $\hat{S}$. Operating on a wave function ψ, we have $\hat{S}\psi = \lambda\psi$, and we call λ the eigenvalue of the symmetry operator $\hat{S}$. Since the eigenvalue of $\hat{C}_\varphi$ is 1 (the orbitals are symmetric for rotation) the functions are σ orbitals. An eigenvalue of +1 for $\hat{\sigma}_v$ is indicated by a +, while −1 is indicated by a −. Further, an eigenvalue of +1 for the operation $\hat{i}$ is indicated with a g (gerade) and an eigenvalue of −1 for $\hat{i}$ with a u (ungerade). The following notation tells the important symmetry properties of ψ_1 and ψ_2:

$$\psi_1 = \sigma_g^+ \qquad \psi_2 = \sigma_u^+ \tag{2-37}$$

Let us now look at $\psi_3 = N(p_z(A) + p_z(B))$ and $\psi_4 = N(p_z(A) - p_z(B))$ (Figure 2-14).

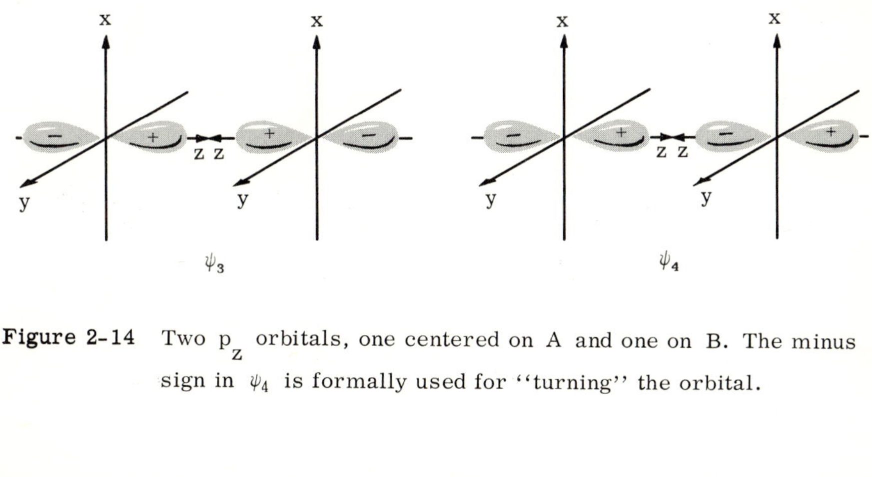

Figure 2-14 Two p_z orbitals, one centered on A and one on B. The minus sign in ψ_4 is formally used for "turning" the orbital.

$$\begin{aligned} \hat{E}\psi_3 &= 1\psi_3 & \hat{E}\psi_4 &= 1\psi_4 \\ \hat{\sigma}_V\psi_3 &= 1\psi_3 & \sigma_V\psi_4 &= 1\psi_4 \\ \hat{C}_\varphi\psi_3 &= 1\psi_3 & C_\varphi\psi_4 &= 1\psi_4 \\ \hat{i}\psi_3 &= 1\psi_3 & \hat{i}\psi_4 &= -1\psi_4 \end{aligned} \tag{2-38}$$

We see that ψ_3 is a σ_g^+ orbital and ψ_4 is a σ_u^+ orbital. We shall now consider π orbitals (Figure 2-15).

$$\psi_5 = N(p_x(A) + p_x(B)) \tag{2-39}$$

$$\psi_6 = N(p_y(A) + p_y(B)) \tag{2-40}$$

We see immediately that rotating ψ_5 by an angle 90° around the line AB will transform ψ_5 into ψ_6. Thus, if we choose φ in $\hat{C}_\varphi$ to be 90° such that x goes to $-$y, we have, using matrix language and matrix multiplication,

$$\hat{C}_{\varphi=90°}\begin{pmatrix}\psi_5\\ \psi_6\end{pmatrix} = \begin{pmatrix}0 & -1\\ 1 & 0\end{pmatrix}\begin{pmatrix}\psi_5\\ \psi_6\end{pmatrix} \tag{2-41}$$

In general, we represent the ψ's as a column vector $\boldsymbol{\Psi}$. We have then, $\hat{S}\,\boldsymbol{\Psi} = \mathbf{A}\boldsymbol{\Psi}$, where $\mathbf{A}$ is the so-called transformation matrix.

The trace (that is, the sum of the diagonal elements) of the transformation matrix $\begin{pmatrix}0 & -1\\ 1 & 0\end{pmatrix}$ is zero. However, this is not the result for an arbitrary value of φ. For rotation of the coordinate system by φ, we have

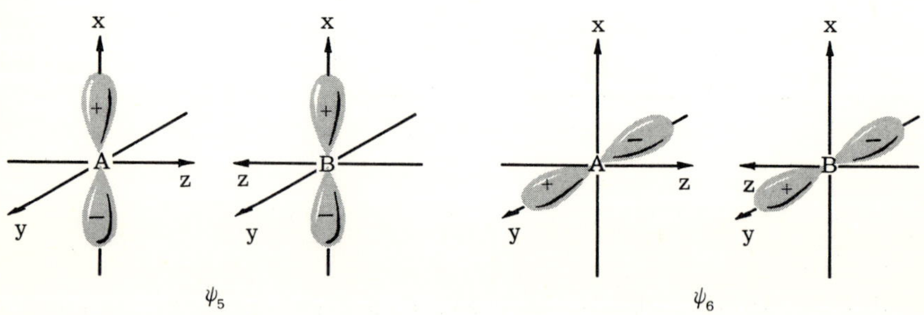

Figure 2-15 p_x and p_y form two π bonding orbitals.

$$\hat{C}_\varphi \begin{pmatrix} x \\ y \end{pmatrix} = \begin{pmatrix} \cos\varphi & -\sin\varphi \\ \sin\varphi & \cos\varphi \end{pmatrix} \begin{pmatrix} x \\ y \end{pmatrix} \qquad (2\text{-}42)$$

Because p_x and p_y behave as x and y, we get

$$\hat{C}_\varphi p_x(A) = \cos\varphi\, p_x(A) - \sin\varphi\, p_y(A) \qquad (2\text{-}43)$$

$$\hat{C}_\varphi p_y(A) = \sin\varphi\, p_x(A) + \cos\varphi\, p_y(A) \qquad (2\text{-}44)$$

with corresponding relationships for $p_x(B)$ and $p_y(B)$. Consequently,

$$\begin{aligned} \hat{C}_\varphi \psi_5 &= \hat{C}_\varphi N(p_x(A) + p_x(B)) \\ &= N(\cos\varphi\, p_x(A) - \sin\varphi\, p_y(A) \\ &\quad + \cos\varphi\, p_x(B) - \sin\varphi\, p_y(B)) \\ &= \cos\varphi\, \psi_5 - \sin\varphi\, \psi_6 \end{aligned} \qquad (2\text{-}45)$$

In the same way, we obtain for $\hat{C}_\varphi \psi_6$

$$\hat{C}_\varphi \psi_6 = \cos\varphi\, \psi_6 + \sin\varphi\, \psi_5 \qquad (2\text{-}46)$$

Written as a matrix equation,

$$\hat{C}_\varphi \begin{pmatrix} \psi_5 \\ \psi_6 \end{pmatrix} = \begin{pmatrix} \cos\varphi & -\sin\varphi \\ \sin\varphi & \cos\varphi \end{pmatrix} \begin{pmatrix} \psi_5 \\ \psi_6 \end{pmatrix} \qquad (2\text{-}47)$$

The trace of the transformation matrix is equal to $2\cos\varphi$. For a rotation $\varphi = \pi/2$, of course, the trace is equal to 0, as found before. Since ψ_5 and ψ_6 are "mixed" under $\hat{C}_\varphi$, we continue the investigation of symmetry properties using matrix notation:

$$\hat{E} \begin{pmatrix} \psi_5 \\ \psi_6 \end{pmatrix} = \begin{pmatrix} 1 & 0 \\ 0 & 1 \end{pmatrix} \begin{pmatrix} \psi_5 \\ \psi_6 \end{pmatrix} \qquad (2\text{-}48)$$

with trace 2. The operation $\hat{\sigma}_v$ gives

$$\hat{\sigma}_v(xz) \begin{pmatrix} \psi_5 \\ \psi_6 \end{pmatrix} = \begin{pmatrix} 1 & 0 \\ 0 & -1 \end{pmatrix} \begin{pmatrix} \psi_5 \\ \psi_6 \end{pmatrix} \qquad (2\text{-}49)$$

with the trace 0.

We could, instead of reflection in the xz plane, have chosen reflection in the yz plane:

$$\hat{\sigma}_v\,(yz)\begin{pmatrix}\psi_5\\ \psi_6\end{pmatrix} = \begin{pmatrix}-1 & 0\\ 0 & 1\end{pmatrix}\begin{pmatrix}\psi_5\\ \psi_6\end{pmatrix} \tag{2-50}$$

again with a transformation matrix with a trace of 0. We see that the transformation matrix is different for $\hat{\sigma}_v\,(xz)$ and $\hat{\sigma}_v\,(yz)$, but the trace is the same. This is true, in general, for equivalent symmetry operations.

Finally, we have for $\hat{i}$,

$$\hat{i}\begin{pmatrix}\psi_5\\ \psi_6\end{pmatrix} = \begin{pmatrix}-1 & 0\\ 0 & -1\end{pmatrix}\begin{pmatrix}\psi_5\\ \psi_6\end{pmatrix} \tag{2-51}$$

with a trace of -2.

We characterize these two degenerate orbitals ψ_5 and ψ_6 by the symbol π_u. The letter π is used to indicate that the orbitals are two-fold-degenerate with the trace $2 \cos \varphi$ under $\hat{C}_\varphi$. As before, the letter u is used to indicate that ψ_5 and ψ_6 change sign for the symmetry operation $\hat{i}$.

Thus, an "accidentally" chosen set of functions is characterized according to its behavior in the molecular framework. As will become apparent, this is of enormous importance in reducing the work of molecular computations. Since we want to characterize the different orbitals according to the traces of their transformation matrices, we use the following *character table,* which applies to all linear molecules with an inversion center. Such molecules are said to belong to the point group $D_{\infty h}$. In a character table the entry under a symmetry operation gives the trace of the appropriate transformation matrix (see Table 2-1). All functions of interest can thus be characterized according to their behavior under the different symmetry operations which transform a molecule "into itself."

Next we shall demonstrate an extraordinarily important concept—that molecular orbitals which have different symmetries cannot be combined. Let us try to calculate the energy of an orbital obtained by combining a function behaving as a σ_g^+ molecular orbital with a function behaving as σ_u^+.

We construct the following linear combination, where ψ_1 transforms as σ_g^+ and ψ_4 transforms as σ_u^+:

$$\psi = a\psi_1 + b\psi_4 \tag{2-52}$$

Table 2-1 Character Table for the Point Group $D_{\infty h}$

	E	C_φ	σ_v	i	iC_φ	$i\sigma_v = C_2$
σ_g^+	1	1	1	1	1	1
σ_u^+	1	1	1	−1	−1	−1
σ_g^-	1	1	−1	1	1	−1
σ_u^-	1	1	−1	−1	−1	1
π_g	2	$2\cos\varphi$	0	2	$2\cos\varphi$	0
π_u	2	$2\cos\varphi$	0	−2	$-2\cos\varphi$	0
Δ_g	2	$2\cos 2\varphi$	0	2	$2\cos 2\varphi$	0
Δ_u	2	$2\cos 2\varphi$	0	−2	$-2\cos 2\varphi$	0
⋮	⋮	⋮	⋮	⋮	⋮	⋮

a and b are the variable parameters. A possible minimum energy is obtained as before by solution of the determinantal equation

$$\begin{vmatrix} H_{11} - W & H_{14} - WS \\ H_{41} - WS & H_{44} - W \end{vmatrix} = 0 \qquad (2\text{-}53)$$

in which $H_{11} = \int \psi_1 \mathcal{H} \psi_1 \, d\tau$, $H_{44} = \int \psi_4 \mathcal{H} \psi_4 \, d\tau$, and $H_{14} = H_{41} = \int \psi_1 \mathcal{H} \psi_4 \, d\tau = \int \psi_4 \mathcal{H} \psi_1 \, d\tau$. We shall now show that $S = H_{41} = H_{14} = 0$, and therefore $W = H_{11}$ or $W = H_{44}$. Thus, the energies are not changed; the orbitals ψ_1 and ψ_4 cannot be combined.

The proof will be done specifically for H_{14}, with quite obvious extension to S.

$$H_{14} = \int \psi_1 \mathcal{H} \psi_4 \, d\tau = \int \sigma_g^+ \mathcal{H} \sigma_u^+ \, d\tau \qquad (2\text{-}54)$$

Since H_{14} is a number that is obtained by evaluation of the definite integral in (2-54), it must not depend on the coordinate system that we choose to calculate this integral. $\mathcal{H}$ is the quantum mechanical expression for the energy; it is in all cases independent of the coordinate system, since the energy of the system cannot depend on how we choose to describe the system. We have then, by using the inversion operator $\hat{i}$,

$$H_{14} = \hat{i}H_{14} = \hat{i}\int \sigma_g^+ \mathcal{H} \sigma_u^+ \, d\tau = \int \left(\hat{i}\sigma_g^+\right)(\hat{i}\mathcal{H})\left(\hat{i}\sigma_u^+\right) d\tau$$

$$= \int \left(\sigma_g^+\right)(\mathcal{H})\left(-\sigma_u^+\right) d\tau = -\int \sigma_g^+ \mathcal{H} \sigma_u^+ \, d\tau = -H_{14} \qquad (2\text{-}55)$$

The only number that is equal to its negative is 0; therefore, $H_{14} = 0$. It follows that the off-diagonal term always must be zero, if ψ's in a linear combination are of different symmetries, for there will, in such a case, always be at least one symmetry operation which gives a different character for the two ψ's, and a proof similar to the above one can be carried through.

On the other hand, if the ψ's have the same symmetry, the phases will disappear, and thus S and H_{12} will be different from zero. This means that two such energy levels will repel each other, and that they

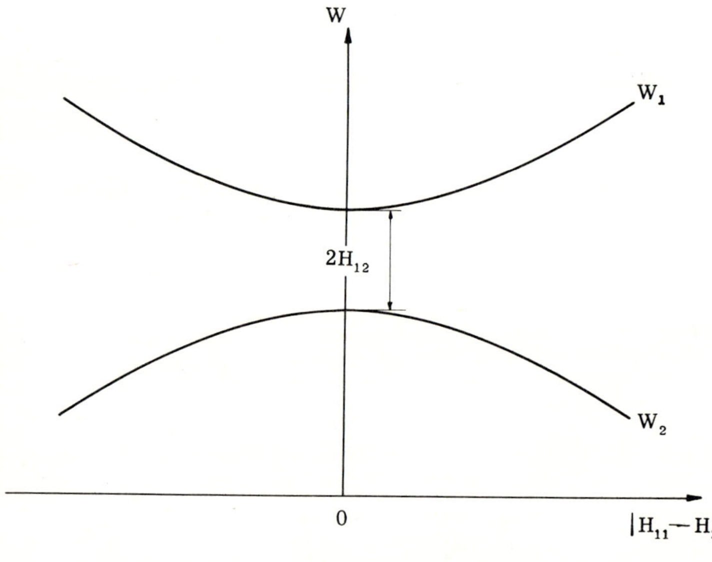

Figure 2-16 The noncrossing rule: two energy levels with the same symmetry properties repel each other and therefore never cross. With $|H_{11}| = |H_{22}|$ and $S = 0$, we have $W = H_{11} \pm H_{12}$; this means that the distance between W_1 and W_2 is $2H_{12}$. In general, from the equation $W^2 - W(H_{11} + H_{22}) - H_{12}^2 + H_{11}H_{12} = 0$, we obtain $W_1 - W_2 = \Delta W = \sqrt{(H_{11} - H_{22})^2 + 4H_{12}^2}$. The two energy levels will thus repel each other and the energy curves for W_1 and W_2 define a hyperbola with $(H_{11} - H_{22})$ as a variable. ΔW is smallest for $H_{11} = H_{22}$.

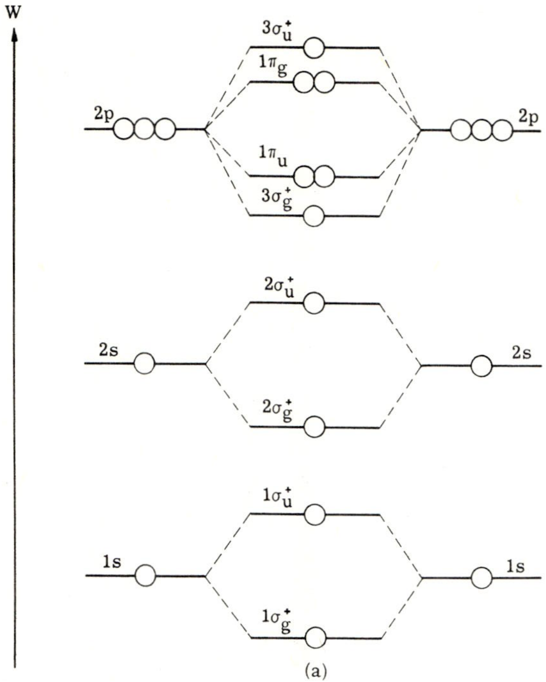

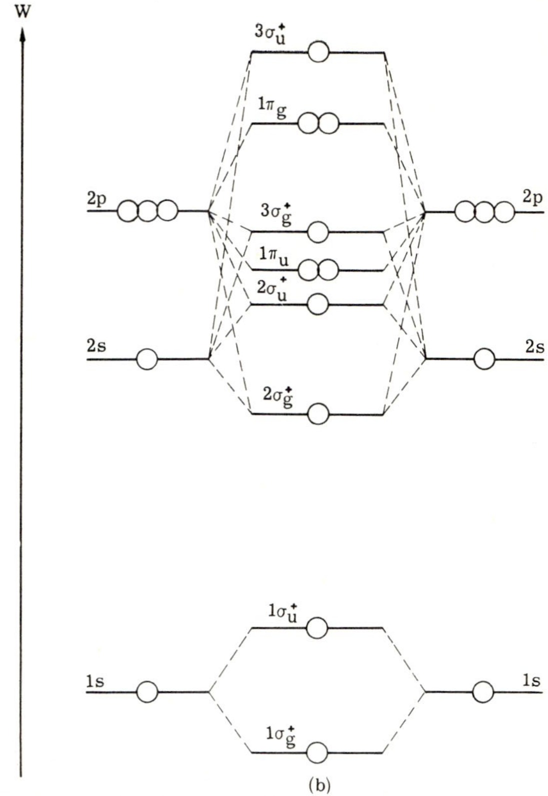

Figure 2-17 Energy diagrams for homonuclear diatomic molecules. The energy-level designations give the symmetries of the corresponding wave functions. The numbering begins at the lowest energy level and goes up, and each symmetry species is numbered separately. Diagram (a) is for no 2s-2p mixing, while (b) represents substantial 2s-2p mixing.

can never cross each other. This important rule can be seen by computing the energy levels using the relevant determinantal equation—this gives $(H_{11} - W)(H_{22} - W) - (H_{12})^2 = 0$ for the simplest case of $S = 0$ (see Figure 2-16). The nearest W_1 and W_2 can approach each other is if $H_{11} = H_{22}$; then $W = H_{11} \pm H_{12}$.

Therefore, it is appropriate to say that, for example, $1\sigma_g^+$ (see Figure 2-17) will be lowered in energy by interaction with all other σ_g^+ orbitals. Such an effect is called *configuration interaction.*

If we only look at the two lowest σ_g^+ orbitals, the secular equation for calculation of the "corrected" energies is

$$\begin{vmatrix} W - W(1\sigma_g^+) & x \\ x & W - W(2\sigma_g^+) \end{vmatrix} = 0 \tag{2-56}$$

with $x = \int \psi(1\sigma_g^+)\mathcal{H}\psi(2\sigma_g^+)\,d\tau$. We have further assumed that $S = \int \psi(1\sigma_g^+)\psi(2\sigma_g^+)\,d\tau$ is equal to zero.

If $W(2\sigma_g^+) \gg W(1\sigma_g^+)$, we have

$$W_1 \approx W(1\sigma_g^+) - \frac{x^2}{W(2\sigma_g^+) - W(1\sigma_g^+)} \tag{2-57}$$

$$W_2 \approx W(2\sigma_g^+) + \frac{x^2}{W(2\sigma_g^+) - W(1\sigma_g^+)} \tag{2-58}$$

One of the energy levels is lowered by the interaction; the other one is raised. These energy changes get *smaller* the larger the energy difference is between the two energy levels which "interact." Of course, if x is very small, we have only small changes in energy in any case.

2-3 DIATOMIC MOLECULES WITH DIFFERENT ATOMIC NUCLEI

We now turn to the diatomic molecules in which the atomic nuclei are different. We still use an LCAO description of the molecular orbitals,

$$\psi = N(\psi_B + \lambda\psi_A) \tag{2-59}$$

The constant λ determines the "polarity" of the orbital—the larger

Table 2-2 Character Table for the Point Group $C_{\infty v}$

	E	C_{φ}	σ_v
σ^+	1	1	1
σ^-	1	1	-1
π	2	$2 \cos \varphi$	0
Δ	2	$2 \cos 2\varphi$	0
⋮	⋮	⋮	⋮

the value of λ the greater the chance of finding the electron around nucleus A. The normalization constant N is given by the condition

$$N^2(1 + \lambda^2 + 2\lambda S) = 1 \tag{2-60}$$

To estimate an energy diagram for the molecular orbitals in such a molecule we first have to examine the symmetry properties of the orbitals.

Since the two atoms of the molecule AB are different, there is no longer a twofold rotation axis, as in an M_2 molecule. Also, the symmetry operation $\hat{i}$, inversion through the center, is not present. Both of these operations would transform A into B (Figure 2-13), and, since A and B are different, the molecule is differently situated after such an operation. Therefore our description of the molecule cannot be independent of the identities of nucleus A and nucleus B.

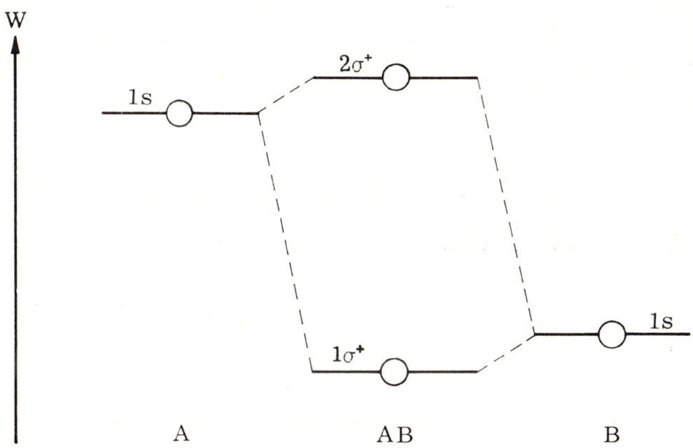

Figure 2-18 A combination of 1s orbitals with different energies.

The symmetry operations $\hat{E}$, $\hat{C}_{\varphi}$, and $\hat{\sigma}_v$ (reflection in a plane that contains the axis A-B) are present. All molecules that possess these symmetry properties have the point-group symmetry $C_{\infty v}$. The orbitals are characterized by symbols similar to those used for a homonuclear diatomic molecule, such as σ^+, π, etc. The character table for $C_{\infty v}$ is given in Table 2-2.

Since the atomic orbitals of the atomic nucleus with larger Z (B) have lower energy than the corresponding orbitals of the other atomic

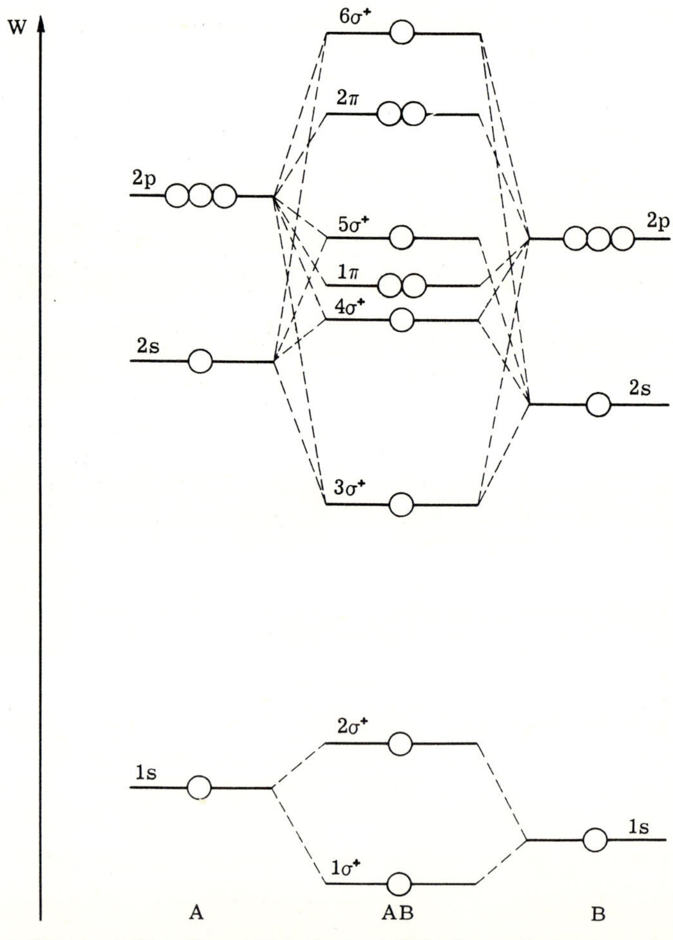

Figure 2-19 The molecular orbitals for a diatomic molecule with different atomic nuclei. The diagram indicates that $4\sigma^+$ repels $5\sigma^+$, making $5\sigma^+$ less stable than the 1π orbitals.

nucleus (A) (the "effective" charge of the nucleus is larger for B than for A), the molecular orbitals that are formed from 1s(A) and 1s(B) are as illustrated in Figure 2-18.

To construct an orbital energy diagram for a diatomic molecule in which the two nuclei are different, we must remember that orbitals with the same symmetry repel each other. As shown earlier, this repulsion is dependent on the energy difference between the molecular orbitals and on the size of the interaction. For diatomic molecules in which the two nuclei are not far from each other in the periodic table, the diagram will be approximately as shown in Figure 2-19.

We now place electrons one after the other into the energy diagram (Figure 2-19). For example, the electronic structure of CO(6 + 8 = 14 electrons) is given as $(1\sigma^+)^2(2\sigma^+)^2(3\sigma^+)^2(1\pi)^4(4\sigma^+)^2(5\sigma^+)^2$. The electronic structure of NO is $(1\sigma^+)^2(2\sigma^+)^2(3\sigma^+)^2(1\pi)^4(4\sigma^+)^2(5\sigma^+)^2(2\pi)^1$. The last electron goes into an antibonding π orbital. The molecule is paramagnetic since it possesses an "unpaired" electron.

On the other hand, if the two nuclei have very different "effective" nuclear charges, as for example the case of LiH, the orbital energies are similar to those indicated in Figure 2-20. Here it is a good approximation to regard the Li(1s) orbital as "nonbonding" and suppose that the chemical bonding between Li and H takes place between the H(1s) and the Li(2s) atomic orbitals. The Li(1s) orbital is, in such an approximation, not a "valence orbital."

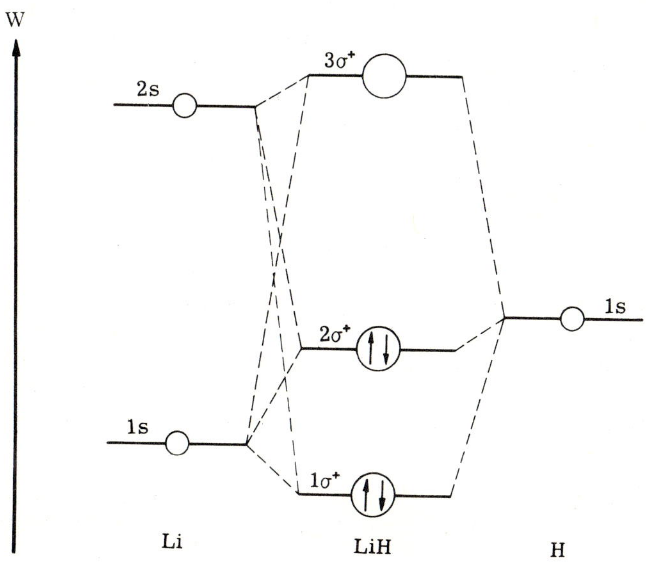

Figure 2-20 Lowest molecular orbitals of LiH.

3 / Electronic States of Molecules

Just as it is possible to characterize the single-electron molecular orbitals using the relevant symmetry operations of the molecule, it is possible to characterize the *total wave function* using the same operations. The total wave function contains *each* and *every* electron coordinate. It is customary to characterize the single molecular orbitals by small letters and the total wave functions by capital letters. For example, we have the single orbital σ_g^+, and the total wave function Σ_g^+.

For systems that contain only one electron there is no difference in the molecular-orbital and the total electronic wave function. For many-electron systems, however, there is *a considerable difference*. It should be noted that for many-electron systems it is only the "symmetry" of the *total* wave function which has physical (and chemical!) significance. This quantity is the only "observable" quantity.†

The ground state and the first and the second excited electronic

† In weighing an object on a balance it is only the result of the weights that is of interest, not how we have chosen to combine the different weights on the other balance pan.

states for the H_2^+ molecule are then (Figure 2-17) $^2\Sigma_g^+$, $^2\Sigma_u^+$, and $^2\Sigma_g^+$. The spin multiplicity, 2S + 1, where S is the value of the spin quantum number (here $S = \frac{1}{2}$), is the left-hand superscript.

The ground state of H_2 is $(1\sigma_g^+)^2$ or, written in determinantal form: $\Psi = |(1\overset{\alpha}{\sigma}{}_g^+)(1\overset{\beta}{\sigma}{}_g^+)|$. Because $\hat{E}\Psi = 1\Psi$, $\hat{C}_\varphi\Psi = 1\Psi$, $\hat{\sigma}_v\Psi = 1\Psi$, $\hat{i}\Psi = 1\Psi$, $\hat{i}\hat{C}_\varphi\Psi = 1\Psi$, and $\hat{i}\hat{\sigma}_v\Psi = 1\Psi$, the ground-state transforms as $^1\Sigma_g^+$, since for this state S = 0, and therefore 2S + 1 = 1. We call the state a "singlet-sigma-g-plus" state. Low-lying excited electronic states for the molecule are obtained by promoting one of the $(1\sigma_g^+)$ electrons to the $(1\sigma_u^+)$ orbital. The configuration $(1\sigma_g^+)(1\sigma_u^+)$ gives, as we shall see, more than one electronic state of the molecule.

Since the two orbitals are different, the two electrons can have the same spin. A total spin quantum number of one (S = 1) is obtained if both electrons have the same spin; on the other hand, a spin quantum number of zero (S = 0) is obtained if the electrons have different spins. With the spin multiplicity equal to 2S + 1, we get both a triplet (S = 1) and a singlet (S = 0) state.

A wave function for the triplet state is

$$\frac{1}{\sqrt{2}}\left[\left(1\overset{\alpha}{\sigma}{}_g^+\right)(1)\left(1\overset{\alpha}{\sigma}{}_u^+\right)(2) - \left(1\overset{\alpha}{\sigma}{}_g^+\right)(2)\left(1\overset{\alpha}{\sigma}{}_u^+\right)(1)\right] \qquad (3\text{-}1)$$

or

$$\Phi = \left|\left(1\overset{\alpha}{\sigma}{}_g^+\right)\left(1\overset{\alpha}{\sigma}{}_u^+\right)\right| \qquad (3\text{-}2)$$

Since $\hat{E}\Phi = 1\Phi$, $\hat{C}_\varphi\Phi = 1\Phi$, $\hat{\sigma}_v\Phi = 1\Phi$, and $\hat{i}\Phi = -1\Phi$, the total wave function is seen to transform as $^3\Sigma_u^+$, a "triplet-sigma-u-plus" state.

The wave function for the singlet state is

$$\frac{1}{2}\Big[\left(1\overset{\alpha}{\sigma}{}_g^+\right)(1)\left(1\overset{\beta}{\sigma}{}_u^+\right)(2) - \left(1\overset{\alpha}{\sigma}{}_g^+\right)(2)\left(1\overset{\beta}{\sigma}{}_u^+\right)(1)$$
$$- \left(1\overset{\beta}{\sigma}{}_g^+\right)(1)\left(1\overset{\alpha}{\sigma}{}_u^+\right)(2) + \left(1\overset{\beta}{\sigma}{}_g^+\right)(2)\left(1\overset{\alpha}{\sigma}{}_u^+\right)(1)\Big] \qquad (3\text{-}3)$$

We have written a combination that is antisymmetric for exchange of both *space* and *spin* coordinates. This singlet wave function can also be written

$$\Phi = \frac{1}{\sqrt{2}}\left[\left|\left(1\overset{\alpha}{\sigma}{}_g^+\right)\left(1\overset{\beta}{\sigma}{}_u^+\right)\right| - \left|\left(1\overset{\beta}{\sigma}{}_g^+\right)\left(1\overset{\alpha}{\sigma}{}_u^+\right)\right|\right] \qquad (3\text{-}4)$$

By performing the symmetry operations we establish that this wave function also transforms as Σ_u^+. Since the spin multiplicity is one, we have a $^1\Sigma_u^+$ state. We might naïvely suppose that both $^1\Sigma_u^+$ and $^3\Sigma_u^+$ have the same energy. But a calculation (Hund's rule!) tells us that the triplet state (in which the spins are parallel) has lower energy than the singlet state.

The next excited configuration occurs if both the $1\sigma_g^+$ electrons are excited into the $1\sigma_u^+$ orbital. The wave function is $|(1\overset{\alpha}{\sigma_u^+})(1\overset{\beta}{\sigma_u^+})|$, which represents a $^1\Sigma_g^+$ state.

The lowest energy levels or *terms* for the hydrogen molecule are given in Figure 3-1.

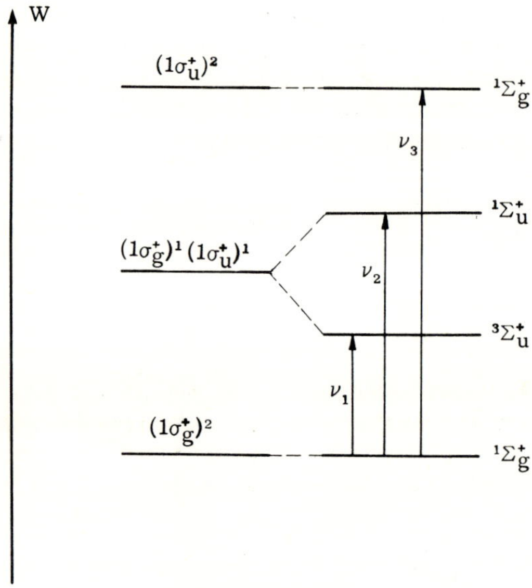

Figure 3-1 Lowest electronic states for H_2. On the left, the electronic configurations are indicated; on the right, the resulting terms. The difference between these terms is given by $h\nu$, where ν is the frequency of light absorbed (in promotion) or emitted (when electrons "fall" to lower levels).

We notice that, in general, filled electron "shells" give a state that is totally symmetric under the symmetry operations. This is due to the fact that there are no "degrees of freedom" in such a case.

Li_2 has 6 electrons, and these 6 electrons fill the three lowest energy levels. The ground state then is $(1\sigma_g^+)^2(1\sigma_u^+)^2(2\sigma_g^+)^2;\,^1\Sigma_g^+$.

N_2 has 14 electrons and the ground state is (Figure 2-17) $(1\sigma_g^+)^2 \times (1\sigma_u^+)^2(2\sigma_g^+)^2(2\sigma_u^+)^2(1\pi_u)^4(3\sigma_g^+)^2;\,^1\Sigma_g^+$. Again, we see that the ground state is $^1\Sigma_g^+$, since all the "shells" are filled.

O_2 has 16 electrons, 2 more than in N_2. These 2 extra electrons go into the $(1\pi_g)$ orbitals. If we write this $(N_2)(1\pi_g)^2$, in which (N_2) stands for the closed nitrogen configuration, we see that different arrangements of the electrons in the $(1\pi_g)$ orbitals are possible. Specifically, the electrons can be in different orbitals with the same or with opposite spin.

The wave functions for O_2 can now be written (we distinguish between the two π orbitals by the letters a and b):

$$\left|\begin{pmatrix}\alpha \\ \pi_g a\end{pmatrix}\begin{pmatrix}\alpha \\ \pi_g b\end{pmatrix}\right| \tag{3-5}$$

This is a component of a spin-triplet state, the three components being

$$\left|(\pi_g{}^a)(\pi_g{}^b)\right|\begin{pmatrix}\alpha\alpha \\ \alpha\beta + \beta\alpha \\ \beta\beta\end{pmatrix}$$

The spin-singlet functions are

$$\frac{1}{\sqrt{2}}\left[\left|\begin{pmatrix}\alpha \\ \pi_g a\end{pmatrix}\begin{pmatrix}\beta \\ \pi_g b\end{pmatrix}\right| - \left|\begin{pmatrix}\beta \\ \pi_g a\end{pmatrix}\begin{pmatrix}\alpha \\ \pi_g b\end{pmatrix}\right|\right] \tag{3-6}$$

$$\left|\begin{pmatrix}\alpha \\ \pi_g a\end{pmatrix}\begin{pmatrix}\beta \\ \pi_g a\end{pmatrix}\right| \tag{3-7}$$

$$\left|\begin{pmatrix}\alpha \\ \pi_g b\end{pmatrix}\begin{pmatrix}\beta \\ \pi_g b\end{pmatrix}\right| \tag{3-8}$$

We now have, with $\pi_g{}^a = (1/\sqrt{2})(p_{x_A} - p_{x_B})$ and $\pi_g{}^b = (1/\sqrt{2}) \times (p_{y_A} - p_{y_B})$:

$$\hat{E}\pi_g{}^a = \pi_g{}^a \tag{3-9}$$

$$\hat{E}\pi_g{}^b = \pi_g{}^b \tag{3-10}$$

$$\hat{C}_{\varphi}\pi_g{}^a = \cos\varphi\ \pi_g{}^a - \sin\varphi\ \pi_g{}^b \tag{3-11}$$

$$\hat{C}_{\varphi}\pi_g{}^b = \sin\varphi\ \pi_g{}^a + \cos\varphi\ \pi_g{}^b \tag{3-12}$$

$$\hat{\sigma}_v\pi_g{}^a = \pi_g{}^a \tag{3-13}$$

$$\hat{\sigma}_v\pi_g{}^b = -\pi_g{}^b \tag{3-14}$$

The molecular wave function $|(\overset{\alpha}{\pi}_g{}^a)(\overset{\alpha}{\pi}_g{}^b)|$, which according to Hund's rule should be the ground state, transforms as follows:

$$\hat{E}\left|\left(\overset{\alpha}{\pi}_g{}^a\right)\left(\overset{\alpha}{\pi}_g{}^b\right)\right| = 1\left|\left(\overset{\alpha}{\pi}_g{}^a\right)\left(\overset{\alpha}{\pi}_g{}^b\right)\right| \tag{3-15}$$

$$\begin{aligned}
\hat{C}_{\varphi}\left|\left(\overset{\alpha}{\pi}_g{}^a\right)\left(\overset{\alpha}{\pi}_g{}^b\right)\right| &= \left|\left(\cos\varphi\left(\overset{\alpha}{\pi}_g{}^a\right) - \sin\varphi\left(\overset{\alpha}{\pi}_g{}^b\right)\right)\right.\\
&\qquad \times \left.\left(\sin\varphi\left(\overset{\alpha}{\pi}_g{}^a\right) + \cos\varphi\left(\overset{\alpha}{\pi}_g{}^b\right)\right)\right|\\
&= \cos\varphi\ \sin\varphi\ \left|\left(\overset{\alpha}{\pi}_g{}^a\right)\left(\overset{\alpha}{\pi}_g{}^a\right)\right|\\
&\qquad + \cos^2\varphi\ \left|\left(\overset{\alpha}{\pi}_g{}^a\right)\left(\overset{\alpha}{\pi}_g{}^b\right)\right| - \sin^2\varphi\ \left|\left(\overset{\alpha}{\pi}_g{}^b\right)\left(\overset{\alpha}{\pi}_g{}^a\right)\right|\\
&\qquad - \sin\varphi\ \cos\varphi\ \left|\left(\overset{\alpha}{\pi}_g{}^b\right)\left(\overset{\alpha}{\pi}_g{}^b\right)\right|
\end{aligned} \tag{3-16}$$

The determinants $|(\overset{\alpha}{\pi}_g{}^a)(\overset{\alpha}{\pi}_g{}^a)|$ and $|(\overset{\alpha}{\pi}_g{}^b)(\overset{\alpha}{\pi}_g{}^b)|$ are zero (the columns are identical). Thus,

$$\begin{aligned}
\hat{C}_{\varphi}\left|\left(\overset{\alpha}{\pi}_g{}^a\right)\left(\overset{\alpha}{\pi}_g{}^b\right)\right| &= \cos^2\varphi\ \left|\left(\overset{\alpha}{\pi}_g{}^a\right)\left(\overset{\alpha}{\pi}_g{}^b\right)\right|\\
&\qquad - \sin^2\varphi\ \left|\left(\overset{\alpha}{\pi}_g{}^b\right)\left(\overset{\alpha}{\pi}_g{}^a\right)\right|
\end{aligned} \tag{3-17}$$

Reversing the columns in the last determinant will change the sign, and with $\sin^2\varphi + \cos^2\varphi = 1$ we have, finally,

$$\hat{C}_{\varphi}\left|\left(\overset{\alpha}{\pi}_g{}^a\right)\left(\overset{\alpha}{\pi}_g{}^b\right)\right| = 1\left|\left(\overset{\alpha}{\pi}_g{}^a\right)\left(\overset{\alpha}{\pi}_g{}^b\right)\right| \tag{3-18}$$

Further,

$$\hat{\sigma}_V \left|\left(\overset{\alpha}{\pi}_g a\right)\left(\overset{\alpha}{\pi}_g b\right)\right| = \left|\left(\overset{\alpha}{\pi}_g a\right)\left(-\overset{\alpha}{\pi}_g b\right)\right| = -1\left|\left(\overset{\alpha}{\pi}_g a\right)\left(\overset{\alpha}{\pi}_g b\right)\right| \tag{3-19}$$

Since both orbitals are symmetric during the inversion $\hat{i}$, their product has to be symmetric. Therefore, referring to Table 2-1, the ground state of O_2 is a $^3\Sigma_g^-$ state. Let us examine the singlet state

$$\Psi_2 = \frac{1}{\sqrt{2}}\left[\left|\left(\overset{\alpha}{\pi}_g a\right)\left(\overset{\beta}{\pi}_g b\right)\right| - \left|\left(\overset{\beta}{\pi}_g a\right)\left(\overset{\alpha}{\pi}_g b\right)\right|\right] \tag{3-20}$$

Here we have

$$E\Psi_2 = 1\Psi_2 \tag{3-21}$$

$$\begin{aligned}\hat{C}_\varphi \Psi_2 &= \frac{1}{\sqrt{2}}\Big[\Big|\left(\cos\varphi\,\overset{\alpha}{\pi}_g a - \sin\varphi\,\overset{\alpha}{\pi}_g b\right)\left(\sin\varphi\,\overset{\beta}{\pi}_g a + \cos\varphi\,\overset{\beta}{\pi}_g b\right)\Big| \\ &\quad - \Big|\left(\cos\varphi\,\overset{\beta}{\pi}_g a - \sin\varphi\,\overset{\beta}{\pi}_g b\right)\left(\sin\varphi\,\overset{\alpha}{\pi}_g a + \cos\varphi\,\overset{\alpha}{\pi}_g b\right)\Big|\Big] \\ &= \cos 2\varphi \frac{1}{\sqrt{2}}\left[\left|\left(\overset{\alpha}{\pi}_g a\right)\left(\overset{\beta}{\pi}_g b\right)\right| - \left|\left(\overset{\beta}{\pi}_g a\right)\left(\overset{\alpha}{\pi}_g b\right)\right|\right] \\ &\quad + \sin 2\varphi \frac{1}{\sqrt{2}}\left[\left|\left(\overset{\alpha}{\pi}_g a\right)\left(\overset{\beta}{\pi}_g a\right)\right| - \left|\left(\overset{\alpha}{\pi}_g b\right)\left(\overset{\beta}{\pi}_g b\right)\right|\right]\end{aligned} \tag{3-22}$$

We notice that Ψ_2 does not go "into itself" during the symmetry operation $\hat{C}_\varphi$ but into a linear combination with another wave function which is itself made up of a linear combination of determinants.

We therefore suspect that Ψ_2 is *one* of the components of a *doubly degenerate set of wave functions*, of which the other component is given by the combination

$$\Psi_3 = \frac{1}{\sqrt{2}}\left[\left|\left(\overset{\alpha}{\pi}_g a\right)\left(\overset{\beta}{\pi}_g a\right)\right| - \left|\left(\overset{\alpha}{\pi}_g b\right)\left(\overset{\beta}{\pi}_g b\right)\right|\right] \tag{3-23}$$

The symmetry operations on the set Ψ_2 and Ψ_3 give

$$\hat{E}\begin{pmatrix}\Psi_2\\ \Psi_3\end{pmatrix} = \begin{pmatrix}1 & 0\\ 0 & 1\end{pmatrix}\begin{pmatrix}\Psi_2\\ \Psi_3\end{pmatrix} \tag{3-24}$$

$$\hat{C}_\varphi\begin{pmatrix}\Psi_2\\ \Psi_3\end{pmatrix} = \begin{pmatrix}\cos 2\varphi & \sin 2\varphi\\ -\sin 2\varphi & \cos 2\varphi\end{pmatrix}\begin{pmatrix}\Psi_2\\ \Psi_3\end{pmatrix} \tag{3-25}$$

and

$$\hat{\sigma}_V \begin{pmatrix} \Psi_2 \\ \Psi_3 \end{pmatrix} = \begin{pmatrix} -1 & 0 \\ 0 & 1 \end{pmatrix} \begin{pmatrix} \Psi_2 \\ \Psi_3 \end{pmatrix} \tag{3-26}$$

The traces of the transformation matrices are, respectively, 2, 2 cos 2φ, and 0. By comparing these with the $D_{\infty h}$ character table (Table 2-1), we find that these two combinations *together* constitute $^1\Delta_g$.

We have not yet considered the linear combination that is orthogonal to Ψ_3. Calling this function Ψ_4 we have

$$\Psi_4 = \frac{1}{\sqrt{2}}\left[\left|\left(\overset{\alpha}{\pi}_g a\right)\left(\overset{\beta}{\pi}_g a\right)\right| + \left|\left(\overset{\alpha}{\pi}_g b\right)\left(\overset{\beta}{\pi}_g b\right)\right|\right] \tag{3-27}$$

The symmetry properties of Ψ_4 follow:

$$\hat{E}\Psi_4 = 1\Psi_4 \tag{3-28}$$

$$\hat{C}_\varphi\Psi_4 = \frac{1}{\sqrt{2}}\left[\left|\left(\cos\varphi\ \overset{\alpha}{\pi}_g a - \sin\varphi\ \overset{\alpha}{\pi}_g b\right)\left(\cos\varphi\ \overset{\beta}{\pi}_g a - \sin\varphi\ \overset{\beta}{\pi}_g b\right)\right|\right.$$

$$\left. + \left|\left(\sin\varphi\ \overset{\alpha}{\pi}_g a + \cos\varphi\ \overset{\alpha}{\pi}_g b\right)\left(\sin\varphi\ \overset{\beta}{\pi}_g a + \cos\varphi\ \overset{\beta}{\pi}_g b\right)\right|\right]$$

$$= 1\Psi_4 \tag{3-29}$$

$$\hat{\sigma}_V\Psi_4 = 1\Psi_4 \tag{3-30}$$

We see that this function transforms as $^1\Sigma_g^+$. The lowest states of O_2 are outlined in Figure 3-2.

It may seem strange on examining Figure 3-2 that all three states have different energies, since they all arise from the same $(\pi_g)^2$ configuration. However, the energy differences are due to the presence of the electron-repulsion term (e^2/r_{12}) in the complete Hamiltonian function. For example, we have for the energy of the ground state,

$$W(^3\Sigma_g^-) = \int \left|\left(\overset{\alpha}{\pi}_g a\right)\left(\overset{\alpha}{\pi}_g b\right)\right| \mathcal{H} \left|\left(\overset{\alpha}{\pi}_g a\right)\left(\overset{\alpha}{\pi}_g b\right)\right| d\tau \tag{3-31}$$

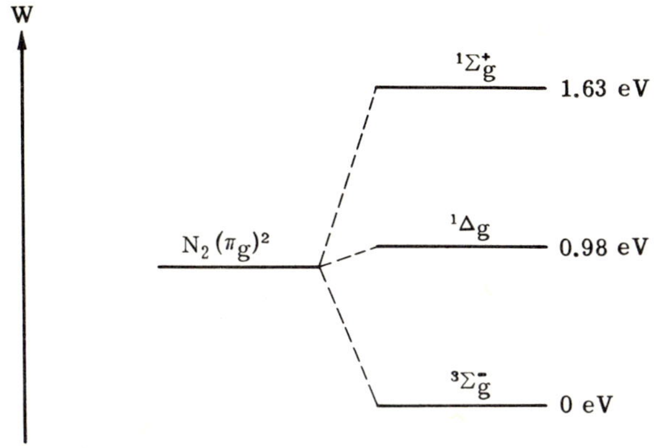

Figure 3-2 Lowest states of the O_2 molecule. The two excited states play a part in the absorption of light in the atmosphere. The energy unit is the electron volt, or eV. The energy of $^3\Sigma_g^-$ is arbitrarily set at zero.

Since

$$\mathcal{H} = \sum_1^2 \mathcal{H}(\text{molecular orbitals}) + e^2/r_{12} \tag{3-32}$$

$$\begin{aligned} W(^3\Sigma_g^-) &= \int \left|\left(\overset{\alpha}{\pi_g} a\right)\left(\overset{\alpha}{\pi_g} b\right)\right| \mathcal{H}\,(\text{molecular orbitals}) \left|\left(\overset{\alpha}{\pi_g} a\right)\left(\overset{\alpha}{\pi_g} b\right)\right| d\tau \\ &\quad + \int \left|\left(\overset{\alpha}{\pi_g} a\right)\left(\overset{\alpha}{\pi_g} b\right)\right| e^2/r_{12} \left|\left(\overset{\alpha}{\pi_g} a\right)\left(\overset{\alpha}{\pi_g} b\right)\right| d\tau \\ &= 2W(\pi_g) + \text{the electron repulsion term} \end{aligned} \tag{3-33}$$

Because the last term is different for each state (the wave functions are different!), the three states have different energies.

4 / Hybridization

The formation of a "chemical bond" will, as we have seen, increase the electronic density between the "bonded" atomic nuclei. We may say to a good approximation that the larger the overlap the stronger the bonding. Therefore, to obtain a strong bond between two atoms, we should take atomic orbitals that point toward each other and have a large overlap.

For example, we can form a good bonding orbital from a linear combination of a 2s and 2p orbital:

$$\Psi = \Psi(2s) \pm \lambda\psi(2p) \tag{4-1}$$

(See Figure 4-1.)

We see that by alternately using a plus and minus in front of the 2p orbital we obtain an electronic density which is concentrated to the right or to the left of the original center of gravity. Forming such a directional orbital, however, requires energy. Figure 4-2 shows this schematically.

The more 2p wave function we take together with the 2s function, the higher the energy of the lowest wave function. For $\lambda = 1$ we have two equivalent orbitals, which have the same energy but which are oriented differently, as shown in Figure 4-1.

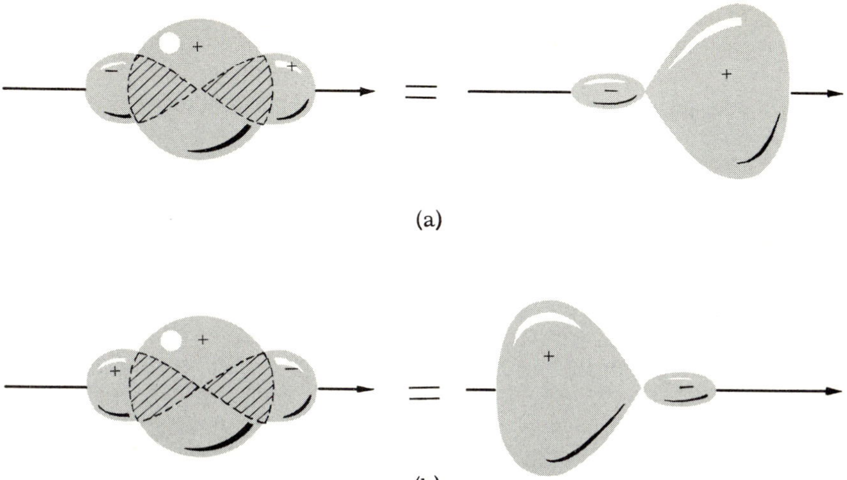

Figure 4-1 "Hybridization" of 2s and 2p orbitals. Boundary surfaces of (a) 2s + 2p, (b) 2s − 2p. In the last case the minus sign is in front of the 2p orbital, effectively turning it around.

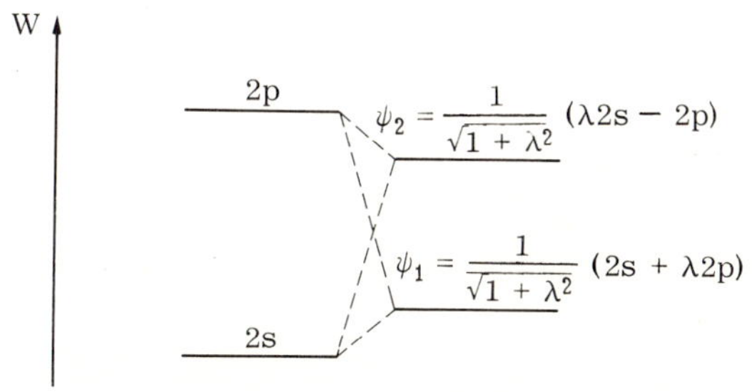

Figure 4-2 The energy changes on hybridizaton of 2s and 2p orbitals. Note that the "hybridized" orbitals are not solutions of the Schrödinger equation for the atom on which they are centered, although their components are solutions. In this way we use the atomic wave functions as basis functions in order to construct trial wave functions which can be used as molecular orbitals. A hybrid orbital always has higher energy than the lowest energy of its atomic components; for example, 2s and 2p are orthogonal to each other and thus the mixture must have an energy W(hybrid), where W(2s) < W(hybrid) < W(2p).

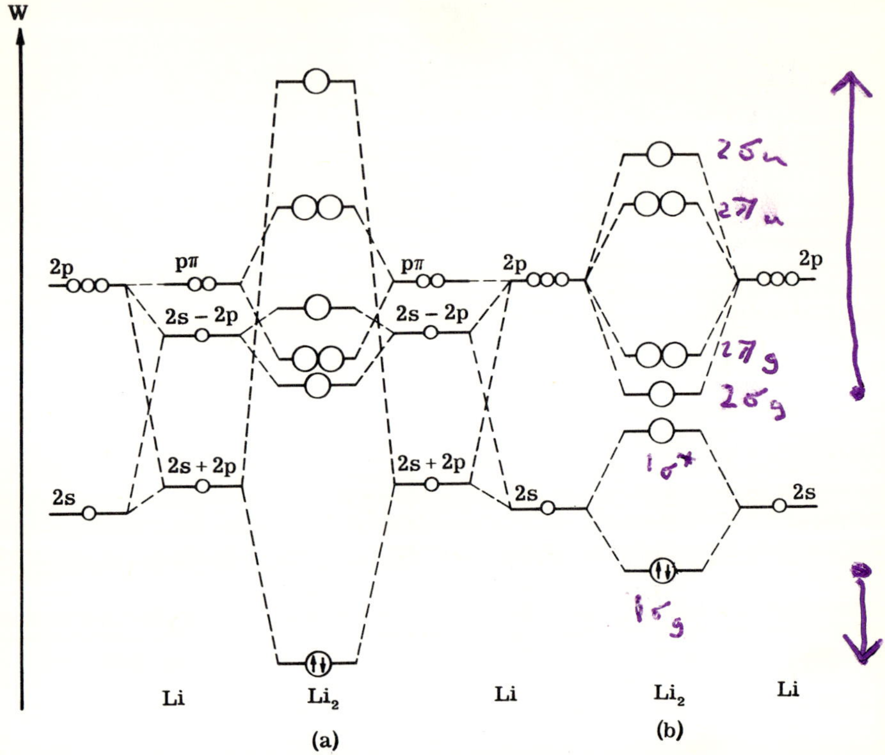

Figure 4-3 Energy diagram for Li_2: (a) with hybridization and (b) without $2s - 2p_z$ hybridization. The (1s) orbitals are left out.

The overlap between two sp hybrid orbitals that are centered on different atomic nuclei is much larger than the overlap of the single 2s or 2p functions. We would then expect a strong bond to be formed between the hybrid orbitals.

In Figure 4-3, (a) and (b) show the situation for Li_2 *with* and *without* hybridization. For the Li_2 molecule a calculation indicates that a substantial increase in the calculated bond energy of the molecule is obtained by mixing about 10 per cent 2p character into the wave function. Thus (see Figure 4-3),

$$\Psi_1 = \frac{1}{\sqrt{1 + \lambda^2}} \psi_{2s} + \frac{\lambda}{\sqrt{1 + \lambda^2}} \psi_{2p} \tag{4-2}$$

$$\frac{\lambda^2}{1 + \lambda^2} = \frac{10}{100} \qquad \lambda = \tfrac{1}{3} \quad \text{or} \quad \Psi_1 = \sqrt{\frac{9}{10}}\, \psi_{2s} + \sqrt{\frac{1}{10}}\, \psi_{2p} \tag{4-3}$$

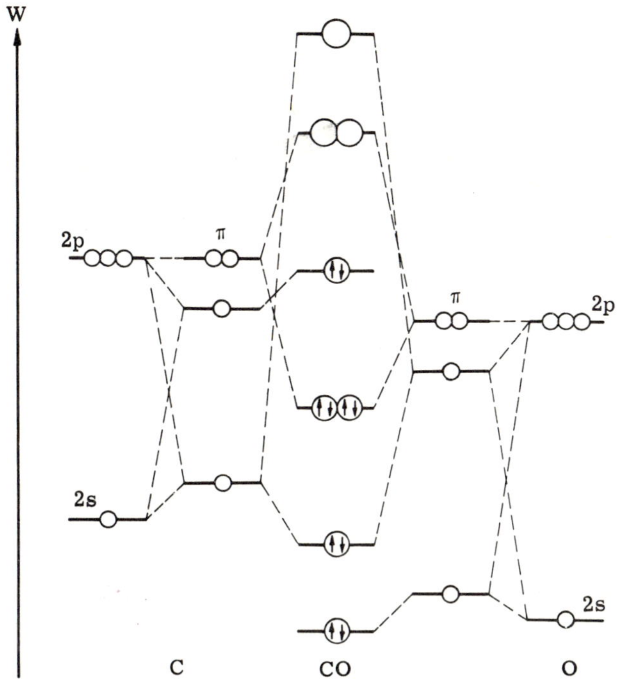

Figure 4-4 Electronic levels of CO with 2s-2p hybridization. The 1s orbitals are not included. A similar diagram can be constructed for the isoelectronic ions, CN^- and NO^+.

As we go down the row to Na_2, K_2, etc., λ gets smaller; the hybrid orbitals are less and less "polar." It has been pointed out by Mulliken that "A little bit of hybridization goes a long way." We use a little energy to construct the hybrid orbitals, but since the bonding is strengthened we obtain a lower energy when the calculation is made. In addition, the Coulomb repulsion of the electrons is reduced, since the orbitals are now directed away from each other. However, we must of course realize that such an accounting is *our* way of keeping track of the total energy of the molecule. Surely the molecule does not care at all which orbitals we use in order to get the strongest bonding!

Let us now consider the CO molecule. Ignoring the $C(1s)^2$ and $O(1s)^2$ levels, we construct the energy diagram shown in Figure 4-4.

We see that this diagram is consistent with the *chemical* properties of CO; the two electrons that have the highest energy are mainly localized on the carbon atomic nucleus. It turns out that the energy of the lone pair level is favorable for bond formation with the valence orbitals of the transition-metal atoms. Our diagram makes it possible to understand why it is the carbon end which coordinates to the metal, and why the M—C—O sequence is linear in metal carbonyls. The maximum σ overlap of the lone-pair s-p function with a metal σ function would be obtained along the —C—O line, at the carbon end.

Because CN^- and NO^+ are isoelectronic with CO, we could make similar remarks concerning their properties as ligands.

5 / Band Intensities

In chemistry it is common to indicate the intensity of a spectral band by stating the maximum molar extinction coefficient ϵ_{max}. Usually the shape of the absorption curve is given as a function of the wave number $\bar{\nu}$, and a curve $\epsilon(\bar{\nu})$ is obtained as given in Figure 5-1.

From a theoretical point of view, the intensity of a transition is given by the *area* under the absorption band. Assuming the band shape to be Gaussian, it is a good approximation to set the intensity equal to (in proper units)

$$f = 4.60 \times 10^{-9} \epsilon_{max} \bar{\nu}_{1/2} \tag{5-1}$$

where $\bar{\nu}_{1/2}$ is the so-called half-width of the band in cm^{-1} (the width of the band where $\epsilon = \frac{1}{2}\epsilon_{max}$).

Theoretically, the intensity f for a spectral transition is given by $f = 1.085 \times 10^{-5}\, \bar{\nu}\, |D|^2$, where D, the transition moment, is given by the integral

$$D = \int \psi_1^* R \psi_2 \, d\tau \tag{5-2}$$

Here ψ_1^* is the wave function for initial state, ψ_2 is the wave function for the final state $[(W_2 - W_1)/hc = \bar{\nu}(cm^{-1})]$, and $\mathbf{R}$ is the

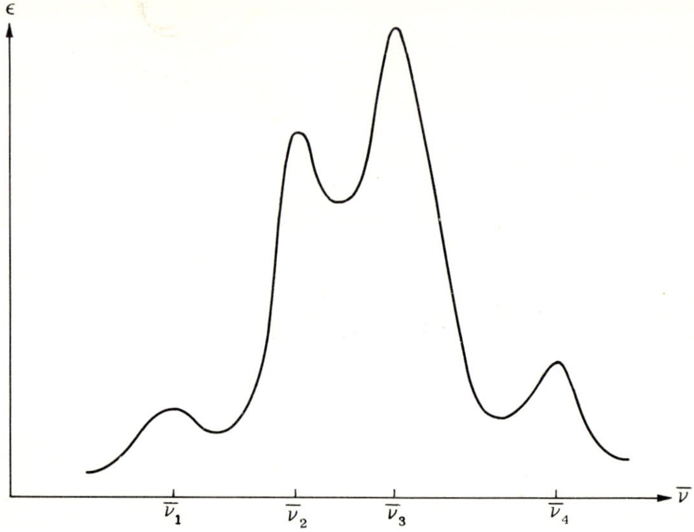

Figure 5-1 Absorption curve $\epsilon(\bar{\nu})$ composed of four absorption bands. Band half-widths are usually between 1000 and 6000 cm^{-1}.

dipole vector $\mathbf{R} = \mathbf{r}_1 + \mathbf{r}_2 + \cdots \mathbf{r}_n$ for all the n electrons included in the wave function.

We note that transitions are allowed only between states which have the same spin quantum number S. If they do not have the same S value, ψ_1 and ψ_2 will be orthogonal to each other, and since R is not a function of the spin, integration over the spin coordinates will give zero. In such a case, we say that the transition is *spin-forbidden.*

Further, if the molecule has an inversion center, transitions can take place only between an even and an odd state. The proof is that R is an odd function (it changes sign on inversion in the center), and since the integral (which represents a *physical* quantity) must not change sign, one of the wave functions, ψ_1 or ψ_2, must change sign during the inversion. This means, of course, that one of the wave functions has to be even (g) and the other one odd (u).

In general, transitions from state 1 to state 2 are allowed only if the integral $\int \psi_1^* \mathbf{R} \psi_2 \, d\tau$ is totally symmetric for all the symmetry operations of the molecule under consideration.

For example, considering the lowest states of H_2 (Figure 3-1), only the transition ${}^1\Sigma_g^+ \rightarrow {}^1\Sigma_u^+$ is allowed. The other ones are forbidden, since ${}^1\Sigma_g^+ \rightarrow {}^3\Sigma_u^+$ is spin-forbidden ($S = 0 \nrightarrow S = 1$) and ${}^1\Sigma_g^+ \rightarrow {}^1\Sigma_g^+$ is parity-forbidden ($g \nrightarrow g$).

Now, to prove that ${}^1\Sigma_g^+ \rightarrow {}^1\Sigma_u^+$ is allowed, we write down the transition moment integral

$$D = \int {}^1\Sigma_g^+ (\mathbf{i}X + \mathbf{j}Y + \mathbf{k}Z)\, {}^1\Sigma_u^+ \, d\tau \tag{5-3}$$

where **i**, **j**, and **k** are unit vectors. Since

$$\hat{C}_\varphi D = \int {}^1\Sigma_g^+ [\mathbf{i}(\cos\varphi X - \sin\varphi Y) + \mathbf{j}(\sin\varphi X + \cos\varphi Y) + \mathbf{k}Z] \times {}^1\Sigma_u^+ \, d\tau \tag{5-4}$$

we see that the X and Y components of the transition moment depend on φ. Since these components are *physical* quantities, the integral consequently cannot depend on our chosen coordinate system; the X and Y components must be zero. We shift our attention to the Z component and get

$$\hat{\sigma}_v D_Z = \int {}^1\Sigma_g^+ (\mathbf{k}Z)\, {}^1\Sigma_u^+ \, d\tau = D_Z \tag{5-5}$$

and

$$\hat{i} D_Z = \int {}^1\Sigma_g^+ (-\mathbf{k}Z)\left(-{}^1\Sigma_u^+\right) d\tau = D_Z \tag{5-6}$$

Thus, the integral is totally symmetric under the relevant symmetry operations and the transition ${}^1\Sigma_g^+ \rightarrow {}^1\Sigma_u^+$ is allowed. We have also established that it will be polarized along the Z axis. In other words, the transition takes place only when the electric vector of light is parallel to the molecular axis.

Ostensibly, only "allowed" transitions should be observed experimentally. In many cases, however, transitions are observed which formally are forbidden. This is not as disastrous as it would appear. Usually it is our model of the molecular structure which is wrong; we assume a static molecular skeleton and forget that vibrations can change this "firm" geometry and allow the molecule to have other structures. These other structures have different "symmetry" elements from those we worked with and give new and different selection rules. For example, we could destroy the inversion center and remove the parity restriction.

Also the selection rule $\Delta S = 0$ is eliminated (especially if the molecule contains heavy atomic nuclei) by the so-called spin-orbit coupling.

One may thus wonder if the selection rules are any use at all. However, a general rule is that the "forbidden" transitions have a *considerably smaller intensity* than the "allowed" ones. The completely allowed transitions have an $\epsilon_{max} \approx 10^3 - 10^5$, parity-forbidden transitions have $\epsilon \approx 10^0 - 10^2$, and spin-forbidden transitions have $\epsilon \approx 10^{-3} - 10^0$. These estimates are necessarily approximate, but they do give the orders of magnitude involved.

As indicated in Figure 5-1, the observed bands are sometimes rather broad. This may be surprising, since it seems to indicate that the electronic energies are poorly defined. However, the explanation is that the wave function of the molecule is a function not only of the electronic motions but the rotational and vibrational motions as well. Assuming the Born-Oppenheimer approximation, we have for the total wave function,

$$\Psi = \psi_{el} \psi_{rot} \psi_{vib}$$

Since $\Delta E_{el} \approx 10^5 \text{ cm}^{-1}$, $\Delta E_{vib} \approx 10^3 \text{ cm}^{-1}$, and $\Delta E_{rot} \approx 1 \text{ cm}^{-1}$, we have $\Delta E_{el} > E_{vib} > \Delta E_{rot}$, and we draw the potential-energy curve for an electronic state as pictured in Figure 5-2.

Neglecting the rotational states, we now have that the transition moment connecting one state to another is

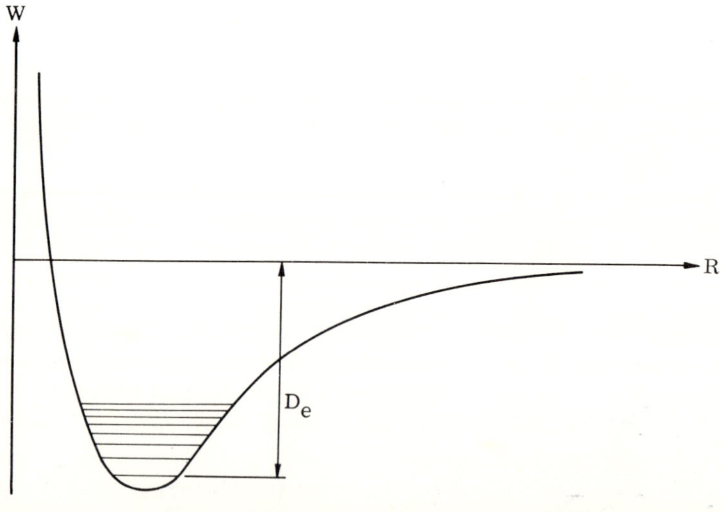

Figure 5-2 Potential surface for an electronic state. D_e is the dissociation energy, and the horizontal lines represent the allowed vibrational states. The rotational states have been omitted.

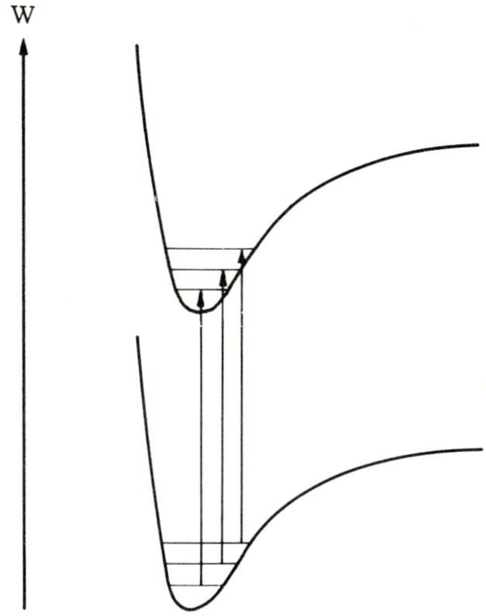

Figure 5-3 Transitions between different electronic states. In this example no change takes place in the vibrational state of the molecule, nor in the equilibrium configuration.

$$D = \int \Psi_1^* \mathbf{R} \Psi_2 \, d\tau$$

$$= \int \psi_1^*(\text{el})\psi_1^*(\text{vib})(\mathbf{R}_{\text{el}} + \mathbf{R}_{\text{vib}})\psi_2(\text{el})\psi_2(\text{vib}) \, d\tau_{\text{el}} \, d\tau_{\text{vib}}$$

which because $\Psi_1(\text{el})$ and $\Psi_2(\text{el})$ are orthogonal to each other reduces to

$$D = \int \psi_1^*(\text{vib})\psi_2(\text{vib}) \, d\tau_{\text{vib}} \int \psi_1^*(\text{el})\mathbf{R}_{\text{el}} \psi_2(\text{el}) \, d\tau_{\text{el}}$$

We notice that the electronic transition moment has been multiplied with a ''vibrational-overlap'' integral. In the solution of the vibrational problem, the vibrational wave functions will depend only upon the geometry and the force constants of the molecule. Therefore, only if all these parameters are identical in the two electronic states 1 and 2 will the two sets of vibrational wave functions be the solutions to the same Schrödinger equation.

In that case one vibrational function will be orthogonal to all the others, and transitions can only take place between two electronic states which have the same vibrational state, as shown in Figure 5-3. However, if these parameters are not completely identical, the

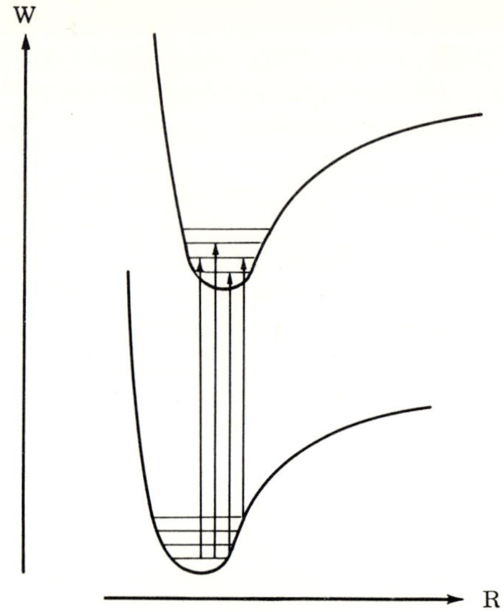

Figure 5-4 Transitions between different electronic states with no requirements on the change in vibrational level.

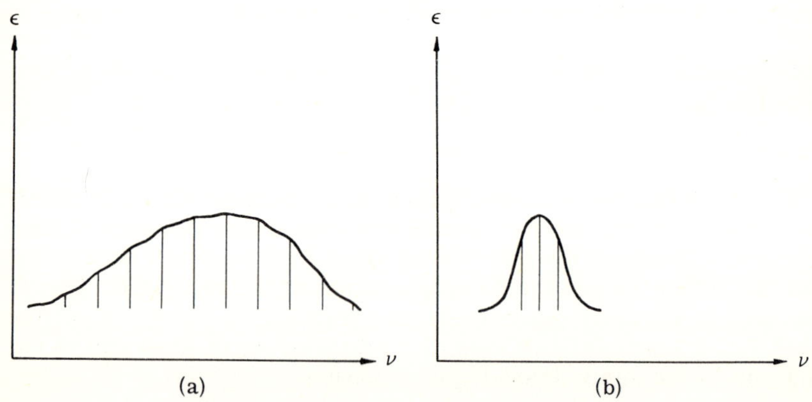

Figure 5-5 Vibrational lines composing an electronic absorption band: (a) Many lines give a broad absorption band; (b) fewer lines give a less complex band.

vibrational wave functions are not orthogonal, and we obtain a situation as pictured in Figure 5-4. The magnitude of a given transition is thus given by the vibrational-overlap integral.

Remembering that the vibrational levels of the ground electronic state of an ensemble of molecules (such as found in the sample we measure) will be populated according to the Boltzmann distribution, we see that a broad electronic absorption band will be composed of many absorption lines. As better and better spectral resolution is obtained, more and more details appear. Cooling the sample down greatly increases the number of molecules in the lowest vibrational state, thereby diminishing the number of possible transitions having "measurable" vibrational overlap. Thus, the number of lines decreases, and the spectrum is easier to interpret. This is illustrated in Figure 5-5.

6 / Triatomic Molecules

The same principles that we have used for the description of diatomic molecules will now be used in a description of the electronic structures of triatomic molecules. However, let it be clear that in using molecular-orbital theory with any hope of success, we first have to know the molecular geometry. Only in very rare cases is it possible from qualitative molecular-orbital considerations to *predict* the geometry of a given molecule. Usually this can only be discussed *after* a thorough calculation has been carried out.

6-1 THE CO_2 MOLECULE

As the first example of a triatomic molecule we choose CO_2. From electron-diffraction measurements and from infrared-absorption spectra we know that in the ground state the molecule is linear (Figure 6-1).

There are 16 valence electrons—6 electrons $(2s)^2(2p)^4$ from each oxygen atom and 4 electrons $(2s)^2(2p)^2$ from the carbon atom. The molecule is linear and has a center of symmetry. Thus, we characterize the orbitals according to the symmetry operations which are found in $D_{\infty h}$. We have for the carbon orbitals,

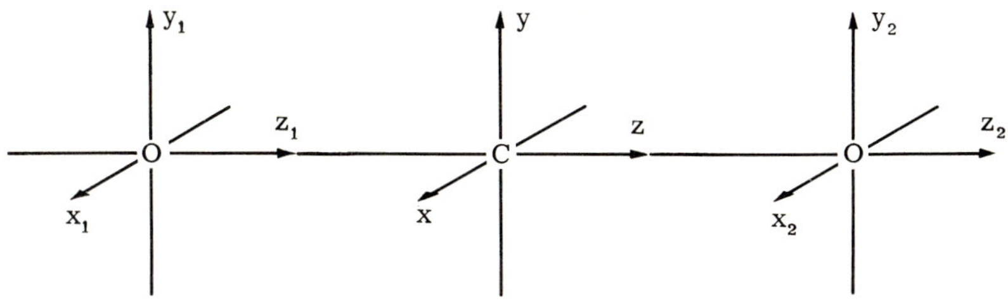

Figure 6-1 Coordinate system for the ground state of the CO_2 molecule.

$$\begin{array}{ll} 2s(C) & \sigma_g^+ \\ 2p_z(C) & \sigma_u^+ \\ \left.\begin{array}{l} 2p_x(C) \\ 2p_y(C) \end{array}\right] & \pi_u \end{array} \tag{6-1}$$

For the oxygen orbitals it is practical to form first the possible linear combinations, and then find the transformation properties of these combinations:

$$\begin{array}{ll} \frac{1}{\sqrt{2}}(2s_1 + 2s_2) & \sigma_g^+ \\ \frac{1}{\sqrt{2}}(2s_1 - 2s_2) & \sigma_u^+ \\ \frac{1}{\sqrt{2}}\left(2p_{z_1} + 2p_{z_2}\right) & \sigma_u^+ \\ \frac{1}{\sqrt{2}}\left(2p_{z_1} - 2p_{z_2}\right) & \sigma_g^+ \\ \left.\begin{array}{l} \frac{1}{\sqrt{2}}\left(2p_{x_1} + 2p_{x_2}\right) \\ \frac{1}{\sqrt{2}}\left(2p_{y_1} + 2p_{y_2}\right) \end{array}\right] & \pi_u \\ \left.\begin{array}{l} \frac{1}{\sqrt{2}}\left(2p_{x_1} - 2p_{x_2}\right) \\ \frac{1}{\sqrt{2}}\left(2p_{y_1} - 2p_{y_2}\right) \end{array}\right] & \pi_g \end{array} \tag{6-2}$$

We now combine the oxygen orbitals with the appropriate carbon orbitals to obtain the molecular orbitals for CO_2. Remember that only orbitals which transform alike can be combined together. We have, for example, the π_u orbitals $\Psi(\pi_u) = \alpha\psi C(\pi_u) + \beta\psi[O \cdots O](\pi_u)$. The coefficients α and β have the same sign in the bonding orbital and different signs in the antibonding orbital. The final σ_g^+, σ_u^+, and π_g combinations are obtained in the same way. The energy-level diagram for CO_2 is given in Figure 6-2.

The 16 valence electrons occupy the lowest orbitals in the ground state, giving the configuration $(1\sigma_g)^2(1\sigma_u)^2(2\sigma_g)^2(2\sigma_u)^2(1\pi_u)^4(1\pi_g)^4$. The filled "shell" configuration for the ground state transforms as $^1\Sigma_g^+$.

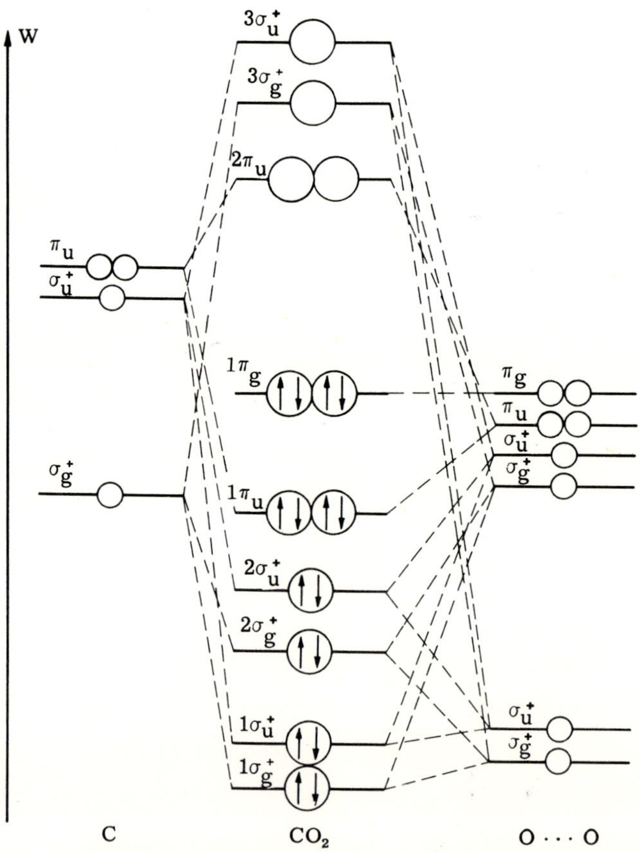

Figure 6-2 Bonding scheme for CO_2.

Low-energy excited states occur by promoting an electron from the $1\pi_g$ orbital to the $2\pi_u$ orbital. The electronic configuration $(\pi_g)^3(\pi_u)^1$ gives rise to states with S = 0 and S = 1. We shall concentrate on the singlet (S = 0) terms, since only transitions from the ground state to singlet states will be "spin-allowed."

The linear combinations of atomic orbitals which transform as π_u and π_g are

$$\pi_u \begin{cases} \left(p_{x_1} + p_{x_2}\right)\frac{1}{\sqrt{2}} \approx (x_1 + x_2);\ \pi_u{}^a \\ \left(p_{y_1} + p_{y_2}\right)\frac{1}{\sqrt{2}} \approx (y_1 + y_2);\ \pi_u{}^b \end{cases}$$
$$\pi_g \begin{cases} \left(p_{x_1} - p_{x_2}\right)\frac{1}{\sqrt{2}} \approx (x_1 - x_2);\ \pi_g{}^a \\ \left(p_{y_1} - p_{y_2}\right)\frac{1}{\sqrt{2}} \approx (y_1 - y_2);\ \pi_g{}^b \end{cases} \tag{6-3}$$

The wave functions for the excited states are given by the possible products $(\pi_g)^3(\pi_u)^1$. Since the distribution of a "hole" is the same as the distribution of an electron in the π_g orbitals (both can be distributed in either $\pi_g{}^a$ or $\pi_g{}^b$), we see that the possible distribution of the 4 electrons in the configuration $(\pi_g)^3(\pi_u)^1$ is equal to the distribution in $(\pi_g)^1(\pi_u)^1$. Such a distribution gives four wave functions:

$$\psi_1 = \left(\pi_g{}^a\right)\left(\pi_u{}^a\right) = (x_1 - x_2)(x_1 + x_2) = x_1{}^2 - x_2{}^2 \tag{6-4}$$

$$\psi_2 = \left(\pi_g{}^a\right)\left(\pi_u{}^b\right) = (x_1 - x_2)(y_1 + y_2)$$
$$= x_1 y_1 + x_1 y_2 - x_2 y_1 - x_2 y_2 \tag{6-5}$$

$$\psi_3 = \left(\pi_g{}^b\right)\left(\pi_u{}^a\right) = (y_1 - y_2)(x_1 + x_2)$$
$$= x_1 y_1 - x_1 y_2 + x_2 y_1 - x_2 y_2 \tag{6-6}$$

$$\psi_4 = \left(\pi_g{}^b\right)\left(\pi_u{}^b\right) = (y_1 - y_2)(y_1 + y_2) = y_1{}^2 - y_2{}^2 \tag{6-7}$$

These four wave functions, or linear combinations of them, represent the lowest excited orbital states for the molecule. They can be used both as spin-singlet and spin-triplet functions since the single orbitals are different.

We now examine how these wave functions transform for the symmetry operations ($\hat{E}$, $\hat{C}_\varphi$, $\hat{\sigma}_V$, $\hat{i}\hat{C}_\varphi$, $\hat{i}\hat{\sigma}_V$). First, note that they are odd—that is, they change sign on inversion in the center. This is a consequence of the fact that the product of an even and an odd function is odd.

The trace for $\hat{E}$ is four. The trace for $\hat{\sigma}_V$ (see Figure 6-1, reflection in the yz plane) is

$$\hat{\sigma}_V\begin{pmatrix}\psi_1\\ \psi_2\\ \psi_3\\ \psi_4\end{pmatrix} = \begin{pmatrix}1 & 0 & 0 & 0\\ 0 & -1 & 0 & 0\\ 0 & 0 & -1 & 0\\ 0 & 0 & 0 & 1\end{pmatrix}\begin{pmatrix}\psi_1\\ \psi_2\\ \psi_3\\ \psi_4\end{pmatrix} \tag{6-8}$$

The transformation matrix is seen to be diagonal. Thus in forming linear combinations of ψ_1, ψ_2, ψ_3, and ψ_4 which are to have a diagonal transformation matrix, we must combine ψ_1 with ψ_4 or ψ_2 with ψ_3.

It is convenient to make these linear combinations, in order to examine how the wave functions ψ_1 to ψ_4 transform under $\hat{C}_\varphi$. We form the new combinations

$$\psi_1 \pm \psi_4 \qquad \text{and} \qquad \psi_2 \pm \psi_3 \tag{6-9}$$

$$\psi_5 = \tfrac{1}{2}(\psi_1 + \psi_4) = \tfrac{1}{2}(x_1^2 + y_1^2) - \tfrac{1}{2}(x_2^2 + y_2^2) \tag{6-10}$$

$$\psi_6 = \tfrac{1}{2}(\psi_1 - \psi_4) = \tfrac{1}{2}(x_1^2 - y_1^2) - \tfrac{1}{2}(x_2^2 - y_2^2) \tag{6-11}$$

$$\psi_7 = \tfrac{1}{2}(\psi_2 + \psi_3) = x_1 y_1 - x_2 y_2 \tag{6-12}$$

$$\psi_8 = \tfrac{1}{2}(\psi_2 - \psi_3) = x_1 y_2 - x_2 y_1 \tag{6-13}$$

A rotation of the coordinate systems 1 and 2 through an angle φ around the connection line O—C—O transforms the coordinates to

$$\hat{C}_\varphi\begin{pmatrix}x\\ y\end{pmatrix} = \begin{pmatrix}\cos\varphi & -\sin\varphi\\ \sin\varphi & \cos\varphi\end{pmatrix}\begin{pmatrix}x\\ y\end{pmatrix} \tag{6-14}$$

On substitution and calculation,

$$\hat{C}_{\varphi}\begin{pmatrix}\psi_5\\ \psi_6\\ \psi_7\\ \psi_8\end{pmatrix} = \begin{pmatrix}1 & 0 & 0 & 0\\ 0 & \cos 2\varphi & -\sin 2\varphi & 0\\ 0 & \sin 2\varphi & \cos 2\varphi & 0\\ 0 & 0 & 0 & 1\end{pmatrix}\begin{pmatrix}\psi_5\\ \psi_6\\ \psi_7\\ \psi_8\end{pmatrix} \qquad (6\text{-}15)$$

From the above, it follows that

$$\hat{\sigma}_{v}\begin{pmatrix}\psi_5\\ \psi_6\\ \psi_7\\ \psi_8\end{pmatrix} = \begin{pmatrix}1 & 0 & 0 & 0\\ 0 & 1 & 0 & 0\\ 0 & 0 & -1 & 0\\ 0 & 0 & 0 & -1\end{pmatrix}\begin{pmatrix}\psi_5\\ \psi_6\\ \psi_7\\ \psi_8\end{pmatrix} \qquad (6\text{-}16)$$

We see that ψ_6 and ψ_7 are "mixed" by the symmetry operation $\hat{C}_{\varphi}$, while ψ_5 and ψ_8 are not "mixed" by any symmetry operation. Finally, we write the transformation matrix for the $\hat{E}$ operation:

$$\hat{E}\begin{pmatrix}\psi_5\\ \psi_6\\ \psi_7\\ \psi_8\end{pmatrix} = \begin{pmatrix}1 & 0 & 0 & 0\\ 0 & 1 & 0 & 0\\ 0 & 0 & 1 & 0\\ 0 & 0 & 0 & 1\end{pmatrix}\begin{pmatrix}\psi_5\\ \psi_6\\ \psi_7\\ \psi_8\end{pmatrix} \qquad (6\text{-}17)$$

Using the character table for a molecule with $D_{\infty h}$ symmetry, we find that ψ_6 and ψ_7 transform as Δ_u, ψ_5 as Σ_u^+ and ψ_8 as Σ_u^-.

The lowest states for the linear CO_2 molecule are given schematically in Figure 6-3a.

The removal of the fourfold degeneracy of the $(\pi_g)^3(\pi_u)^1$ configuration is as usual caused by the repulsion term (e^2/r_{12}) between the electrons. As we have seen, we find out beforehand that there will be three energy levels from such a configuration [six in all, if we consider triplet states (S = 1) also]. However, it is only by more quantitative calculations or by experiments that we can find the energies of the states.

Electronic transitions can take place from the ground state to the three excited states. But these transitions are not all orbitally allowed. By (as for the H_2 molecule) looking at the integrals $\int \Sigma_g^+ RX \, d\tau$, in which X is an excited state, we find that only the transition $\Sigma_g^+ \to \Sigma_u^+$ is allowed. The first intense absorption band for CO_2 has a maximum at 1335A (75,000 cm^{-1}). This band is assigned to the transition ${}^1\Sigma_g^+ \to {}^1\Sigma_u^+$.

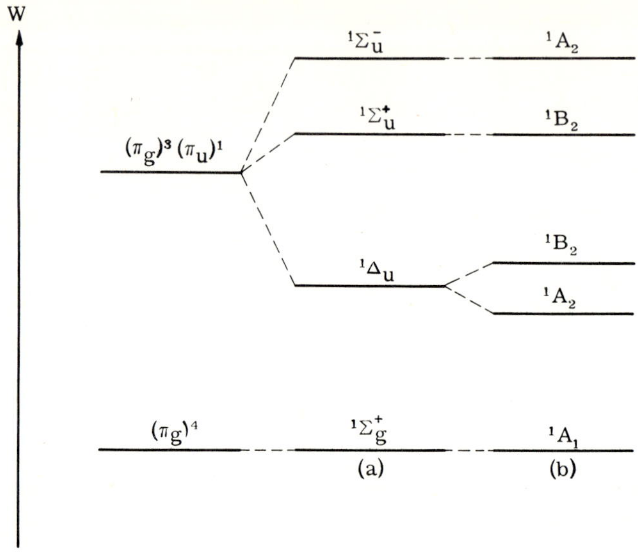

Figure 6-3 The lowest electronic singlets (S = 0) for the CO_2 molecule with (a) a linear configuration in both the ground state and the excited states and (b) a bent configuration.

In the above assignment we assumed that the molecule is linear in the excited state. This merits closer investigation. If we look at Figure 6-2, we see that the π_g orbital is a nonbonding orbital, and with $\pi_u{}^b$ filled we have used up all the available π-bonding character in the ground state. If we twist the O's and make an angle O–C–O (bent), some of the π bonding is destroyed. Thus, the ground state should prefer a linear structure, but this is not necessarily the case if an electron is excited into the antibonding π_u orbital.

Table 6-1 Character Table for C_{2v}

	E	$C_2(y)$	$\sigma_v(xy)$	$\sigma_v(yz)$
A_1	1	1	1	1
A_2	1	1	−1	−1
B_1	1	−1	1	−1
B_2	1	−1	−1	1

Table 6-2 Comparison of Orbital Symmetry Properties in $D_{\infty h}$ and C_{2v}

Orbitals in $D_{\infty h}$	E	$C_2(y)$	$\sigma_v(xy)$	$\sigma_v(yz)$	Symmetry in C_{2v}
$\sigma_g^+ = (s_1 + s_2)$	1	1	1	1	a_1
$\sigma_u^+ = (s_1 - s_2)$	1	−1	−1	1	b_2
$\pi_u = px^C$	1	−1	1	−1	b_1
$\pi_u = py^C$	1	1	1	1	a_1
$\pi_g^a = (x_1 - x_2)$	1	1	−1	−1	a_2
$\pi_g^b = (y_1 - y_2)$	1	−1	−1	1	b_2

The angular molecule has the following symmetry operations (symmetry C_{2v}):

$$\hat{E} \quad \hat{C}_2(y) \quad \hat{\sigma}_v(xy) \quad \hat{\sigma}_v(yz)$$

The characters for the representations in C_{2v} are given in Table 6-1.

We place the bent CO_2 molecule in the coordinate system shown in Figure 6-4 and examine how the previously constructed molecular orbitals transform in the new molecular geometry. We obtain the results shown in Table 6-2. Note that in the C_{2v} symmetry there

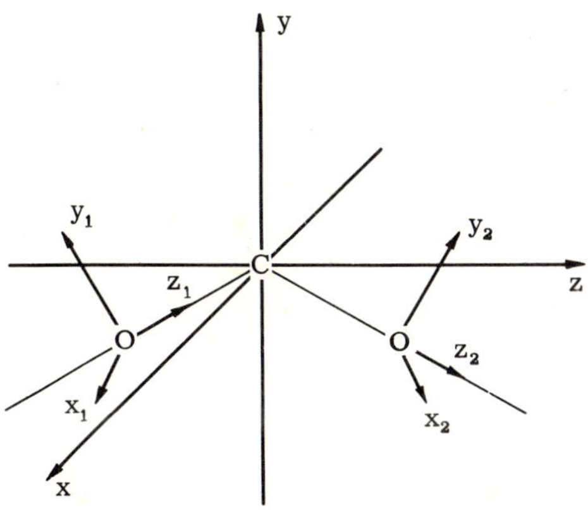

Figure 6-4 The CO_2 molecule lies in the yz plane.

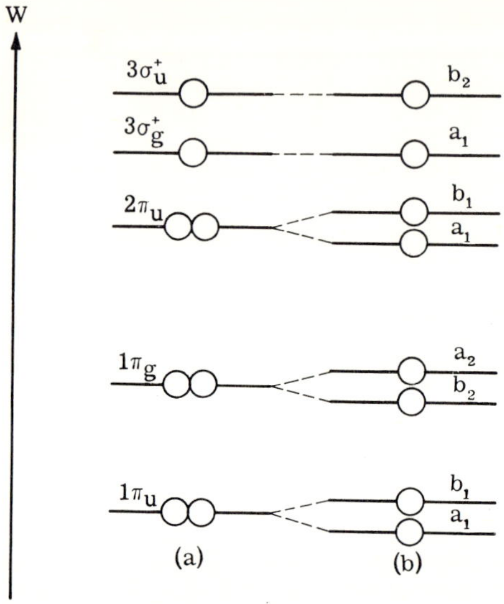

Figure 6-5 Molecular orbitals for (a) a linear CO_2 molecule and (b) a bent CO_2 molecule.

are no doubly degenerate sets and that, consequently, the degeneracy of the π orbitals is removed.

In Figure 6-5 the molecular-orbital energy levels for the linear CO_2 molecule (Figure 6-2) are displayed on the left-hand side. By bending the O—C—O angle a little, we have the energy levels shown on the right-hand side.

In the first approximation, the splitting of the π levels will be symmetrical; one component will be as much above the original level as the other one will be below. Since $1\pi_g$ and $1\pi_u$ in the ground state are filled with electrons, and since all the orbitals more stable than $1\pi_u$ are filled, there is no net gain in stability in the ground state by bending the molecule.

If we consider the splitting of $2\pi_u$, we see that one of the new a_1 levels can interact with $a_1\,(3\sigma_g^+)$, since they have the same symmetry and are close to each other. Thus, $a_1\,(2\pi_u)$ will be strongly stabilized at the expense of $a_1\,(3\sigma_g^+)$. The other decomposed component, $b_1\,(2\pi_u)$, has no nearby orbital of the same symmetry with which it can interact. Thus, the b_1 orbital is rather unaffected by the bending (Figure 6-6). All this means that if we excite an electron from $1\pi_g \rightarrow 2\pi_u$ (components b_1, a_1), the CO_2 molecule will bend if the electron occupies $a_1\,(2\pi_u)$ but will stay linear if it is excited to $b_1\,(2\pi_u)$.

The ground state is linear with a total wave function which transforms as $^1\Sigma_g^+$. Of the four "product functions" $(b_2)^2(a_2)^1(a_1)^1$, $(b_2)^2 \times (a_2)^1(b_1)^1$, $(a_2)^2(b_2)^1(a_1)^1$, and $(a_2)^2(b_2)^1(b_1)^1$, which are wave functions for the low-energy excited states, the first and third will be "bent" states and the second and fourth will have a linear configuration.

The total symmetries for these four states are

$$\begin{aligned} &(b_2)^2(a_2)^1(a_1)^1:\ {}^1A_2 \quad &&\text{(bent)} \\ &(b_2)^2(a_2)^1(b_1)^1:\ {}^1B_2 \quad &&\text{(linear)} \\ &(a_2)^2(b_2)^1(a_1)^1:\ {}^1B_2 \quad &&\text{(bent)} \\ &(a_2)^2(b_2)^1(b_1)^1:\ {}^1A_2 \quad &&\text{(linear)} \end{aligned} \tag{6-18}$$

Figure 6-3 shows the correlation of the excited states in the bent molecule to the states in the linear CO_2. From the single orbitals we see that on bending we have $^1\Sigma_u^- \rightarrow {}^1A_2$, etc.

The absorption spectrum of CO_2 has three bands, as shown in Figure 6-7. The absorption at 75,000 cm^{-1} is still interpreted as the "allowed" transition $^1\Sigma_g^+ \rightarrow {}^1\Sigma_u^+({}^1B_2)$. The molecule is almost

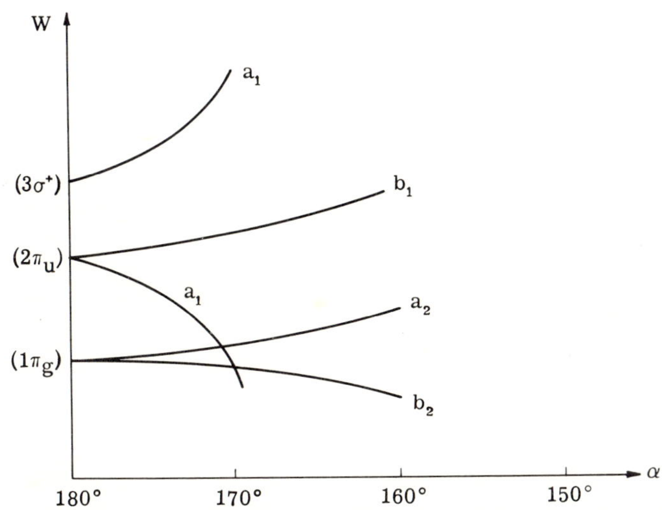

Figure 6-6 Variation of the molecular-orbital energy levels in CO_2 as a function of α, the angle O—C—O.

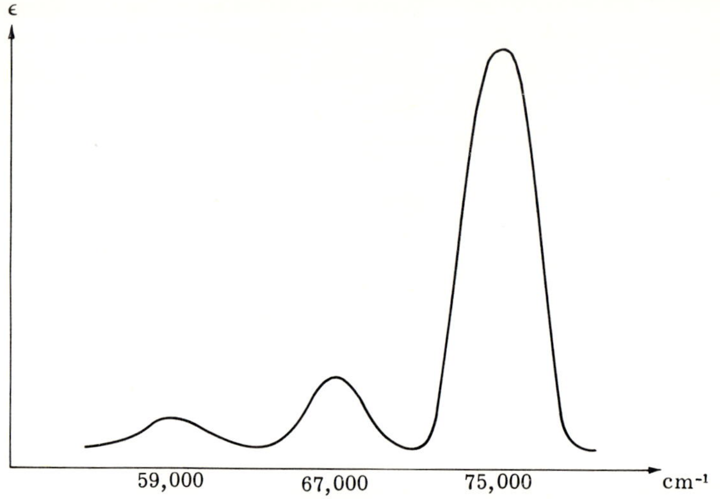

Figure 6-7 The absorption spectrum of the CO_2 molecule.

linear in the $^1\Sigma_u^+$ excited state, as predicted. The weaker bands at 59,000 cm^{-1} and 67,000 cm^{-1} are assigned, respectively, $^1\Sigma_g^+ \rightarrow {}^1A_2$ $(^1\Delta_u)$ and $^1\Sigma_g^+ \rightarrow {}^1B_2\,(^1\Delta_u)$. These two excited states are associated with a bent molecular configuration.

6-2 THE H_2O MOLECULE

As the second example of a triatomic molecule, we shall consider the bonding in the water molecule. Water in its ground state has an angular configuration in which the H—O—H angle is about 105°. This angle is often explained as the angle of 90° between two 2p oxygen orbitals, expanded a little due to the mutual repulsion of the two protons (Figure 6-8). One ignores in this description the two "lone pairs" of electrons on oxygen $(2s)^2(2p_z)^2$. However, we must remember that the "lone pairs" of electrons repel each other strongly, and calculations have shown that in H_2O this is a very important effect. The eight valence electrons in the H_2O molecule will naturally try to stay as far apart from each other as possible. This is accomplished by hybridizing the 2s and the three 2p orbitals of the oxygen to four equivalent orbitals, which are directed toward the corners of a regular tetrahedron (Figure 6-9).

The ion O^{2-} will have its four lone pairs distributed in the four

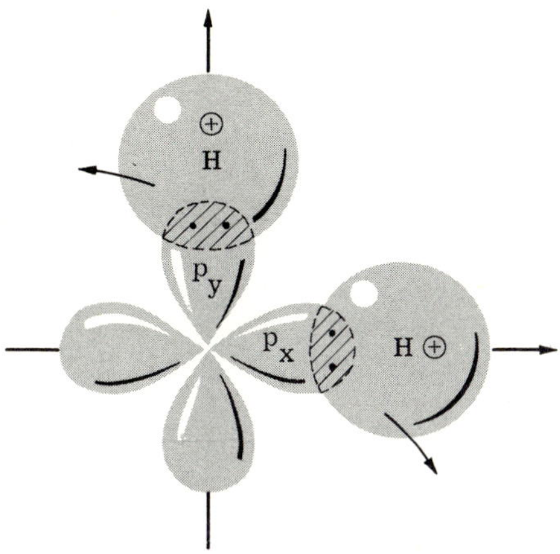

Figure 6-8 Naïve picture of H_2O's ground state.

tetrahedral orbitals. Two protons use two of these orbitals, which destroys a little of the hybridization. The two H^+'s draw electron density away from the oxygen. Thus, the two orbitals which the protons share have a little more 2p "character." This is the localized description of the two O—H bonds, with the two lone pairs in the remaining two orbitals.

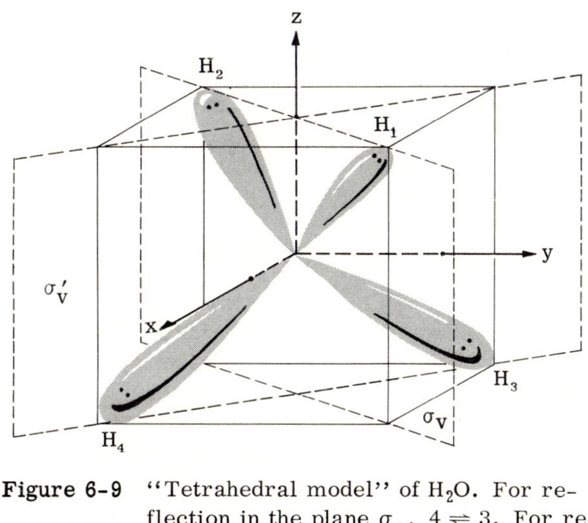

Figure 6-9 "Tetrahedral model" of H_2O. For reflection in the plane σ_v, $4 \rightleftharpoons 3$. For reflection in σ_v', $1 \rightleftharpoons 2$.

Ideally one can, therefore, consider the H—O—H angle as derived from the tetrahedral angle of 109° 28′. This angle is smaller in the H_2O molecule (about 105°), owing to the expansion of the electron cloud by the two protons. This idea is consistent with the fact that the F—O—F angle in OF_2 is only 102°, since we expect F^+ to draw much more electron density than H^+. The result is that the pure sp^3 hybridization is disturbed even more. The 2p character in the F—O bond is larger and the angle is closer to 90°.

Let us now construct four equivalent orbitals directed out to the corners of a cube, as in Figure 6-9. As our basis functions we have 2s, $2p_x$, $2p_y$, and $2p_z$. An orbital directed toward 1 in the first octant must be made up of equivalent amounts of $2p_x$, $2p_y$, and $2p_z$, and with 2s distributed equally among the four hybrids we have

$$\psi_1 = \tfrac{1}{2}(2s) + \tfrac{1}{2}(2p_x) + \tfrac{1}{2}(2p_y) + \tfrac{1}{2}(2p_z) \tag{6-19}$$

Similarly, for ψ_2, ψ_3, and ψ_4,

$$\psi_2 = \tfrac{1}{2}(2s) - \tfrac{1}{2}(2p_x) - \tfrac{1}{2}(2p_y) + \tfrac{1}{2}(2p_z) \tag{6-20}$$

$$\psi_3 = \tfrac{1}{2}(2s) - \tfrac{1}{2}(2p_x) + \tfrac{1}{2}(2p_y) - \tfrac{1}{2}(2p_z) \tag{6-21}$$

$$\psi_4 = \tfrac{1}{2}(2s) + \tfrac{1}{2}(2p_x) - \tfrac{1}{2}(2p_y) - \tfrac{1}{2}(2p_z) \tag{6-22}$$

Notice that the four hybrid orbitals all are orthogonal to each other. They are also normalized to 1; the coefficients of $\frac{1}{2}$ are needed for that purpose. The four hybrid orbitals are called sp^3 hybrids.

Using two of the sp^3 hybrids to form bonds with the two hydrogen atoms, e.g., ψ_1 and ψ_2, we have a set of LCAO—MO's,

$$\varphi_1 = \alpha\psi_1 + \beta\psi_1\,\sigma(H) \tag{6-23}$$

$$\varphi_2 = \alpha\psi_2 + \beta\psi_2\,\sigma(H) \tag{6-24}$$

The energies of the molecular orbitals will be, in order of increasing energy,

$$W^b(\varphi_1) = W^b(\varphi_2) < W(\psi_3) = W(\psi_4) < W^*(\varphi_1)$$
$$= W^*(\varphi_2) \tag{6-25}$$

With six electrons from oxygen and two electrons from the hydrogens, the electronic ground state of water in this formulation is

$${}^1A_1:\ (\varphi_1^b)^2(\varphi_2^b)^2(\psi_3)^2(\psi_4)^2$$

It is clear that all the bonding orbitals and all the nonbonding orbitals are occupied.

This description is then a *localized* description, because we consider one O—H bond as completely independent of the other one. Let us now consider a *delocalized* description of the bonding in water. First we investigate the transformation properties of the various atomic orbitals, because we want to form as general a set of molecular orbitals as possible.

The symmetry operations of the water molecule are collected in Table 6-3, together with the transformation properties of the sundry orbitals classified under C_{2v}, the point group of water (see Figure 6-9).

The most general molecular orbitals which can be constructed are thus of the type

$$\psi_1 = \lambda_1 (s) + \lambda_2 \left(p_z\right) + \lambda_3 (\sigma_1 + \sigma_2) \tag{6-26}$$

$$\psi_2 = \lambda_4 \left(p_x + p_y\right) + \lambda_5 (\sigma_1 - \sigma_2) \tag{6-27}$$

$$\psi_3 = \left(p_x - p_y\right) \tag{6-28}$$

The bonding scheme is set out in Figure 6-10.

The ground state is ${}^1A_1\,(a_1^b)^2\,(b_1^b)^2\,(a_1)^2\,(b_2)^2$. Notice that the 3×3 determinantal equation for the a_1 orbital gives three roots: one strongly bonding, one nearly nonbonding, and one strongly antibonding.

By making the linear combinations $\psi_1 \pm \psi_2$ and putting $\lambda_1 = \lambda_2 = \lambda_4 = \alpha$ and $\lambda_3 = \lambda_5 = \frac{1}{2}\beta$, we obtain

$$\varphi_1 = \psi_1 + \psi_2 = \alpha\left((s) + \left(p_x\right) + \left(p_y\right) + \left(p_z\right)\right) + \beta\sigma_1 \tag{6-29}$$

$$\varphi_2 = \psi_1 - \psi_2 = \alpha\left((s) - \left(p_x\right) - \left(p_y\right) + \left(p_z\right)\right) + \beta\sigma_2 \tag{6-30}$$

These new combinations are just the localized "hybrid" orbitals for

Table 6-3 Symmetry Orbitals for H_2O

	E	$C_2(z)$	σ_v	σ_v'	Transformation properties of the orbitals
A_1	1	1	1	1	s, p_z, $(\sigma_1 + \sigma_2)$
A_2	1	1	−1	−1	
B_1	1	−1	1	−1	$(p_x + p_y)$, $(\sigma_1 - \sigma_2)$
B_2	1	−1	−1	1	$(p_x - p_y)$

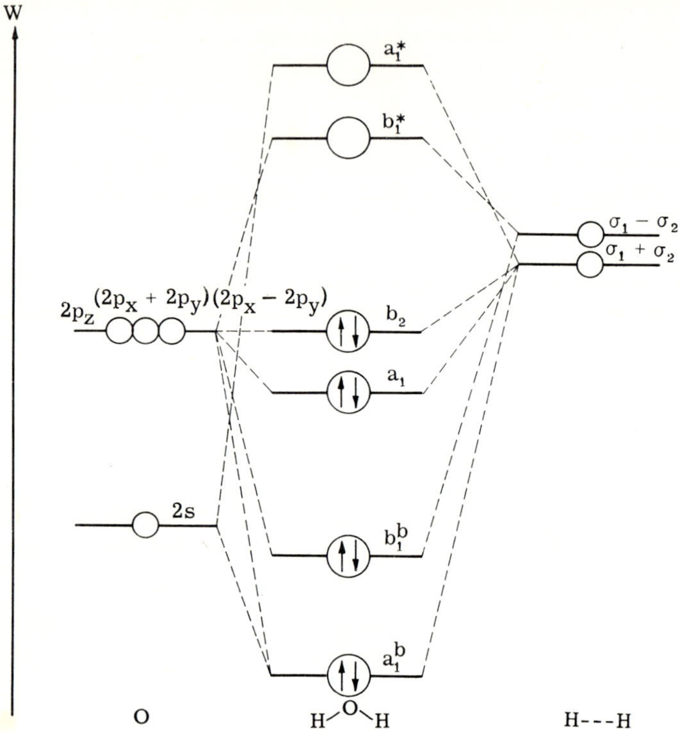

Figure 6-10 Molecular-orbital description of the bonding in water.

the water molecule. The molecular-orbital method gives the most general description of the bonding in the water molecule, and we see that in this description the bonding between the H's and the O is completely delocalized. To convert to the localized description we place restrictions on some of the variational parameters. This means we have a less flexible description. However, we have an added advantage, since we can use our chemical intuition to pinpoint the location of the "bonds" in a molecule.

6-3 NO_2

Nitrogen dioxide has 17 valence electrons. Since this is an odd number, there must be at least 1 unpaired electron. In the ground state the O—N—O angle is approximately 132°. The orbitals of the

Table 6-4 Character Table for C_{2v} Symmetry

	E	$C_2(z)$	$\sigma_v(xz)$	$\sigma_v(yz)$
A_1	1	1	1	1
A_2	1	1	−1	−1
B_1	1	−1	1	−1
B_2	1	−1	−1	1

Table 6-5 Transformation Properties of NO_2 Orbitals

	E	$C_2(z)$	$\sigma_v(xz)$	$\sigma_v(yz)$	Irreducible representation in C_{2v}
N(s)	1	1	1	1	a_1
N(p_x)	1	−1	1	−1	b_1
N(p_y)	1	−1	−1	1	b_2
N(p_z)	1	1	1	1	a_1
O($s_1 + s_2$)	1	1	1	1	a_1
O($s_1 - s_2$)	1	−1	1	−1	b_1
O($p_{x_1} + p_{x_2}$)	1	−1	1	−1	b_1
O($p_{x_1} - p_{x_2}$)	1	1	1	1	a_1
O($p_{y_1} + p_{y_2}$)	1	−1	−1	1	b_2
O($p_{y_1} - p_{y_2}$)	1	1	−1	−1	a_2
O($p_{z_1} + p_{z_2}$)	1	1	1	1	a_1
O($p_{z_1} - p_{z_2}$)	1	−1	1	−1	b_1

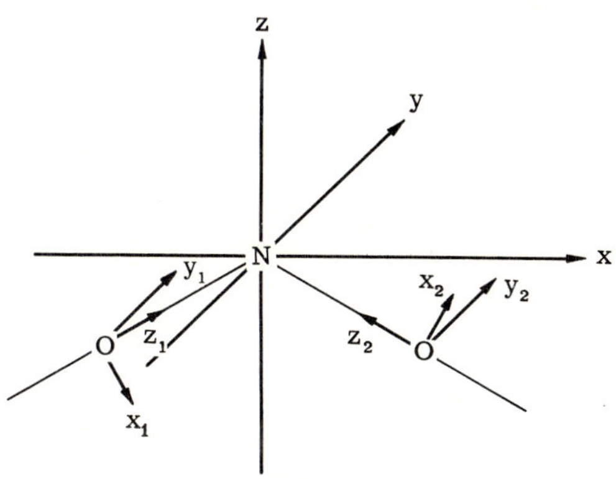

Figure 6-11 The NO_2 molecule is situated in the xz plane. The symmetry operations are E, $C_2(z)$, $\sigma_v(xz)$, and $\sigma_v(yz)$.

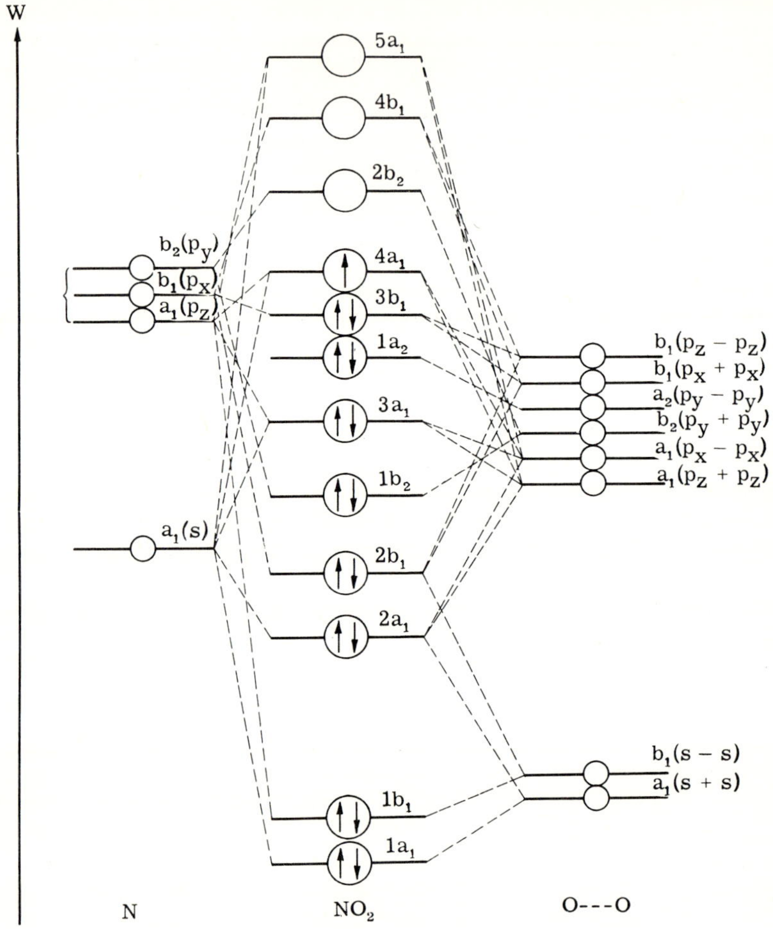

Figure 6-12 Energy-level scheme for NO_2. Some of the dashed lines are omitted in order to simplify the diagram.

molecule must be characterized according to the symmetry operations in the C_{2v} point group. For practical reasons we orient the NO_2 molecule a little differently in the coordinate system than the angular CO_2 molecule. The symmetry operations change accordingly, but are, of course, equivalent in both cases (see Figure 6-11 and Table 6-4).

We now examine the transformation properties of the orbitals on N and on the two O atomic nuclei (see Table 6-5). The energy-level scheme for NO_2 is shown in Figure 6-12. The ground-state configuration is $(1a_1)^2(1b_1)^2(2a_1)^2(2b_1)^2(1b_2)^2(3a_1)^2(1a_2)^2(3b_1)^2(4a_1)^1$, which gives $^2A_1$. The first excited state is $\cdots(1a_2)^2(3b_1)^1(4a_1)^2$, with the total symmetry $^2B_1$. The absorption band in NO_2, which has a maximum at approximately 4400 A (23,000 cm^{-1}), may be tentatively assigned to the $^2A_1 \rightarrow {}^2B_1$ transition.

6-4 O_3 AND SO_2

The O_3 and the SO_2 molecules have 18 valence electrons. They both have angular molecular configurations, with O—O—O and O—S—O angles of about 120°. The energy-level diagram for NO_2 can be used to describe the electronic structures of O_3 and SO_2. There is one more electron to place in the diagram in Figure 6-12. The ground state is then $\cdots(1a_2)^2(3b_1)^2(4a_1)^2;\ {}^1A_1$.

For SO_2 the lowest transitions are

$$\cdots(1a_2)^2(3b_1)^2(4a_1)^2;\ {}^1A_1 \rightarrow (1a_2)^2(3b_1)^2(4a_1)^1(2b_2)^1;\ {}^1B_2$$

$$\cdots(1a_2)^2(3b_1)^2(4a_1)^2;\ {}^1A_1 \rightarrow (1a_2)^1(3b_1)^2(4a_1)^2(2b_2)^1;\ {}^1B_1$$

and

$$\cdots(1a_2)^2(3b_1)^2(4a_1)^2;\ {}^1A_1 \rightarrow (1a_2)^2(3b_1)^1(4a_1)^2(2b_2)^1;\ {}^1A_2 \qquad (6\text{-}31)$$

Since the components (X,Y,Z) of the dipole vector transform as (B_1, B_2, A_1), the transition $^1A_1 \rightarrow {}^1B_2$ is orbitally allowed and polarized at right angles to the plane of the molecule. The $^1A_1 \rightarrow {}^1B_1$ transition is polarized parallel to the plane of the molecule ($A_1 \times B_1 \times B_1 = A_1$), while the transition $^1A_1 \rightarrow {}^1A_2$ is *not* allowed as an electrical-dipole transition ($A_1 \times A_2 = A_2 \neq B_1$, B_2, or A_1).† The spectrum of SO_2 is shown schematically in Figure 6-13.

Experiments show that the band at about 27,000 cm^{-1} is perpendicularly polarized and this transition is assigned $^1A_1 \rightarrow {}^1B_2$. The

---

†Recall that transitions M → N are allowed only if the integral $\int$MRN $d\tau$, in which **R** is the dipole vector, is different from zero. The integral is nonzero provided that the product of the functions M, N, and **R** is invariant to all symmetry operations in the molecule under consideration. With $M = A_1$, the trace of M is +1 for all the symmetry operations. By examining the character table for C_{2v} symmetry, we see that the transition is orbitally allowed only if the electronic excited state transforms in the same way as one of the components of the dipole vector.

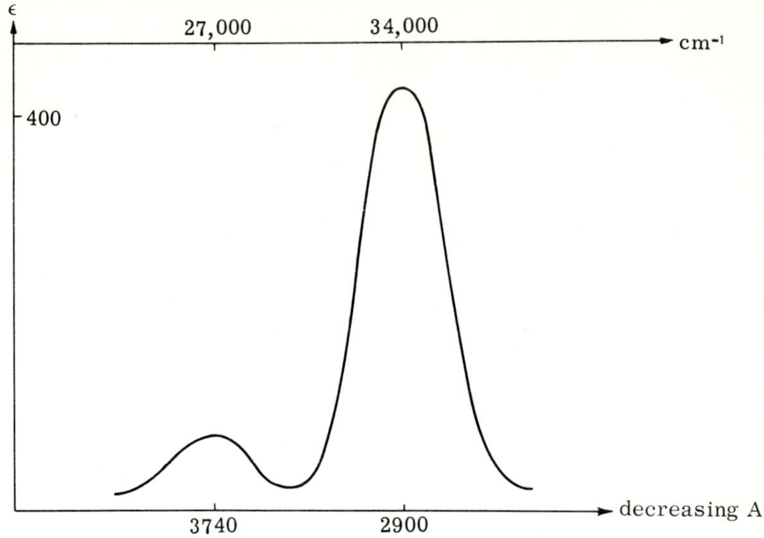

Figure 6-13 The absorption spectrum of the SO_2 molecule.

transition $^1A_1 \rightarrow {}^1B_1$ is found at 34,000 cm^{-1}. The molecular geometry of the excited state is very probably quite different from the geometry in the ground state.

The absorption spectrum of O_3 should be similar to the spectrum of SO_2. The weak band in O_3 at 6000 A (= 16,600 cm^{-1}) with $\epsilon_{max} \approx 1$, can be assigned as the orbitally forbidden $^1A_1 \rightarrow {}^1A_2$ transition.

7 / Selected Molecules with Four or More Atoms

7-1 H_2O_2

The first tetratomic molecule we shall treat is H_2O_2. The structure of H_2O_2, from electron-diffraction measurements, is "helical," as shown in Figure 7-1.

For the O—O bonding, we form $(2p_z + 2s)$ and $(2p_z - 2s)$ hybrids on the two oxygen atomic nuclei (Figure 7-2).

The bonding in the ground state of O_2^{2-} is indicated by the diagram

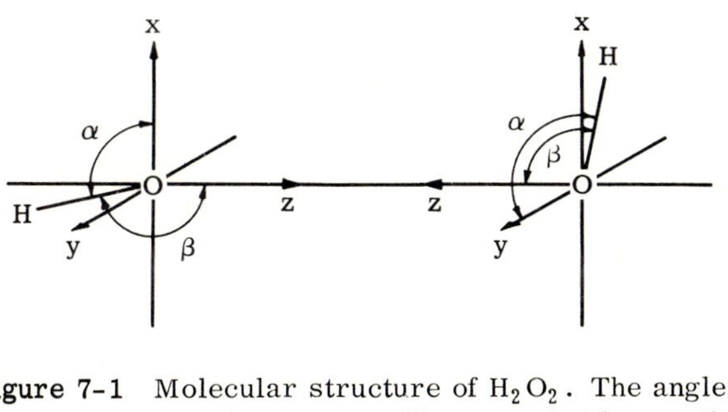

Figure 7-1 Molecular structure of H_2O_2. The angles α and β are actually somewhat larger than 90°.

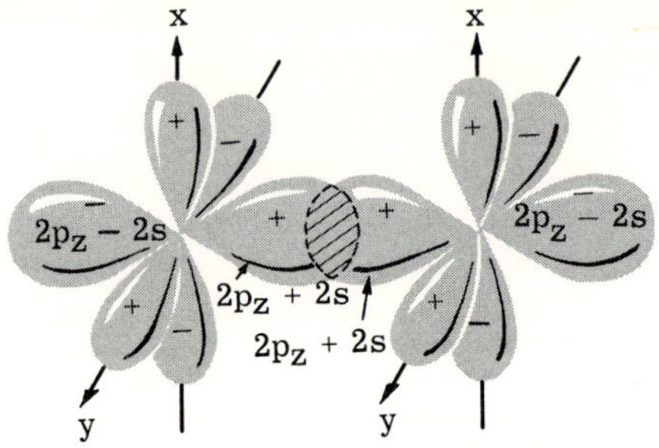

Figure 7-2 Orbitals in O_2^{2-}.

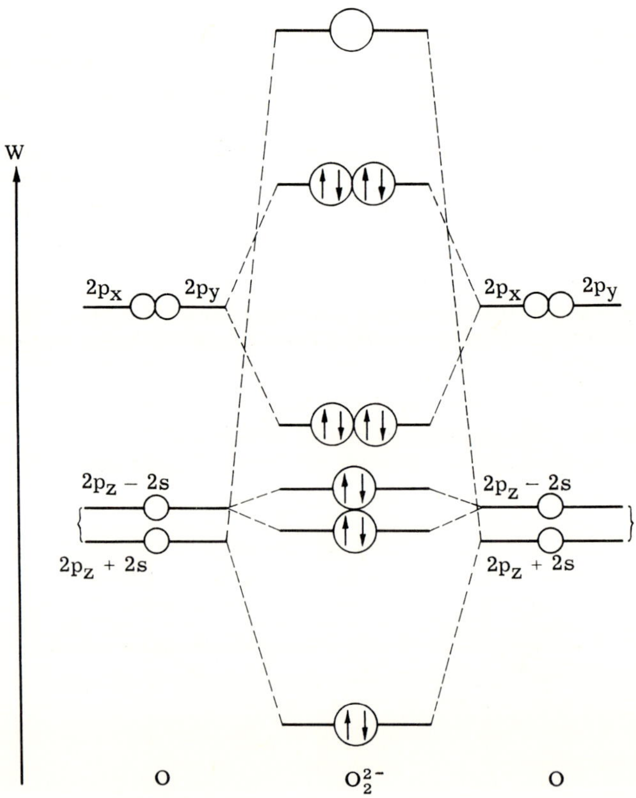

Figure 7-3 A bonding scheme for O_2^{2-}.

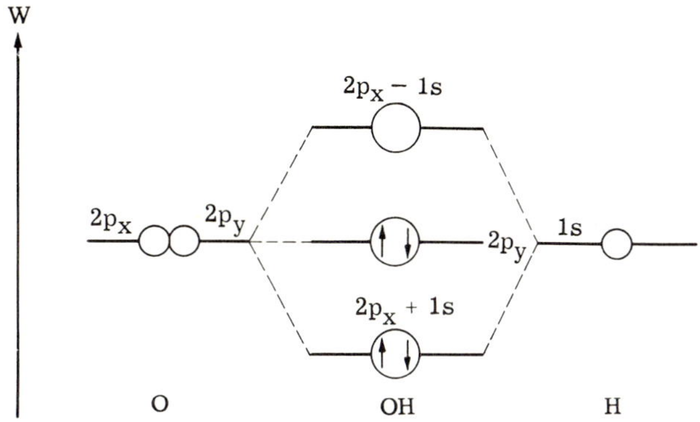

Figure 7-4 Bonding between O($2p_x$) and 1s(H) in H_2O_2.

given in Figure 7-3. Note that we have 14 valence electrons in the energy levels. With 14 electrons the bonding and the antibonding π orbitals are filled, and thus the O_2^{2-} ion is not as stable as O_2. The 4 electrons in ($2p_z$ − 2s) orbitals are the "lone pairs."

If we bond a proton to one of the O's using a $2p_x$ orbital, there is a resulting gain in stability. This is shown by the energy-level diagram in Figure 7-4. The second proton can be bonded to the other oxygen by using either the $2p_x$ or the $2p_y$ orbital. If the $2p_x$ orbital is used, the two $2p_y$ orbitals on the O's are able to combine to give bonding and antibonding π orbitals (Figure 7-5). This is not favorable because both the bonding and the antibonding $\pi(p_y)$ orbitals are fully occupied.

However, by using the $2p_x$ orbital on one of the oxygens and the

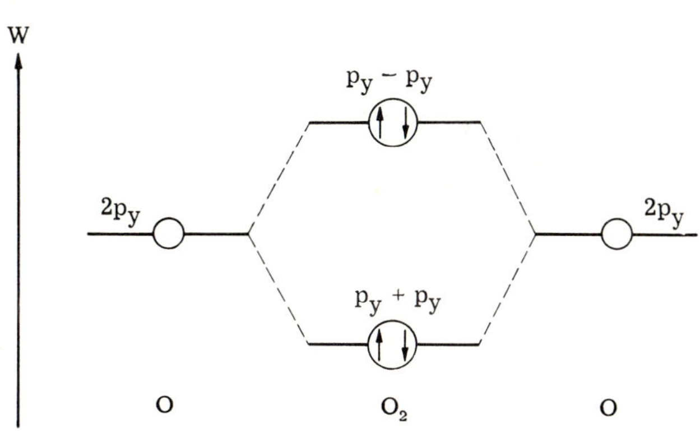

Figure 7-5 Hypothetical π bonding between O(p_y) − O(p_y).

$2p_y$ orbital on the other oxygen for the O—H bonds, the $2p_{y_1}$ and $2p_{x_2}$ orbitals that remain are not able to interact. These occupied orbitals represent two new "lone pairs" of electrons.

This type of argument allows us to rationalize the "staggered" molecular structure exhibited by H_2O_2 (shown in Figure 7-1). The above argument is a fancy way of saying that the lone pairs repel each other strongly, and thus prefer to occupy orbitals that are perpendicular to each other. This simple view is consistent with the observed structure. We could also say that a "slight" sp^3 hybridization (as in H_2O) exists, with the H—O—O angle smaller than the tetrahedral angle, owing to the difference in the bonding and nonbonding electron pairs. This is, of course, in agreement with the experimental structure determinations.

7-2 FORMALDEHYDE, H_2CO

The formaldehyde molecule is planar in its ground state (Figure 7-6). We first construct three strong σ bonds involving the carbon, the two hydrogen, and the oxygen atoms. Since the angles in the plane all are approximately 120°, we construct three equivalent orbitals in the xy plane which are directed from carbon toward H_1, H_2, and O. For this purpose we hybridize the three carbon atomic orbitals 2s, $2p_x$, and $2p_y$. From three orbitals we can construct three linearly independent hybrid orbitals.

Along the x axis we have an orbital $\psi_0 = \alpha(p_x) + \beta(s)$, in which α and β are the "mixing coefficients" in the hybridization. Since we want to construct three equivalent orbitals, each one must have $\frac{1}{3}$ of the C(2s) orbital.

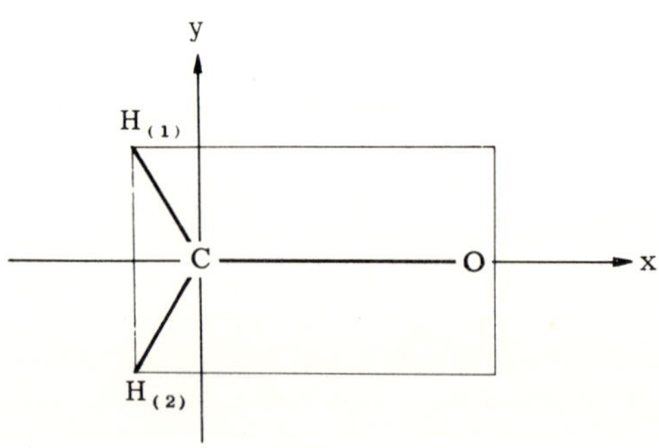

Figure 7-6 The geometry of formaldehyde in the ground state.

Recall that $2p_x$, $2p_y$, $2p_z$, and 2s all are orthogonal to each other. Thus, the condition for the normalization of ψ_0 is

$$\alpha^2 + \beta^2 = 1 \tag{7-1}$$

with $\beta^2 = \frac{1}{3}$, we have $\alpha^2 = \frac{2}{3}$. Finally,

$$\psi_0 = \sqrt{\frac{2}{3}}\,(p_x) + \sqrt{\frac{1}{3}}\,(s) \tag{7-2}$$

To construct an orbital equivalent to ψ_0, we rotate ψ_0 by 120°. Since

$$\hat{C}_3\begin{pmatrix} x \\ y \end{pmatrix} = \begin{pmatrix} \cos 120^\circ & -\sin 120^\circ \\ \sin 120^\circ & \cos 120^\circ \end{pmatrix}\begin{pmatrix} x \\ y \end{pmatrix} \tag{7-3}$$

and p_x transforms as x, we have for the orbitals directed toward $H_{(1)}$

$$\psi_{H_{(1)}} = \hat{C}_3\psi_0 = \sqrt{\frac{2}{3}}\left(-\frac{1}{2}(p_x) - \frac{\sqrt{3}}{2}(p_y)\right) + \sqrt{\frac{1}{3}}\,(s) \tag{7-4}$$

A further rotation of $\psi_{H_{(1)}}$ by 120° gives

$$\psi_{H_{(2)}} = \sqrt{\frac{2}{3}}\left(-\frac{1}{2}(p_x) + \frac{\sqrt{3}}{2}(p_y)\right) + \sqrt{\frac{1}{3}}\,(s) \tag{7-5}$$

which goes back to ψ_0 on a final rotation of 120°. These three orbitals are called the sp^2, or "trigonal," hybrids of carbon. We notice that $C(2p_z)$ has not been used, and that the boundary surfaces of the three hybrids actually form 120° angles with each other.

On the oxygen we hybridize $2p_x$ and 2s as usual (see Figure 7-6) and form $2p_x - 2s$ and $2p_x + 2s$. The σ electronic structure for formaldehyde is shown in Figure 7-7, with a "lone pair" on O directed away from the carbon.

All together we have, ignoring $C(1s)^2$ and $O(1s)^2$ electrons, $2 \times 1(H) + 4(C) + 6(O) = 12$ valence electrons, of which we have accounted for 8. Still available are $C(p_z)$ and $O(2p_y)(2p_z)$. The π-type orbitals situated on C and O are shown in Figure 7-8. We see that $C(p_z)$ and $O(p_z)$ form π molecular orbitals. The $2p_y$ orbital on oxygen is nonbonding. Assuming that the σ bonding orbitals are very stable and therefore that the σ antibonding orbitals have very high energy, we obtain a schematic energy diagram as given in Figure 7-9.

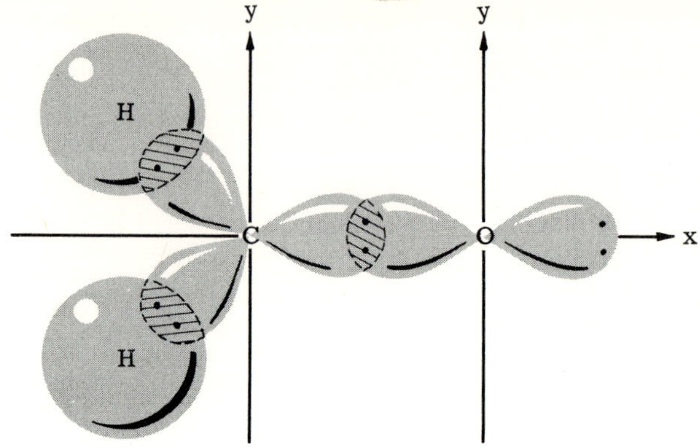

Figure 7-7 The σ electronic structure of formaldehyde.

Formaldehyde has symmetry operations which place it in the point group C_{2v}. The character table for C_{2v} was given in Table 6-1. Since $O(p_y)$ transforms as b_2 and $O(p_z)$ as b_1, the ground state is $\ldots (b_1^b)^2 (b_2)^2$; $^1A_1$.

The lowest electronic excited state occurs on exciting one of the electrons in the b_2 nonbonding orbital to the antibonding π orbital (b_1^*). The electronic configuration is then $\ldots (b_1^b)^2 (b_2)^1 (b_1^*)^1$. Since the two different orbitals (b_2) and (b_1^*) each contain an electron, the excited electronic configuration gives states with $S = 0$ and $S = 1$. Because the product of two functions transforming as b_2 and b_1 transforms as a_2 (see Table 6-1), we obtain $^3A_2$ and $^1A_2$ excited states (Figure 7-10). Hund's rule tells us that the triplet state $^3A_2$ has lower energy than the $^1A_2$ state.

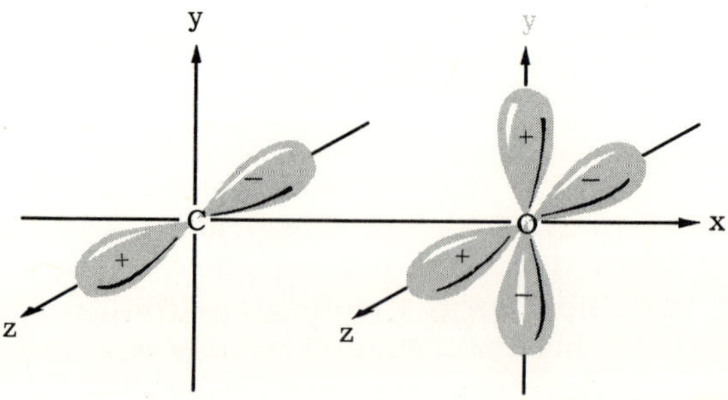

Figure 7-8 π orbitals in formaldehyde.

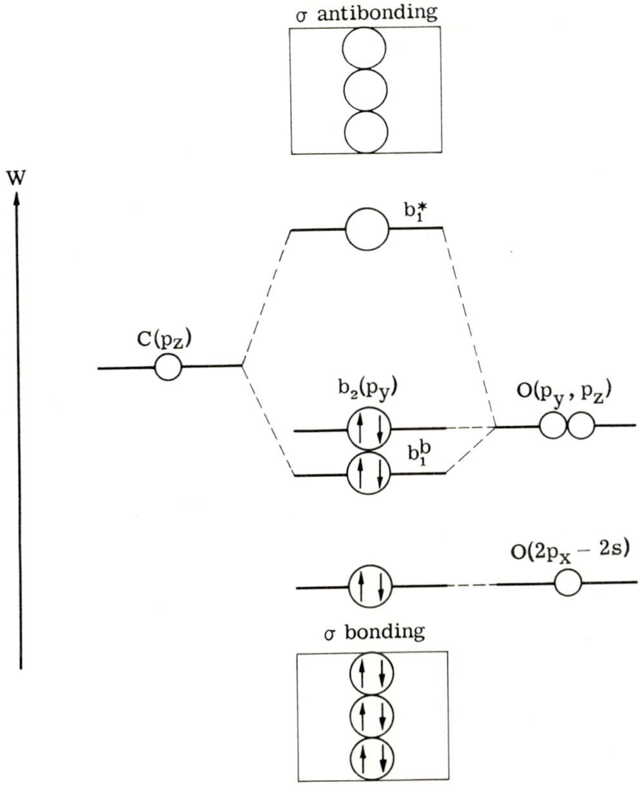

Figure 7-9 Molecular orbitals in H_2CO.

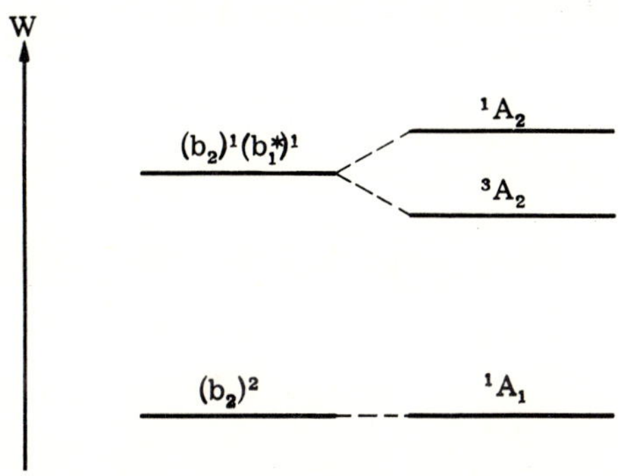

Figure 7-10 The lowest terms for formaldehyde.

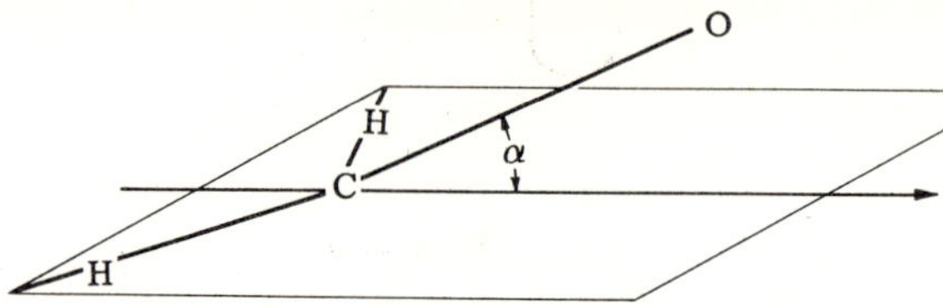

Figure 7-11 The molecular structure of formaldehyde in the excited state $^1A_2$. The barrier for the "umbrella inversion" of O is $\approx 650\ \text{cm}^{-1}$.

Because the components (X, Y, Z) of the dipole vector in C_{2V} transform as (A_1, B_2, B_1) the transition $^1A_1 \rightarrow {}^1A_2$ is orbitally forbidden. The transition $^1A_1 \rightarrow {}^3A_2$ is both orbitally forbidden and spin-forbidden. Although the transitions are theoretically forbidden, they are both observed but with small intensities. The $^1A_1 \rightarrow {}^3A_2$ transition is found between 4000 and 3000 A, and $^1A_1 \rightarrow {}^1A_2$ is found between 3700 and 2300 A. We shall not discuss the mechanisms that make these transitions slightly allowed. The $^1A_1 \rightarrow {}^1A_2$ transition, with a maximum around 3000 A, is a trademark of the carbonyl group.

Both of these transitions transfer a nonbonding oxygen electron into the π antibonding C=O orbital. Both transitions should, therefore, be accompanied by an increase in the C—O bond length. This is verified experimentally. In the excited state the C—O distance is ~1.31 A, but it is only 1.22 A in the ground state. In addition, in the excited states the oxygen atomic nucleus is no longer in the plane formed by H╲C╱H. The out-of-plane bending is about 20° (see Fig. 7-11).

7-3 THE BORON HYDRIDE B_2H_6

The last molecule we shall consider in this chapter is diborane, B_2H_6. Electron diffraction indicates the molecular structure of B_2H_6 as shown in Figure 7-12. H_α, H_β, H_γ, and H_δ lie in the yz plane, while B_2H_2 lies in the xz plane.

If we ignore the $B(1s)^2$ electrons we have $2 \times 3(B) + 6(H) = 12$ valence electrons. We assume that the B 2s and 2p orbitals are sp^3-hybridized. Thus 8 electrons can be placed in four σ bonds from the two borons to H_α, H_β, H_γ, and H_δ. This means that we have 4 electrons left to dispose of in the six orbitals $(\chi_1 \cdots \chi_6)$. Clearly, diborane is an "electron-deficient" molecule. There are more valence orbitals than there are electrons.

We construct the following orbitals in an attempt to explain the

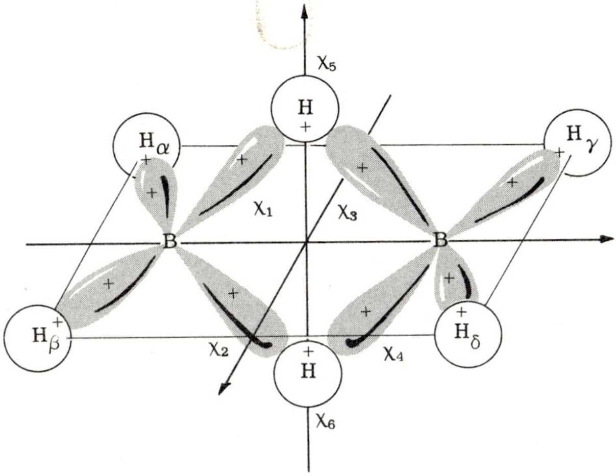

Figure 7-12 The structure of diborane.

bonding in the B_2H_2 part of diborane, in which the two hydrogens bridge the two borons:

$$\frac{1}{2}(\chi_1 + \chi_2 + \chi_3 + \chi_4) = \psi_1 \tag{7-6}$$

$$\frac{1}{2}(\chi_1 - \chi_2 - \chi_3 + \chi_4) = \psi_2 \tag{7-7}$$

$$\frac{1}{2}(\chi_1 + \chi_2 - \chi_3 - \chi_4) = \psi_3 \tag{7-8}$$

$$\frac{1}{2}(\chi_1 - \chi_2 + \chi_3 - \chi_4) = \psi_4 \tag{7-9}$$

$$\frac{1}{\sqrt{2}}(\chi_5 + \chi_6) = \psi_5 \tag{7-10}$$

$$\frac{1}{\sqrt{2}}(\chi_5 - \chi_6) = \psi_6 \tag{7-11}$$

The symmetry operations of the molecule place it in the point group D_{2h}. The D_{2h} character table is given in Table 7-1.

By examining how $\psi_1 \cdots \psi_6$ transform, we find $\psi_1(a_g)$, $\psi_2(B_{2g})$, $\psi_3(b_{1u})$, $\psi_4(b_{3u})$, $\psi_5(a_g)$, and $\psi_6(b_{3u})$. Notice that ψ_1 and ψ_5, and also ψ_4 and ψ_6, can be combined. Thus we obtain the bonding scheme shown in

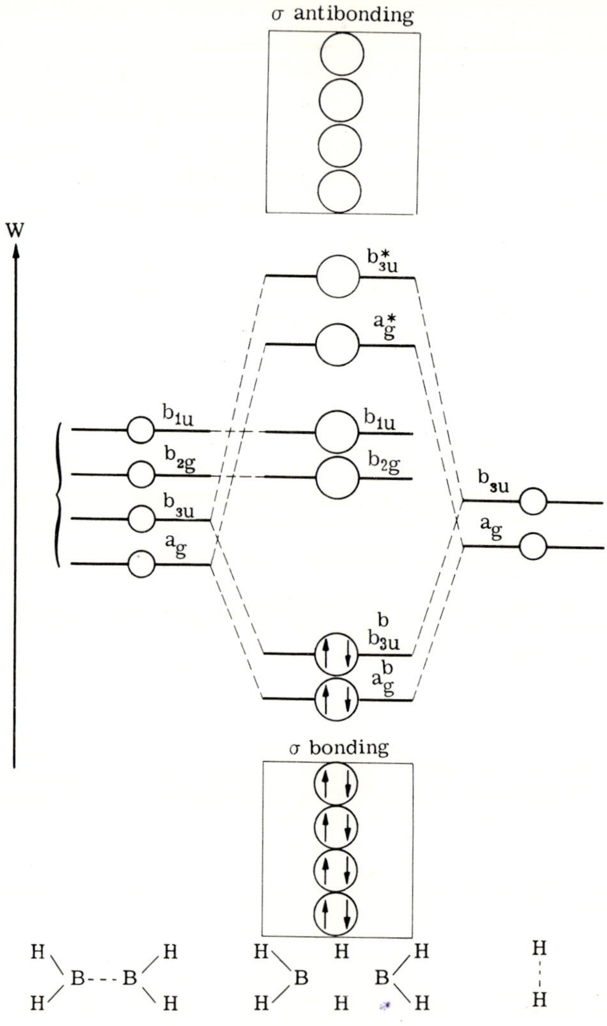

Figure 7-13 Bonding scheme for B_2H_6.

Figure 7-13. The four electrons completely fill the a_g and b_{3u} bonding orbitals. Thus the ground state is ^1A_g.

The exercises we have gone through for a few simple molecules can be carried out no matter how complicated the molecular geometry is. The only condition is that we know this geometry. The placing of *single-electron* energy levels is best accomplished by doing a thorough calculation, but the qualitative level schemes

Table 7-1 Character Table for D_{2h}

	E	$C_2(z)$	$C_2(y)$	$C_2(x)$	i	$\sigma_v(xy)$	$\sigma_v(xz)$	$\sigma_v(yz)$
A_g	1	1	1	1	1	1	1	1
B_{1g}	1	1	−1	−1	1	1	−1	−1
B_{2g}	1	−1	1	−1	1	−1	1	−1
B_{3g}	1	−1	−1	1	1	−1	−1	1
A_u	1	1	1	1	−1	−1	−1	−1
B_{1u}	1	1	−1	−1	−1	−1	1	1
B_{2u}	1	−1	1	−1	−1	1	−1	1
B_{3u}	1	−1	−1	1	−1	1	1	−1

arrived at by the use of symmetry properties are often very useful. The energies of *excited* states for molecules are very difficult to calculate by any theoretical method—therefore, we almost always rely on experiments for this information.

In the next chapter we shall discuss molecular orbital theory as applied to transition metal complexes. Since we shall be dealing with d orbitals, the power of the symmetry methods developed up to this point will be clearly shown.

8 / Molecular Orbitals Involving d Valence Orbitals

8-1 GENERAL CONSIDERATIONS

The transition metal ions possess a very stable set of d orbitals, and it is likely that d orbitals are involved in bonding in all transition metal complexes, regardless of structure. The common structures that use d valence orbitals for forming σ bonding molecular orbitals are square-planar, tetrahedral, and octahedral. Examples of these structures are given in Figure 8-1.

In addition to a set of nd valence orbitals, the transition metal ions have available $(n + 1)s$ and $(n + 1)p$ valence orbitals. Since the transition metal ion is centrally located in most complexes, the molecular orbitals are conveniently written in the form

$$\psi = N\left(\psi_M + \lambda\Phi_{lig}\right) \tag{8-1}$$

where ψ_M is the metal orbital, Φ_{lig} is a normalized combination of ligand orbitals (ligands are the groups attached to the metal in a complex), λ is the mixing coefficient, and N is a normalizing constant:

$$N^2(1 + \lambda^2 + 2\lambda G) = 1 \tag{8-2}$$

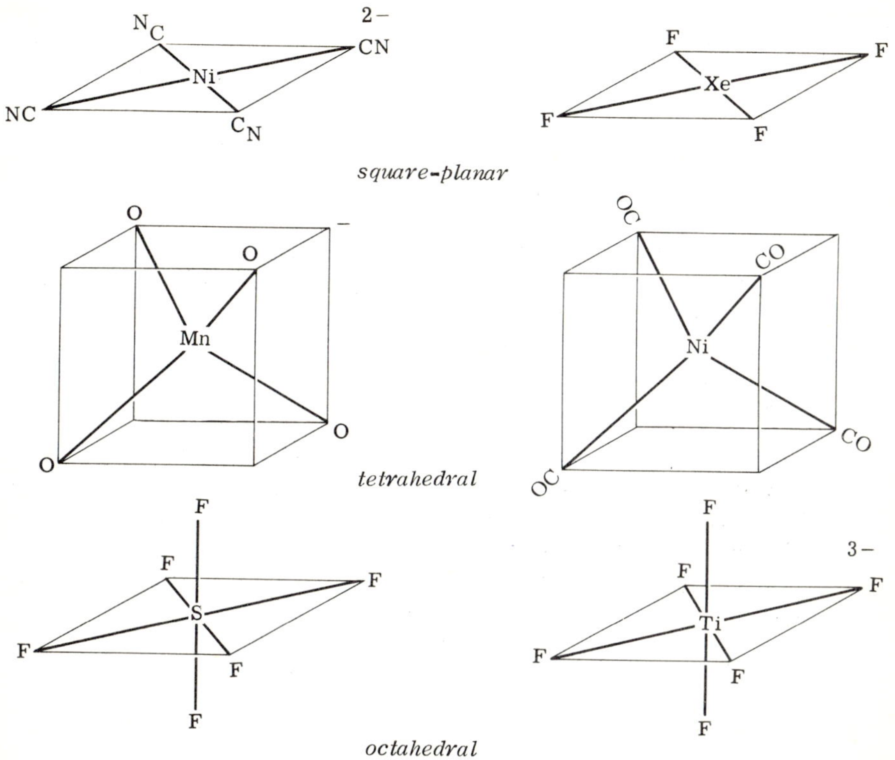

Figure 8-1 Structures that use d orbitals in bonding.

In Eq. (8-2) G is the total overlap of the metal orbital with the linear combination of ligand orbitals,

$$G = \int \psi_M \cdot \Phi_{lig} \, d\tau \qquad (8\text{-}3)$$

A quantity of some usefulness in the discussion of the electronic structures of complexes is the "fraction of electronic charge" found on the central ion. Consider, for instance, a complex ML_6^{Z+}, containing ligands which are neutral molecules. In an ionic model the charge distribution would be $M^{Z+}(L_6)^0$. It is, however, much more likely that the positive charge is "smeared out" over the complex, with the charge on the metal close to zero, and each ligand having a positive charge of approximately Z/6.

Examining the molecular orbital $\psi = N(\psi_M + \lambda\Phi_{lig})$, it is apparent that the fraction of electronic charge "belonging" to the metal atom is N^2. However, to keep the formal charges straight we must also give to the metal atom half of $2N^2\lambda G$, which is the "overlap population." Thus, this fraction for each singly-occupied orbital is given as

$$N^2(1 + \lambda G) = \frac{1 + \lambda G}{1 + 2\lambda G + \lambda^2} \tag{8-4}$$

We notice that for $\lambda = 1$, the fraction is $\frac{1}{2}$, as it should be for symmetry reasons; each center shares the electron equally. The fractional positive charge on the metal atom in a complex ML_n^{+Z} is thus

$$Z - \sum_i \frac{1 + \lambda_i G_i}{1 + 2\lambda_i G_i + \lambda_i^2} \tag{8-5}$$

where the summation is taken over all the valence electrons.

The importance of the concept of fractional charge is the fact that it enables us to make rough estimates of the values of certain molecular integrals. Consider an LCAO, $\psi = N(\psi_i + \lambda\Phi_j)$. In a variational treatment we have $H_{ii} = \int \psi_i \mathcal{H} \psi_i \, d\tau$, where $\mathcal{H}$ is the Hamiltonian of the molecule. It is now possible to estimate H_{ii} as

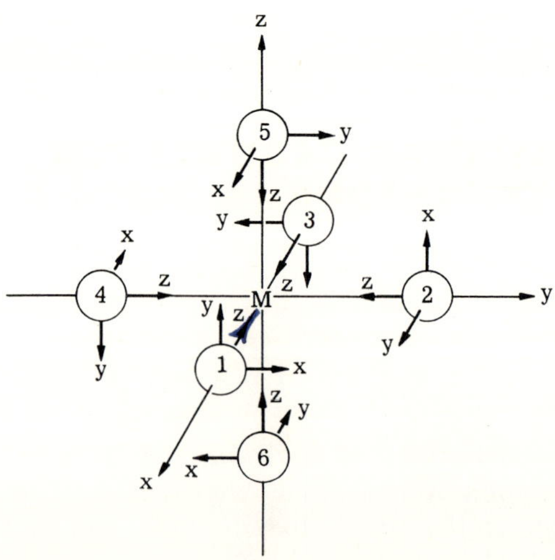

Figure 8-2 Coordinate system for an octahedral complex.

Table 8-1 Character Table for O_h

O_h	E	$8C_3$	$3C_2$	$6C_4$	$6C_2'$	i	$8S_6$	$3\sigma_h$	$6S_4$	$6\sigma_d$
A_{1g}	1	1	1	1	1	1	1	1	1	1
A_{2g}	1	1	1	−1	−1	1	1	1	−1	−1
E_g	2	−1	2	0	0	2	−1	2	0	0
T_{1g}	3	0	−1	1	−1	3	0	−1	1	−1
T_{2g}	3	0	−1	−1	1	3	0	−1	−1	1
A_{1u}	1	1	1	1	1	−1	−1	−1	−1	−1
A_{2u}	1	1	1	−1	−1	−1	−1	−1	1	1
E_u	2	−1	2	0	0	−2	1	−2	0	0
T_{1u}	3	0	−1	1	−1	−3	0	1	−1	1
T_{2u}	3	0	−1	−1	1	−3	0	1	1	−1

the ionization potential of an electron from an atom with a charge equal to the fractional charge on the center i.

Let us now consider in some detail the molecular orbitals for an octahedral complex containing a first-row transition metal ion. The orbitals that will be used in the bonding scheme are the 3d, 4s, and 4p orbitals of the central atom and the ns and np orbitals of the ligands. The coordinate system that is convenient for the construction of σ and π MO's is shown in Figure 8-2. The character table for the O_h symmetry is given in Table 8-1.

8-2 MOLECULAR ORBITALS FOR AN OCTAHEDRAL MOLECULE

We shall now find the transformation properties of the 3d, 4s, and 4p orbitals of the central atom in octahedral symmetry. The symmetry operations are given in Figure 8-3.

We have, by looking at one of the C_3 symmetry operations, that $\hat{C}_3 \begin{pmatrix} x \\ y \\ z \end{pmatrix} = \begin{pmatrix} y \\ z \\ x \end{pmatrix}$. Using this operation,

$$\hat{C}_3 (s) = (s) \tag{8-6}$$

$$\hat{C}_3 \begin{pmatrix} p_x \\ p_y \\ p_z \end{pmatrix} = \begin{pmatrix} 0 & 1 & 0 \\ 0 & 0 & 1 \\ 1 & 0 & 0 \end{pmatrix} \begin{pmatrix} p_x \\ p_y \\ p_z \end{pmatrix} \tag{8-7}$$

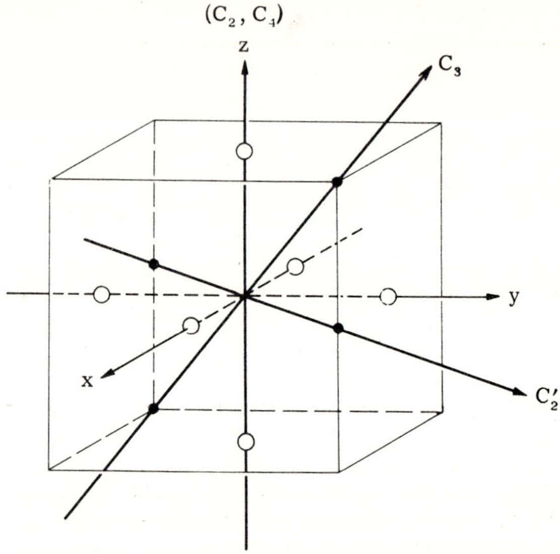

Figure 8-3 Symmetry operations in O_h.

and

$$\hat{C}_3 \begin{pmatrix} d_{xz} \\ d_{yz} \\ d_{xy} \\ d_{z^2} \\ d_{x^2-y^2} \end{pmatrix} = \begin{pmatrix} 0 & 0 & 1 & 0 & 0 \\ 1 & 0 & 0 & 0 & 0 \\ 0 & 1 & 0 & 0 & 0 \\ 0 & 0 & 0 & -\frac{1}{2} & \frac{\sqrt{3}}{2} \\ 0 & 0 & 0 & -\frac{\sqrt{3}}{2} & -\frac{1}{2} \end{pmatrix} \begin{pmatrix} d_{xz} \\ d_{yz} \\ d_{xy} \\ d_{z^2} \\ d_{x^2-y^2} \end{pmatrix} \tag{8-8}$$

It is evident that the sets (s), (p_x, p_y, p_z), $(d_{z^2}, d_{x^2-y^2})$, and

Table 8-2 Traces of Transformation Matrices

	E	C_3	C_2	C_4	C_2'	i	...
(s)	1	1	1	1	1	1	
(p_x, p_y, p_z)	3	0	−1	1	−1	−3	
$(d_{x^2-y^2}, d_{z^2})$	2	−1	2	0	0	2	
(d_{xy}, d_{xz}, d_{yz})	3	0	−1	−1	1	3	

(d_{xz}, d_{yz}, d_{xy}) are mixed together under this symmetry operation. By performing the other symmetry operations, we find that the traces of the transformation matrices are as given in Table 8-2. By comparison with the character table of O_h we see that (s) transforms as a_{1g}, (p_x, p_y, p_z) as t_{1u}, $(d_{x^2-y^2}, d_{z^2})$ as e_g, and (d_{xy}, d_{xz}, d_{yz}) as t_{2g}.

8-3 LIGAND-ORBITAL REPRESENTATIONS

In the next step we find the linear combinations of the ligand ns and np orbitals which can be used for bonding. The σ orbitals are (Figure 8-2) ns and np_z and the π orbitals are np_x and np_y. Since ns and np_z will transform in exactly the same way, we shall adopt a linear combination of the two for the σ valence orbital furnished by the ligands, i.e.,

$$\sigma_{(lig)} = \alpha\psi(s) + \sqrt{1 - \alpha^2}\,\psi(p_z) \tag{8-9}$$

First we construct six linearly independent molecular orbitals from the six ligand σ functions. The trace of the six σ functions under the symmetry operations of O_h is

E	C_3	C_2	C_4	C_2'	i	S_6	σ_h	S_4	σ_d
6	0	2	2	0	0	0	4	0	2

From the character table 8-1 we see that the six orbitals behave as the combination a_{1g}, t_{1u}, and e_g.

The totally symmetric $\sigma(a_{1g})$ orbital can be written by inspection:

$$\sigma(a_{1g}) = \frac{1}{\sqrt{6}}(\sigma_1 + \sigma_2 + \sigma_3 + \sigma_4 + \sigma_5 + \sigma_6) \tag{8-10}$$

This ligand combination is normalized, neglecting overlap between the six σ orbitals. Thus, we have for the complete a_{1g} molecular wave function,

$$\psi(a_{1g}) = c_1(4s) + c_2\frac{1}{\sqrt{6}}(\sigma_1 + \sigma_2 + \sigma_3 + \sigma_4 + \sigma_5 + \sigma_6) \tag{8-11}$$

with

$$c_1^2 + c_2^2 + 2c_1c_2G = 1 \tag{8-12}$$

Remembering that p orbitals change sign on inversion in the

center, the "extension" of the metal p orbitals to the ligands leads to the functions (Figure 8-2)

$$t_{1u}:\begin{cases}\psi(p_x) = c_3(4p_x) + c_4\dfrac{1}{\sqrt{2}}(\sigma_1 - \sigma_3)\\ \psi(p_y) = c_3(4p_y) + c_4\dfrac{1}{\sqrt{2}}(\sigma_2 - \sigma_4)\\ \psi(p_z) = c_3(4p_z) + c_4\dfrac{1}{\sqrt{2}}(\sigma_5 - \sigma_6)\end{cases} \qquad (8\text{-}13)$$

Again, we see that the p orbitals and the linear combinations of σ orbitals "follow" each other under the various symmetry operations.

With $d_{x^2-y^2}$ and d_{z^2} we construct the σ orbitals as follows: The "extension" of the $d_{x^2-y^2}$ orbital to the ligand orbitals yields

$$\psi(d_{x^2-y^2}) = c_5(3d_{x^2-y^2}) + c_6\tfrac{1}{2}(\sigma_1 - \sigma_2 + \sigma_3 - \sigma_4) \qquad (8\text{-}14)$$

Let us now rotate this orbital, using the threefold axis (see Figure 8-2):

$$\begin{aligned}\hat{C}_3\psi(d_{x^2-y^2}) &= c_5\left(-\frac{1}{2}d_{x^2-y^2} - \frac{\sqrt{3}}{2}d_{z^2}\right) + c_6\frac{1}{2}(\sigma_2 - \sigma_5 + \sigma_4 - \sigma_6)\\ &= -\tfrac{1}{2}\left(c_5 d_{x^2-y^2} + c_6\tfrac{1}{2}(\sigma_1 - \sigma_2 + \sigma_3 - \sigma_4)\right)\\ &\quad - \frac{\sqrt{3}}{2}\left(c_5 d_{z^2} + c_6\frac{1}{2\sqrt{3}}(2\sigma_5 + 2\sigma_6 - \sigma_1 - \sigma_2 - \sigma_3 - \sigma_4)\right)\end{aligned} \qquad (8\text{-}15)$$

Hence these two linear combinations are the σ molecular orbitals transforming as e_g. Note that we have now used up all the σ valence orbitals; e.g., let us add up all the σ_1 parts:

$$\sigma_1:\ \frac{1}{6}(s) + \frac{1}{2}(p_x) + \frac{1}{4}(d_{x^2-y^2}) + \frac{1}{12}(d_{z^2}) = 1 \qquad (8\text{-}16)$$

We now turn our attention to the π ligand orbitals. We see that the proper wave function with (d_{xz}) is (see Figure 8-2)

$$\psi(d_{xz}) = c_6(d_{xz}) + c_7\tfrac{1}{2}(y_1 + x_5 + x_3 + y_6) \qquad (8\text{-}17)$$

and $\psi(d_{yz})$ and $\psi(d_{xy})$ can be obtained by use of the $\hat{C}_3$ operator. All in all, by matching the metal-orbital lobes in sign and magnitude we get the results shown in Table 8-3.

Table 8-3 Proper Ligand Orbital Combinations for an Octahedral Complex

Representation	Ligand combination must match	Ligand combination
a_{1g}	s	$\sigma_1 + \sigma_2 + \sigma_3 + \sigma_4 + \sigma_5 + \sigma_6$
e_g	$2z^2 - x^2 - y^2$	$2\sigma_5 + 2\sigma_6 - \sigma_1 - \sigma_2 - \sigma_3 - \sigma_4$
	$x^2 - y^2$	$\sigma_1 - \sigma_2 + \sigma_3 - \sigma_4$
t_{2g}	xz	$y_1 + x_5 + x_3 + y_6$
	yz	$x_2 + y_5 + y_4 + x_6$
	xy	$x_1 + y_2 + y_3 + x_4$
t_{1u}	x	$\sigma_1 - \sigma_3$
		$y_2 + x_5 - x_4 - y_6$
	y	$\sigma_2 - \sigma_4$
		$x_1 + y_5 - y_3 - x_6$
	z	$\sigma_5 - \sigma_6$
		$y_1 + x_2 - x_3 - y_4$

The possible combinations of ligand π orbitals are found by looking at the trace of all the π ligand orbitals under the various symmetry operations of the octahedron:

E	C_3	C_2	C_4	C_2'	i	S_6	σ_h	S_4	σ_d
12	0	−4	0	0	0	0	0	0	0

The traces of the transformation matrices are made up of t_{1g}, t_{2g}, t_{1u}, and t_{2u} sets of functions. We can, as shown above, easily find the combinations which transform like t_{2g} and t_{1u}.

We have yet to find the t_{1g} and t_{2u} ligand combinations. There are no metal orbitals with these symmetries. Therefore, we know t_{1g} and t_{2u} will be nonbonding in the metal complex. The t_{1g} and t_{2u} combinations are found by recalling from the O_h character table that the character under C_2 is -1 for T_1 and 1 for T_2. This means that the t_{2g} combinations are converted to t_{1g} by changing every other sign,

$$t_{1g}\begin{cases} y_1 - x_5 + x_3 - y_6 \\ x_2 - y_5 + y_4 - x_6 \\ x_1 - y_2 + y_3 - x_4 \end{cases} \qquad (8\text{-}18)$$

Similary, t_{2u} is obtained from t_{1u} by alternate sign changes,

Table 8-4 Metal and Ligand Orbitals for the Molecular Orbitals of an Octahedral Complex

Representation	Metal orbital	Ligand orbitals σ	Ligand orbitals π
a_{1g}	4s	$\frac{1}{\sqrt{6}}(\sigma_1 + \sigma_2 + \sigma_3 + \sigma_4 + \sigma_5 + \sigma_6)$	
e_g	$3d_{x^2-y^2}$	$\frac{1}{2}(\sigma_1 - \sigma_2 + \sigma_3 - \sigma_4)$	
	$3d_{z^2}$	$\frac{1}{2\sqrt{3}}(2\sigma_5 + 2\sigma_6 - \sigma_1 - \sigma_2 - \sigma_3 - \sigma_4)$	
t_{1u}	$4p_x$	$\frac{1}{\sqrt{2}}(\sigma_1 - \sigma_3)$	$\frac{1}{2}(y_2 + x_5 - x_4 - y_6)$
	$4p_y$	$\frac{1}{\sqrt{2}}(\sigma_2 - \sigma_4)$	$\frac{1}{2}(x_1 + y_5 - y_3 - x_6)$
	$4p_z$	$\frac{1}{\sqrt{2}}(\sigma_5 - \sigma_6)$	$\frac{1}{2}(y_1 + x_2 - x_3 - y_4)$
t_{2g}	$3d_{xz}$ $3d_{yz}$ $3d_{xy}$		$\frac{1}{2}(y_1 + x_5 + x_3 + y_6)$ $\frac{1}{2}(x_2 + y_5 + y_4 + x_6)$ $\frac{1}{2}(x_1 + y_2 + y_3 + x_4)$
t_{1g}			$\frac{1}{2}(y_1 - x_5 + x_3 - y_6)$ $\frac{1}{2}(x_2 - y_5 + y_4 - x_6)$ $\frac{1}{2}(x_1 - y_2 + y_3 - x_4)$
t_{2u}			$\frac{1}{2}(y_2 - x_5 - x_4 + y_6)$ $\frac{1}{2}(x_1 - y_5 - y_3 + x_6)$ $\frac{1}{2}(y_1 - x_2 - x_3 + y_4)$

$$t_{2u}\begin{cases} y_2 - x_5 - x_4 + y_6 \\ x_1 - y_5 - y_3 + x_6 \\ y_1 - x_2 - x_3 + y_4 \end{cases} \tag{8-19}$$

The metal and normalized ligand combinations for an octahedral complex are grouped in σ and π classifications and summarized in Table 8-4.

8-4 GROUP OVERLAP OF METAL AND LIGAND ORBITALS

The total overlap of a metal orbital and a linear combination of ligand orbitals is called the group overlap G. That is,

$$G = \int \psi_M \Phi_L \, d\tau \tag{8-20}$$

where ψ_M is a normalized metal orbital and Φ_L is a normalized combination of ligand orbitals.

The G's are usually expressed in terms of the two-atom overlap integrals,

$$S(a,b) = \int \Phi_a \Phi_b \, d\tau \tag{8-21}$$

The standard two-atom overlaps are shown in Figure 8-4. For an octahedral complex, we have the following group overlaps:

$$G[a_{1g}(\sigma)] = \int 4s \frac{1}{\sqrt{6}} (\sigma_1 + \sigma_2 + \sigma_3 + \sigma_4 + \sigma_5 + \sigma_6) \, d\tau$$

$$= \frac{6}{\sqrt{6}} S(\sigma, 4s) = \sqrt{6}\, S(\sigma, 4s) \tag{8-22}$$

$$G[e_g(\sigma)] = \int 3d_{z^2} \frac{1}{2\sqrt{3}} (2\sigma_5 + 2\sigma_6 - \sigma_1 - \sigma_2 - \sigma_3 - \sigma_4) \, d\tau$$

$$= \frac{1}{2\sqrt{3}} (2 + 2 + 2)\, S(\sigma, 3d_\sigma) = \sqrt{3}\, S(\sigma, 3d_\sigma)$$

$$= \int 3d_{x^2-y^2} \tfrac{1}{2} (\sigma_1 - \sigma_2 + \sigma_3 - \sigma_4) \, d\tau \tag{8-23}$$

$$G[t_{1u}(\sigma)] = \int 4p_x \frac{1}{\sqrt{2}} (\sigma_1 - \sigma_3) \, d\tau = \sqrt{2}\, S(\sigma, 4p_\sigma)$$

$$= \int 4p_y \frac{1}{\sqrt{2}} (\sigma_2 - \sigma_4) \, d\tau = \int 4p_z \frac{1}{\sqrt{2}} (\sigma_5 - \sigma_6) \, d\tau \tag{8-24}$$

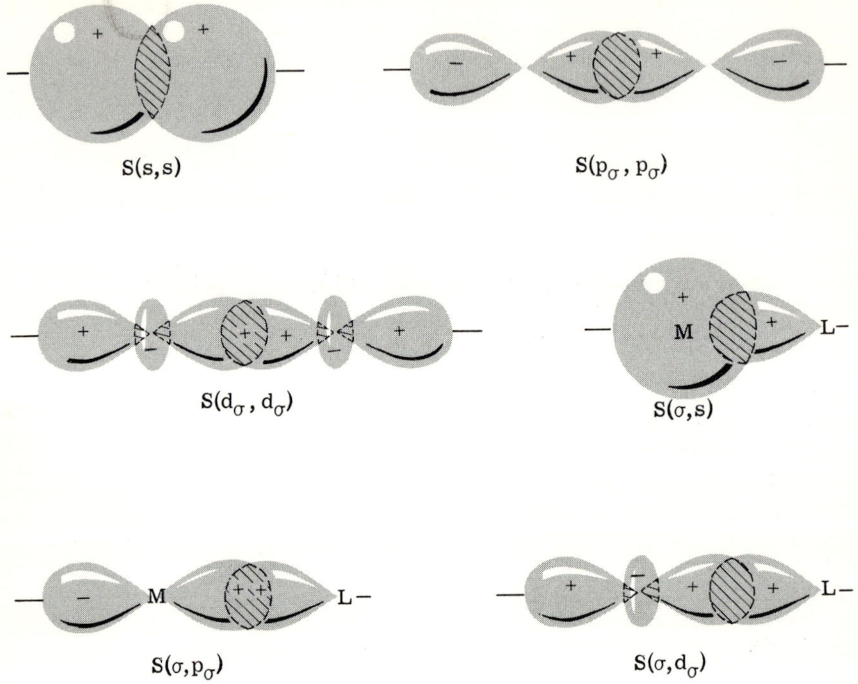

Figure 8-4 Standard two-atom overlaps.

$$G[t_{2g}(\pi)] = \int 3d_{xz}\,\tfrac{1}{2}(y_1 + x_5 + x_3 + y_6)\,d\tau = 2S(p_\pi, 3d_\pi)$$
$$= \int 3d_{yz}\,\tfrac{1}{2}(x_2 + y_5 + y_4 + x_6)\,d\tau$$
$$= \int 3d_{xy}\,\tfrac{1}{2}(x_1 + y_2 + y_3 + x_4)\,d\tau \qquad (8\text{-}25)$$

$$G[t_{1u}(\pi)] = \int 4p_x\,\tfrac{1}{2}(y_2 + x_5 - x_4 - y_6)\,d\tau = 2S(p_\pi, 4p_\pi)$$
$$= \int 4p_y\,\tfrac{1}{2}(x_1 + y_5 - y_3 - x_6)\,d\tau$$
$$= \int 4p_z\,\tfrac{1}{2}(y_1 + x_2 - x_3 - y_4)\,d\tau \qquad (8\text{-}26)$$

In Appendix 8A is given an example calculation of $G[e_g(\sigma)]$.

8-5 ENERGY CALCULATIONS

Approximate energies of the molecular orbitals are obtained by solving the secular equation $|H_{ij} - WG_{ij}| = 0$. There is one secular equation to solve for each type of orbital. Thus, we have a_{1g}, e_g, t_{2g}, and t_{1u} secular equations in O_h symmetry.

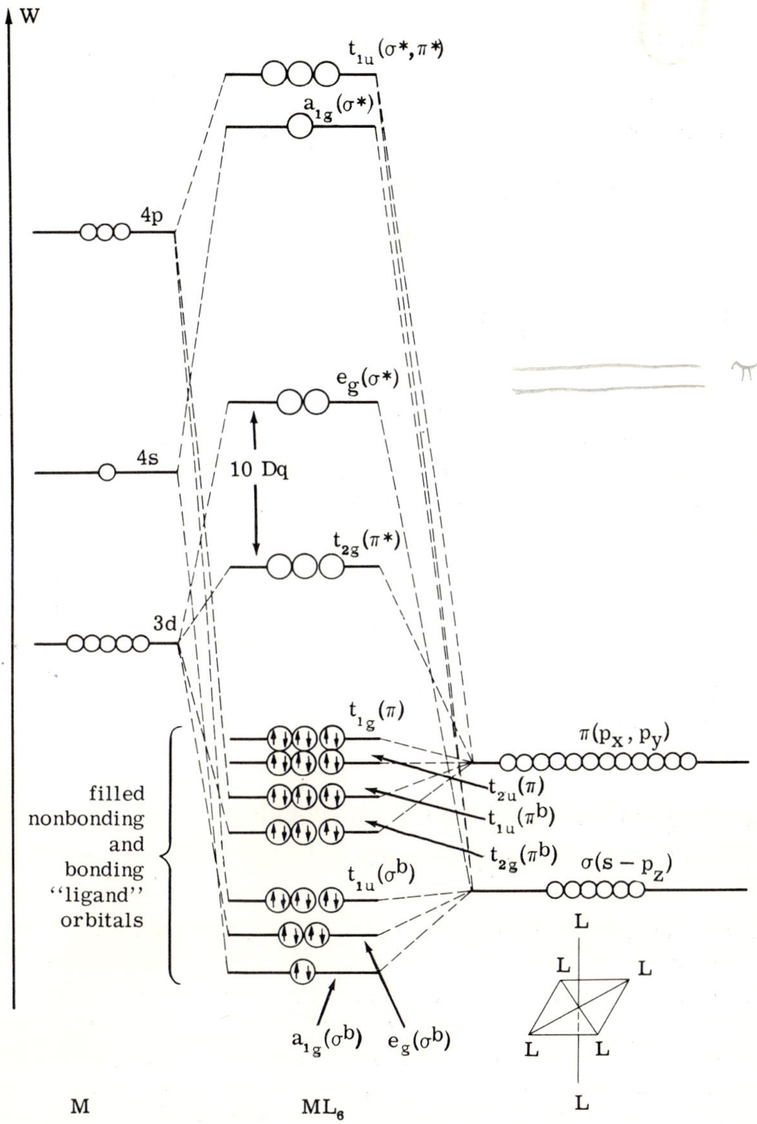

Figure 8-5 Molecular-orbital energy-level scheme for an octahedral complex.

The problem is in estimating the Coulomb (H_{ii}) and exchange (H_{ij}) integrals. A simple method estimates the Coulomb integrals as atomic ionization potentials. Since single-electron ionization potentials are functions of charge and electronic configuration, it is important to iterate until the Coulomb integrals taken are appropriate for the final charge distribution and electronic configuration calculated for the complex.

The exchange integrals are assumed to be proportional to the overlap integrals since $\psi_M \Phi_L \approx \frac{1}{2} G_{ML} (\psi_M \psi_M + \Phi_L \Phi_L)$:

$$H_{ij} = k\left(\frac{H_{ii} + H_{jj}}{2}\right) \left(G_{ij}\right) \tag{8-28}$$

or

$$H_{ij} = -k \sqrt{H_{ii} H_{jj}}\ G_{ij} \tag{8-27}$$

A value of k of about 2 has been used in most cases.

The energies of the molecular orbitals for an octahedral complex ML_6 are shown in Figure 8-15. There are 36 + n electrons to place in the molecular orbitals (6 from each ligand and n from the central atom). As a simple example, we find that the ground state of TiF_6^{3-} is

$$\left[a_{1g}(\sigma^b)\right]^2 \left[e_g(\sigma^b)\right]^4 \left[t_{1u}(\sigma^b)\right]^6 \left[t_{2g}(\pi^b)\right]^6 \left[t_{1u}(\pi^b)\right]^6 \left[t_{2u}(\pi)\right]^6$$
$$\times \left[t_{1g}(\pi)\right]^6 \left[t_{2g}(\pi^*)\right]^1 = {}^2T_{2g}$$

since Ti^{3+} contributes 1 electron. The details of a calculation of CrF_6^{3-} are given in Appendix 8B.

8-6 ELECTRONIC SPECTRA OF METAL COMPLEXES

A low-lying excited state in TiF_6^{3-} occurs upon excitation of an electron from $t_{2g}(\pi^*) \rightarrow e_g(\sigma^*)$. It is now established that the colors of many transition metal complexes are due to such "d-d" transitions. The energy separation between $e_g(\sigma^*)$ and $t_{2g}(\pi^*)$ is known as Δ, or 10 Dq, and we see that the molecular-orbital theory predicts a positive Δ because of the difference in the covalent interaction of the d_σ and d_π valence orbitals. The value of 10 Dq in TiF_6^{3-} is obtained from the weak-absorption-band system in the 5000 to 7000 A region of the spectrum, which is due to the parity-forbidden transition ${}^2T_{2g} \rightarrow {}^2E_g$. Then $h\bar{\nu} = 10\ Dq \approx 17{,}000\ cm^{-1}$ for the complex TiF_6^{3-}. The value of 10 Dq in all cases of interest is obtained from experiment. It is found that for trivalent ions of the first transition series $Dq \approx 2000\ cm^{-1}$ and for divalent ions of the first transition series $Dq \approx 1000\ cm^{-1}$.

Higher electronic excited states arise on excitation of an electron from a bonding or nonbonding MO into $t_{2g}(\pi^*)$ or $e_g(\sigma^*)$. Since the bonding and nonbonding MO's are mainly localized around the ligands, and $t_{2g}(\pi^*)$ and $e_g(\sigma^*)$ are mainly located on the metal, this type of transition is known as ligand-to-metal *charge transfer* (abbreviated $L \rightarrow M$). The first such band occurs higher than 50,000 cm^{-1} in TiF_6^{3-}.

In certain complexes, low-lying excited states arise by the transfer of an electron from an orbital based on the metal to an orbital based on the ligands. This is called metal-to-ligand ($M \rightarrow L$) charge transfer, and it is exhibited by complexes containing ligands such as NO, CO, and CN^-, which have relatively stable π^* orbitals.

APPENDIX 8A/Evaluation of $G[e_g(\sigma)]$

Recall from Eq. (8-23) that

$$G[e_g(\sigma)] = \int (3d_{z^2}) \frac{1}{2\sqrt{3}} (2\sigma_5 + 2\sigma_6 - \sigma_1 - \sigma_2 - \sigma_3 - \sigma_4)\, d\tau \tag{8-29}$$

Expanding, we have

$$\begin{aligned} &\int (3d_{z^2}) \frac{1}{2\sqrt{3}} (2\sigma_5 + 2\sigma_6 - \sigma_1 - \sigma_2 - \sigma_3 - \sigma_4)\, d\tau \\ &= \frac{1}{2\sqrt{3}} \Big[\int (3d_{z^2})(2\sigma_5)\, d\tau + \int (3d_{z^2})(2\sigma_6)\, d\tau \\ &\quad - \int (3d_{z^2})(\sigma_1 + \sigma_2 + \sigma_3 + \sigma_4)\, d\tau \Big] \end{aligned} \tag{8-30}$$

The integrals involving σ_5 and σ_6 are just the two-atom overlaps $S(\sigma, d_\sigma)$ shown in Figure 8-1. Thus we have

$$\int (3d_{z^2})(2\sigma_5)\, d\tau = \int (3d_{z^2})(2\sigma_6)\, d\tau = 2S(\sigma, 3d_\sigma) \tag{8-31}$$

The integral involving σ_1, σ_2, σ_3, and σ_4 is transformed into the

Table 8-5 Coordinate Transformations

M to ①	M to ②	M to ③	M to ④
$z \to y$	$z \to x$	$z \to -x$	$z \to -y$
$x \to -z$	$x \to y$	$x \to z$	$x \to -x$
$y \to x$	$y \to -z$	$y \to -y$	$y \to z$

standard two-atom overlap integral $S(\sigma,3d_\sigma)$ by rotating the metal coordinate system to coincide with the coordinate systems of ligands ①, ②, ③, and ④. Referring back to Figure 7-12 we see that the rotation of metal coordinates leads to the transformations of Table 8-5. Since the angular part of $3d_{z^2} = c(3z^2 - r^2)$, we find

$$-(3d_{z^2})(\sigma_1) = -c(3y^2 - r^2)(\sigma) \tag{8-32}$$

$$-(3d_{z^2})(\sigma_2) = -c(3x^2 - r^2)(\sigma) \tag{8-33}$$

$$-(3d_{z^2})(\sigma_3) = -c(3x^2 - r^2)(\sigma) \tag{8-34}$$

$$-(3d_{z^2})(\sigma_4) = -c(3y^2 - r^2)(\sigma) \tag{8-35}$$

Adding equations (8-32), (8-33), (8-34), and (8-35), we have

$$\begin{aligned}-\int (3d_{z^2})(\sigma_1 + \sigma_2 + \sigma_3 + \sigma_4)\,d\tau &= -c\int (6x^2 + 6y^2 - 4r^2)\sigma\,d\tau \\ &= -c\int (2x^2 + 2y^2 - 4z^2)\sigma\,d\tau = 2\int c(3z^2 - r^2)\sigma\,d\tau \\ &= 2\int (3d_{z^2})(\sigma)\,d\tau = 2S(\sigma,3d_\sigma)\end{aligned} \tag{8-36}$$

Thus, from Eq. (8-30),

$$\begin{aligned}G[e_g(\sigma)] &= \frac{1}{2\sqrt{3}}\left[2S(\sigma,3d_\sigma) + 2S(\sigma,3d_\sigma) + 2S(\sigma,3d_\sigma)\right] \\ &= \sqrt{3}\,S(\sigma,3d_\sigma)\end{aligned} \tag{8-37}$$

APPENDIX 8B/ Example Calculations

8-7 BASIS FUNCTIONS

The valence orbitals taken for a molecular-orbital calculation of a transition metal complex are the nd, (n + 1)s, and (n + 1)p metal orbitals and appropriate σ and π functions of the ligands. Many of these valence orbitals are not individually basis functions for an irreducible representation in the symmetry under consideration. Symmetry basis functions transforming properly must be constructed, by methods analogous to those used throughout this volume. The results for a number of important symmetries are tabulated in this volume in various places, as follows:

$D_{\infty h}$: Eqs. (6-1) and (6-2)
C_{2v}: Table 6-5
O_h: Table 8-4
T_d: Table 8-6
C_{4v}: Table I, Ballhausen-Gray reprint
D_{4h}: Table II, Gray-Ballhausen reprint

8-8 NORMALIZATION INCLUDING LIGAND-LIGAND OVERLAP

The basis functions referred to in Section 8-7 are normalized assuming zero ligand-ligand overlap. In reality, of course, the ligand valence orbitals overlap, and this should be taken into consideration when normalizing the basis functions. As an example, consider one of the $T_2(\sigma_s)$ functions for a tetrahedral molecule (see Table 8-6). For σ_s of T_2, row 1:

$$\begin{aligned} &\tfrac{1}{4}\int (s_1 - s_2 + s_3 - s_4)^2\, d\tau \\ &\quad = \tfrac{1}{4}\int (s_1^2 + s_2^2 + s_3^2 + s_4^2 - 2s_1s_2 + 2s_1s_3 - 2s_1s_4 \\ &\qquad - 2s_2s_3 + 2s_2s_4 - 2s_3s_4)\, d\tau \\ &\quad = \tfrac{1}{4}[4 - 4S(s,s)] = [1 - S(s,s)] \end{aligned} \tag{8-38}$$

Thus,

$$\sigma_s(t_2^{(1)}) = \frac{1}{2[1 - S(s,s)]^{1/2}}(s_1 - s_2 + s_3 - s_4) \tag{8-39}$$

Table 8-6 Basis Functions for T_d Molecules[a]

Irreducible representation	Row	Metal orbitals	Ligand orbitals
A_1	1	s	$\frac{1}{2}(s_1 + s_2 + s_3 + s_4)$, $\frac{1}{2}\left(p_{z_1} + p_{z_2} + p_{z_3} + p_{z_4}\right)$
E	1	d_{z^2}	$\frac{1}{2}\left(p_{x_1} - p_{x_2} - p_{x_3} + p_{x_4}\right)$
	2	$d_{x^2-y^2}$	$\frac{1}{2}\left(p_{y_1} - p_{y_2} - p_{y_3} + p_{y_4}\right)$
T_2	1	p_x, d_{yz}	$\frac{1}{2}\left(p_{z_1} - p_{z_2} + p_{z_3} - p_{z_4}\right)$, $\frac{1}{2}(s_1 - s_2 + s_3 - s_4)$, $\frac{1}{4}\left[p_{x_1} + p_{x_2} - p_{x_3} - p_{x_4} + \sqrt{3}\left(-p_{y_1} - p_{y_2} + p_{y_3} + p_{y_4}\right)\right]$
	2	p_y, d_{xz}	$\frac{1}{2}\left(p_{z_1} + p_{z_2} - p_{z_3} - p_{z_4}\right)$, $\frac{1}{2}(s_1 + s_2 - s_3 - s_4)$, $\frac{1}{4}\left[p_{x_1} - p_{x_2} + p_{x_3} - p_{x_4} + \sqrt{3}\left(p_{y_1} - p_{y_2} + p_{y_3} - p_{y_4}\right)\right]$
	3	p_z, d_{xy}	$\frac{1}{2}\left(p_{z_1} - p_{z_2} - p_{z_3} + p_{z_4}\right)$, $\frac{1}{2}(s_1 - s_2 - s_3 + s_4)$, $-\frac{1}{2}\left(p_{x_1} + p_{x_2} + p_{x_3} + p_{x_4}\right)$
T_1	1		$\frac{1}{4}\left[\sqrt{3}\left(p_{x_1} + p_{x_2} - p_{x_3} - p_{x_4}\right) + p_{y_1} + p_{y_2} - p_{y_3} - p_{y_4}\right]$
	2		$\frac{1}{4}\left[\sqrt{3}\left(p_{x_1} - p_{x_2} + p_{x_3} - p_{x_4}\right) - p_{y_1} + p_{y_2} - p_{y_3} + p_{y_4}\right]$
	3		$\frac{1}{2}\left(p_{y_1} + p_{y_2} + p_{y_3} + p_{y_4}\right)$

[a] Referred to the coordinate system shown in Figure 8-6.

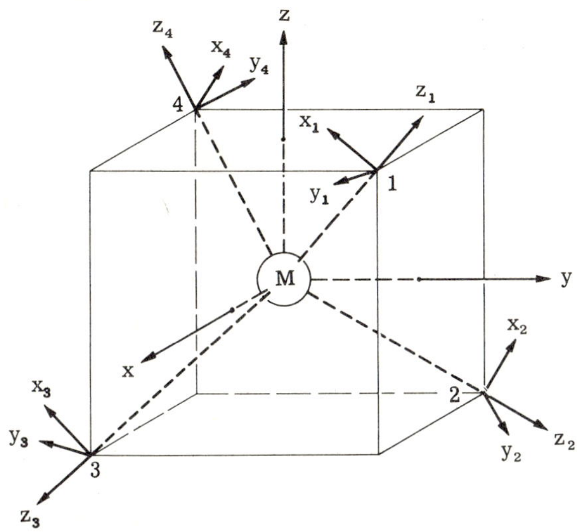

Figure 8-6 Coordinate system for a tetrahedral complex.

Table 8-7 Correction for Ligand-Ligand Overlap—Tetrahedral Geometry[a]

Irreducible representation	basis function Ψ_i	Correction factor N_i
A_1	σ_s	$\left[1 + 3S\left(s_L, s_L\right)\right]^{-1/2}$
	σ_p	$\left[1 + 2S\left(p_{\sigma L}, p_{\sigma L}\right) + S\left(p_{\pi L}, p_{\pi L}\right)\right]^{-1/2}$
E	π	$\left[1 + \frac{1}{2}S\left(p_{\sigma L}, p_{\sigma L}\right) - \frac{1}{2}S\left(p_{\pi L}, p_{\pi L}\right)\right]^{-1/2}$
T_2	σ_p	$\left[1 - \frac{2}{3}S\left(p_{\sigma L}, p_{\sigma L}\right) - \frac{1}{3}S\left(p_{\pi L}, p_{\pi L}\right)\right]^{-1/2}$
	σ_s	$\left[1 - S\left(s_L, s_L\right)\right]^{-1/2}$
	π	$\left[1 + \frac{1}{6}S\left(p_{\sigma L}, p_{\sigma L}\right) + \frac{11}{6}S\left(p_{\pi L}, p_{\pi L}\right)\right]^{-1/2}$
T_1	π	$\left[1 - \frac{1}{2}S\left(p_{\sigma L}, p_{\sigma L}\right) - \frac{3}{2}S\left(p_{\pi L}, p_{\pi L}\right)\right]^{-1/2}$

[a] The subscripts σ and π denote the type of overlap involved; L means ligand orbital.

Table 8-8 Correction for Ligand-Ligand Overlap—Octahedral Geometry[a]

Irreducible representation	Ligand function	Correction factor N_i
A_{1g}	$z_1 + z_2 + z_3 + z_4 + z_5 + z_6$	$\left[1 + S\left(\sigma_L, \sigma_L; 2R\right) + 2S\left(\sigma_L, \sigma_L; \sqrt{2}R\right) + 2S\left(\pi_L, \pi_L; \sqrt{2}R\right)\right]^{-1/2}$
E_g	$z_1 - z_2 + z_3 - z_4$	$\left[1 + S\left(\sigma_L, \sigma_L; 2R\right) - S\left(\sigma_L, \sigma_L; \sqrt{2}R\right) - S\left(\pi_L, \pi_L; \sqrt{2}R\right)\right]^{-1/2}$
T_{1g}	$x_2 - y_5 + y_4 - x_6$	$\left[1 - S\left(\sigma_L, \sigma_L; 2R\right) - S\left(\sigma_L, \sigma_L; \sqrt{2}R\right) - S\left(\pi_L, \pi_L; \sqrt{2}R\right)\right]^{-1/2}$
T_{2g}	$y_1 + x_5 + x_3 + y_6$	$\left[1 - S\left(\pi_L, \pi_L; 2R\right) + S\left(\sigma_L, \sigma_L; \sqrt{2}R\right) + S\left(\pi_L, \pi_L; \sqrt{2}R\right)\right]^{-1/2}$
T_{1u}	$z_1 - z_3$	$\left[1 - S\left(\sigma_L, \sigma_L; 2R\right)\right]^{-1/2}$
	$y_2 + x_5 - x_4 - y_6$	$\left[1 + S\left(\pi_L, \pi_L; 2R\right) + 2S\left(\pi_L, \pi_L; \sqrt{2}R\right)\right]^{-1/2}$
T_{2u}	$y_2 - x_5 - x_4 + y_6$	$\left[1 + S\left(\pi_L, \pi_L; 2R\right) - 2S\left(\pi_L, \pi_L; \sqrt{2}R\right)\right]^{-1/2}$

[a] $S\left(\sigma_L, \sigma_L; 2R\right)$ means the overlap between two ligand σ functions at distance 2R, where R is the metal-ligand internuclear separation.

By similar procedures, it is found that the basis functions previously tabulated must be multiplied by correction factors in order to be correctly normalized including ligand-ligand overlap. For tetrahedral and octahedral geometries, correction factors are given in Tables 8-7 and 8-8.

8-9 EVALUATION OF A GROUP OVERLAP INTEGRAL: $G_{T_2}(d,\sigma_s)$

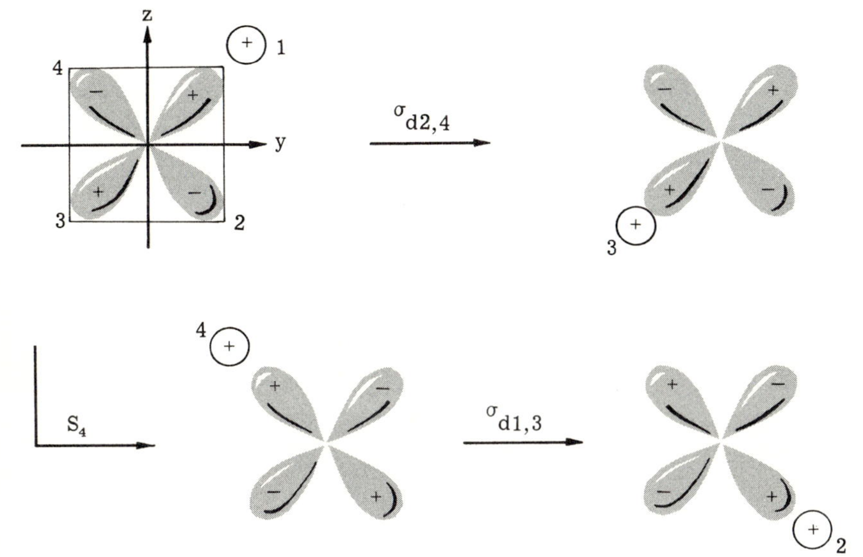

So

$$S(d_{yz}, s_1) = S(d_{yz}, s_3) = -S(d_{yz}, s_4) = -S(d_{yz}, s_2) \tag{8-40}$$

$$\begin{aligned} G_{T_2}(d,\sigma_S) &= \int d_{yz} \frac{(s_1 - s_2 + s_3 - s_4)}{2[1 - S(s,s)]^{1/2}}\, d\tau \\ &= \frac{1}{2[1 - S(s,s)]^{1/2}} [S(d_{yz}, s_1) - S(d_{yz}, s_2) \\ &\quad + S(d_{yz}, s_3) - S(d_{yz}, s_4)] \\ &= \frac{1}{2[1 - S(s,s)]^{1/2}} [4S(d_{yz}, s_1)] \end{aligned} \tag{8-41}$$

$$\boxed{G_{T_2}(d,\sigma_S) = \frac{2S(d_{yz}, s_1)}{[1 - S(s,s)]^{1/2}}} \tag{8-42}$$

To evaluate $S(d_{yz}, s_1)$ we may resolve d_{yz} into d orbitals based on a new coordinate system x′, y′, z′, with the z′ axis coincident with the z_1 axis of atom 1, the x' axis parallel to the x_1 axis, and the y' axis parallel to the y_1 axis. This is shown in Figure 8-7. The new coordinate system x', y', z' is related to the old coordinate system x, y, z by the following Euler angles[17]:

$$\Phi = -\frac{\pi}{4} \qquad \theta = -\left(\frac{\pi}{2} - b\right) \tag{8-43}$$

where $\tan b = \dfrac{1}{\sqrt{2}}$, $\sin b = \dfrac{1}{\sqrt{3}}$, and $\cos b = \dfrac{\sqrt{2}}{\sqrt{3}}$.

$$\psi = -\frac{\pi}{2}$$

$$\underset{\text{new}}{\begin{pmatrix} x' \\ y' \\ z' \end{pmatrix}} = \begin{pmatrix} \cos\psi & \sin\psi & 0 \\ -\sin\psi & \cos\psi & 0 \\ 0 & 0 & 1 \end{pmatrix} \begin{pmatrix} 1 & 0 & 0 \\ 0 & \cos\theta & \sin\theta \\ 0 & -\sin\theta & \cos\theta \end{pmatrix} \begin{pmatrix} \cos\Phi & \sin\Phi & 0 \\ -\sin\Phi & \cos\Phi & 0 \\ 0 & 0 & 1 \end{pmatrix} \underset{\text{old}}{\begin{pmatrix} x \\ y \\ z \end{pmatrix}}$$

$$\begin{pmatrix} x' \\ y' \\ z' \end{pmatrix} = \begin{pmatrix} 0 & -1 & 0 \\ 1 & 0 & 0 \\ 0 & 0 & 1 \end{pmatrix} \begin{pmatrix} 1 & 0 & 0 \\ 0 & \frac{1}{\sqrt{3}} & -\frac{\sqrt{6}}{3} \\ 0 & \frac{\sqrt{6}}{3} & \frac{1}{\sqrt{3}} \end{pmatrix} \begin{pmatrix} \frac{1}{\sqrt{2}} & -\frac{1}{\sqrt{2}} & 0 \\ \frac{1}{\sqrt{2}} & \frac{1}{\sqrt{2}} & 0 \\ 0 & 0 & 1 \end{pmatrix}$$

$$\begin{pmatrix} x' \\ y' \\ z' \end{pmatrix} = \begin{pmatrix} -\frac{1}{\sqrt{6}} & -\frac{1}{\sqrt{6}} & \frac{\sqrt{6}}{3} \\ \frac{1}{\sqrt{2}} & -\frac{1}{\sqrt{2}} & 0 \\ \frac{1}{\sqrt{3}} & \frac{1}{\sqrt{3}} & \frac{1}{\sqrt{3}} \end{pmatrix} \begin{pmatrix} x \\ y \\ z \end{pmatrix} \tag{8-44}$$

$$\underset{\text{old}}{\begin{pmatrix} x \\ y \\ z \end{pmatrix}} = \begin{pmatrix} -\frac{1}{\sqrt{6}} & \frac{1}{\sqrt{2}} & \frac{1}{\sqrt{3}} \\ -\frac{1}{\sqrt{6}} & -\frac{1}{\sqrt{2}} & \frac{1}{\sqrt{3}} \\ \frac{\sqrt{6}}{3} & 0 & \frac{1}{\sqrt{3}} \end{pmatrix} \underset{\text{new}}{\begin{pmatrix} x' \\ y' \\ z' \end{pmatrix}}$$

Some useful relationships involving the normalized angular parts of the d orbitals are as follows:

$$\left(\frac{15}{4\pi}\right)^{1/2} \frac{xz}{r^2} = d_{xz}$$

$$\left(\frac{15}{4\pi}\right)^{1/2} \frac{yz}{r^2} = d_{yz}$$

$$\left(\frac{15}{4\pi}\right)^{1/2} \frac{xy}{r^2} = d_{xy}$$

$$\left(\frac{5}{16\pi}\right)^{1/2} \left(\frac{3z^2 - r^2}{r^2}\right) = d_{z^2} \qquad (8\text{-}45)$$

$$\left(\frac{5}{16\pi}\right)^{1/2} \left(\frac{x^2 - y^2}{r^2}\right) = \frac{\sqrt{3}}{3} d_{x^2-y^2}$$

$$\left(\frac{5}{16\pi}\right)^{1/2} \left(\frac{y^2 - z^2}{r^2}\right) = -\frac{1}{2} d_{z^2} - \frac{\sqrt{3}}{6} d_{x^2-y^2}$$

$$\left(\frac{5}{16\pi}\right)^{1/2} \left(\frac{z^2 - x^2}{r^2}\right) = \frac{1}{2} d_{z^2} - \frac{\sqrt{3}}{6} d_{x^2-y^2}$$

We can now carry out the resolution of the old d_{yz} orbital in the new (primed) coordinate system.

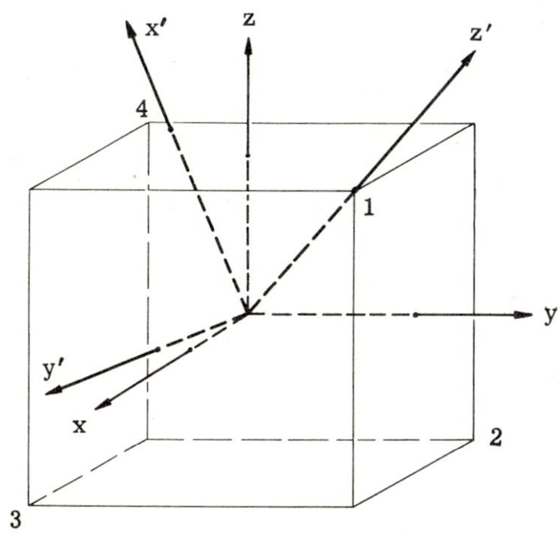

Figure 8-7 Transformed coordinate system for evaluation of an overlap integral, $S(d_{yz}, s_1)$.

Table 8-9 Group Overlap Integrals—Tetrahedral Geometry

A_1	$G_{A_1}(s, \sigma_s)$	$= \dfrac{2S(s_M, s_L)}{\left[1 + 3S(s_L, s_L)\right]^{1/2}}$
	$G_{A_1}(s, \sigma_p)$	$= \dfrac{-2S(s_M, p_{\sigma L})}{\left[1 + 2S(p_{\sigma L}, p_{\sigma L}) + S(p_{\pi L}, p_{\pi L})\right]^{1/2}}$
	$G_{A_1}(\sigma_s, \sigma_p)$	$= \dfrac{-\sqrt{6}\, S(s_L, p_{\sigma L})}{\left[1 + 3S(s_L, s_L)\right]^{1/2} \left[1 + 2S(p_{\sigma L}, p_{\sigma L}) + S(p_{\pi L}, p_{\pi L})\right]^{1/2}}$
E	$G_E(d, \pi)$	$= \dfrac{2\sqrt{6}\, S(d_{\pi M}, p_{\pi L})}{3\left[1 + \frac{1}{2}S(p_{\sigma L}, p_{\sigma L}) - \frac{1}{2}S(p_{\pi L}, p_{\pi L})\right]^{1/2}}$
T_2	$G_{T_2}(p, \sigma_p)$	$= \dfrac{-2\sqrt{3}\, S(p_{\sigma M}, p_{\sigma L})}{3\left[1 - \frac{2}{3}S(p_{\sigma L}, p_{\sigma L}) - \frac{1}{3}S(p_{\pi L}, p_{\pi L})\right]^{1/2}}$
	$G_{T_2}(p, d)$	$= 0$
	$G_{T_2}(p, \sigma_s)$	$= \dfrac{2\sqrt{3}\, S(p_{\pi M}, p_{\pi L})}{3\left[1 - S(s_L, s_L)\right]^{1/2}}$

$$G_{T_2}(p,\pi) = \frac{-2\sqrt{6}\, S\left(p_{\pi M}, p_{\pi L}\right)}{3\left[1 + \frac{1}{6} S\left(p_{\sigma L}, p_{\sigma L}\right) + \frac{11}{6} S\left(p_{\pi L}, p_{\pi L}\right)\right]^{1/2}}$$

$$G_{T_2}\left(d, \sigma_p\right) = \frac{-2\sqrt{3}\, S\left(d_{\sigma M}, p_{\sigma L}\right)}{3\left[1 - \frac{2}{3} S\left(p_{\sigma L}, p_{\sigma L}\right) - \frac{1}{3} S\left(p_{\pi L}, p_{\pi L}\right)\right]^{1/2}}$$

$$G_{T_2}\left(\sigma_p, \sigma_s\right) = \frac{\sqrt{6}\, S\left(p_{\sigma L}, s_L\right)}{3\left[1 - \frac{2}{3} S\left(p_{\sigma L}, p_{\sigma L}\right) - \frac{1}{3} S\left(p_{\pi L}, p_{\pi L}\right)\right]^{1/2}\left[1 - S\left(s_L, s_L\right)\right]^{1/2}}$$

$$G_{T_2}\left(\sigma_p, \pi\right) = \frac{2\sqrt{2}\, S\left(p_{\sigma L}, p_{\sigma L}\right) - S\left(p_{\pi L}, p_{\pi L}\right)}{3\left[1 - \frac{2}{3} S\left(p_{\sigma L}, p_{\sigma L}\right) - \frac{1}{3} S\left(p_{\pi L}, p_{\pi L}\right)\right]^{1/2}\left[1 + \frac{1}{6} S\left(p_{\sigma L}, p_{\sigma L}\right) + \frac{11}{6} S\left(p_{\pi L}, p_{\pi L}\right)\right]^{1/2}}$$

$$G_{T_2}\left(d, \sigma_s\right) = \frac{2\sqrt{3}\, S\left(d_{\sigma M}, s_L\right)}{3\left[1 - S\left(s_L, s_L\right)\right]^{1/2}}$$

$$G_{T_2}(d,\pi) = \frac{2\sqrt{2}\, S\left(d_{\pi M}, p_{\pi L}\right)}{3\left[1 + \frac{1}{6} S\left(p_{\sigma L}, p_{\sigma L}\right) + \frac{11}{6} S\left(p_{\pi L}, p_{\pi L}\right)\right]^{1/2}}$$

$$G_{T_2}\left(\sigma_s, \pi\right) = \frac{-2\sqrt{3}\, S\left(s_L, p_{\sigma L}\right)}{3\left[1 - S\left(s_L, s_L\right)\right]^{1/2}\left[1 + \frac{1}{6} S\left(p_{\sigma L}, p_{\sigma L}\right) + \frac{11}{6} S\left(p_{\pi L}, p_{\pi L}\right)\right]^{1/2}}$$

$$d_{yz} = \left(\frac{15}{4\pi}\right)^{1/2} \frac{yz}{r^2}$$

$$= \left(\frac{15}{4\pi}\right)^{1/2} \left[\frac{\left(-\frac{1}{\sqrt{6}} x' - \frac{1}{\sqrt{2}} y' + \frac{1}{\sqrt{3}} z'\right)\left(\frac{\sqrt{6}}{3} x' + \frac{1}{\sqrt{3}} z'\right)}{r^2} \right]$$

$$= \frac{2\sqrt{3}}{3}\left(\frac{5}{16\pi}\right)^{1/2}\left(\frac{z'^2 - x'^2}{r^2}\right) \qquad (8\text{-}46)$$

$$+ \left(\frac{15}{4\pi}\right)^{1/2}\left(\frac{\sqrt{2}}{6}\frac{x'z'}{r^2} - \frac{\sqrt{3}}{3}\frac{x'y'}{r^2} - \frac{\sqrt{6}}{6}\frac{y'z'}{r^2}\right)$$

$$= \frac{\sqrt{3}}{3} d_{z'^2} + \frac{\sqrt{2}}{6} d_{x'z'} - \frac{\sqrt{6}}{6} d_{y'z'} - \frac{1}{3} d_{x'^2-y'^2} - \frac{\sqrt{3}}{3} d_{x'y'}$$

A check of the normalization gives $\frac{1}{3} + \frac{1}{18} + \frac{1}{6} + \frac{1}{9} + \frac{1}{3} = 1$. Next we can evaluate the overlap integral $S(d_{yz}, s_1)$. Since the z' axis coincides with the internuclear axis, d_{yz} has been resolved into components which are σ, π, or δ with respect to the internuclear axis, whereas s_1 is, of course, σ with respect to this axis.

$$S(d_{yz}, s_1) = \int d_{yz} s_1 \, d\tau$$

$$= \frac{\sqrt{3}}{3}\int d_{z'^2} s_1 \, d\tau + \frac{\sqrt{2}}{6}\int d_{x'z'} s_1 \, d\tau - \frac{\sqrt{6}}{6}\int d_{y'z'} s_1 \, d\tau$$

$$- \frac{1}{3}\int d_{x'^2-y'^2} s_1 \, d\tau - \frac{\sqrt{3}}{3}\int d_{x'y'} s_1 \, d\tau \qquad (8\text{-}47)$$

$$= \frac{\sqrt{3}}{3} S(d_{z'^2}, s_1) + 0 + 0 + 0 + 0 = \frac{\sqrt{3}}{3} S(d_\sigma, s)$$

$$G_{T_2}(d, \sigma_S) = \frac{2S(d_{yz}, s_1)}{[1 - S(s_L, s_L)]^{1/2}}$$

$$\boxed{G_{T_2}(d, \sigma_S) = \frac{2\sqrt{3}}{3} \frac{S(d_{\sigma M}, s_L)}{[1 - S(s_L, s_L)]^{1/2}}} \qquad (8\text{-}48)$$

where M and L denote metal and ligand, respectively.

Table 8-10 Group Overlap Integrals—Octahedral Geometry[a]

A_{1g}	$G_{A_{1g}}(s,\sigma_L) = 6N_{A_{1g}(\sigma_L)}[S(s_M,\sigma_L;R)]$
E_g	$G_{E_g}(d_\sigma,\sigma_L) = 6N^{z^2}_{E_g(\sigma_L)}[S(d_\sigma,\sigma_L;R)]$
T_{2g}	$G_{T_{2g}}(d_\pi,\pi_L) = 4N_{T_{2g}(\pi_L)}[S(d_\pi,\pi_L;R)]$
T_{1u}	$G_{T_{1u}}(p_\sigma,\sigma_L) = 2N_{T_{1u}(\sigma_L)}[S(p_\sigma,\sigma_L;R)]$
	$G_{T_{1u}}(p_\pi,\pi_L) = 4N_{T_{1u}(\pi_L)}[S(p_\pi,\pi_L;R)]$

[a] $N_{A_{1g}(\sigma_L)}$, for example, is the normalization constant for the $a_{1g}\,\sigma_L$ combination, including ligand-ligand overlap.

Similar procedures can be used to evaluate other group overlap integrals. The results for tetrahedral and octahedral geometries are given in Tables 8-9 and 8-10.

8-10 RADIAL FUNCTIONS

The radial functions for atoms through argon tabulated by Clementi[1] are particularly useful for simple MO calculations. Other sources are Watson[2] and Richardson et al.[3] for transition metal atoms, and Slater[4] for simple single-exponential-term functions.

8-11 BOND DISTANCES

Bond distances are needed for overlap integral calculations. Standard sources are Sutton,[5] Pauling,[6] and Wells.[7]

8-12 OVERLAP INTEGRALS

Evaluation of two-atom overlap integrals has been discussed by Mulliken et al.[8] Tables of overlap integrals are given in Mulliken et al.,[8] Jaffe et al.,[9] Leifer et al.,[10] and Craig et al.[11] Lofthus gives some additional master formulas.[12]

8-13 COULOMB INTEGRALS (H_{ii})

The Coulomb integrals H_{ii} which appear in the secular equation are approximated as valence-state ionization energies (VSIE's) corrected for ligand-ligand overlap, if any. In general, each VSIE is a function of the charge and the orbital configuration of the atom in the molecule in question. Arlen Viste and Harold Basch at Columbia have calculated the average energies of all the terms in the important orbital configurations for the atoms hydrogen through krypton. These average term energies were calculated for neutral, singly ionized, and doubly ionized atoms, and in some instances for higher ionized species; the one-electron ionization energies of various orbital configurations are given in Table 8-11 for the first-row transition metal atoms. One-electron ionization energies for some important ligand atoms are given in Table 8-12.

In Sections 8-16 and 8-17 the example calculations of MnO_4^- and CrF_6^{3-} outline in some detail the use of the VSIE's given in Tables 8-11 and 8-12.

8-14 EXCHANGE INTEGRALS (H_{ij})

The exchange integral between two orbitals on different atoms is assumed to be a function of the overlap of the two orbitals as well as their stabilities. Wolfsberg and Helmholz suggested the approximation[13]

$$H_{ij} = FG_{ij}\left(\frac{H'_{ii} + H'_{jj}}{2}\right) \tag{8-49}$$

with F approximately two. We have used the very similar approximation[14]

$$H_{ij} = -2G_{ij}\left(H'_{ii} \cdot H'_{jj}\right)^{1/2} \tag{8-50}$$

which is, of course, approximately equal to the Wolfsberg-Helmholz formula when $F \approx 2$ and $H'_{ii} \approx H'_{jj}$.

8-15 OVERLAP CORRECTION FOR H_{ii} OF LIGANDS

For the ligand functions we have

$$\psi_i = \sum_\alpha a_{i\alpha}\Phi_{1\alpha} \tag{8-51}$$

$$\int \psi_i^2 \, d\tau = \frac{1}{N_i^2} = \sum_\alpha a_{i\alpha}^2 + \sum\sum_{\alpha \neq \beta} a_{i\alpha} a_{i\beta} S(i\alpha, i\beta) \quad (8\text{-}52)$$

The basic assumption is that for two different atomic orbitals (two orbitals on different atoms, or different orbitals on the same atom), Φ_k and Φ_n,

$$H_{kn} = -2.00 \left(H_{kk} \cdot H_{nn}\right)^{1/2} S(k,n) \quad (8\text{-}53)$$

Using this assumption,

$$\begin{aligned} H_{ii} &= \int (N_i \psi_i) H (N_i \psi_i) \, d\tau \\ &= N_i^2 H'_{ii} \left[\sum a_{i\alpha}^2 + 2 \sum\sum_{\alpha \neq \beta} a_{i\alpha} a_{i\beta} S(i\alpha, i\beta)\right] \end{aligned} \quad (8\text{-}54)$$

$H_{i\alpha i\alpha} = H'_{ii}$, the diagonal element uncorrected for ligand-ligand overlap. If ψ_i has already been normalized neglecting ligand-ligand overlap, then N_i is the correction factor for the normalization, which was tabulated previously. In this case, $\sum_\alpha a_{i\alpha}^2 = 1$.

Let

$$\begin{aligned} X_i &= \sum\sum_{\alpha \neq \beta} a_{i\alpha} a_{i\beta} S(i\alpha, i\beta) \\ H_{ii} &= H'_{ii} \left(\frac{1 + 2X_i}{1 + X_i}\right) \end{aligned} \quad (8\text{-}55)$$

For the central metal-atom orbitals, of course, $H_{ii} = H'_{ii}$. That the off-diagonal elements H_{ij} are to be computed from the uncorrected H'_{ii} rather than the corrected H_{ii} can be shown as follows:

$$\begin{aligned} G_{ij} &= \int (N_i \psi_i)(N_j \psi_j) \, d\tau = N_i N_j \sum_\alpha \sum_\mu a_{i\alpha} a_{j\mu} S(i\alpha, j\mu) \\ H_{ij} &= N_i N_j \sum_\alpha \sum_u a_{i\alpha} a_{j\mu} \left[-2.00 \left(H_{i\alpha i\alpha} H_{j\mu j\mu}\right)^{1/2} S(i\alpha, j\mu)\right] \end{aligned} \quad (8\text{-}56)$$

(The assumption $\Phi_{i\alpha} \neq \Phi_{j\mu}$ is a reasonable one, since ψ_i and ψ_j are symmetry basis functions belonging to the same row of the same irreducible representation.)

$$H_{ij} = -2.00 \left(H'_{ii} \cdot H'_{jj}\right)^{1/2} G_{ij} \quad (8\text{-}57)$$

Table 8-11 VSIE Results—First-Row Transition Metals

Valence state ionization energies (VSIE's) are tabulated here. The VSIE's are obtained by appropriately combining the values of W_{av} for the two configurations under consideration, together with the appropriate ionization potential. W_{av} is the weighted mean of the energies of the terms arising from a configuration, relative to the ground state of the atom or ion in question. The weighting factor is equal to the total degeneracy (spin × orbital) of the term. If necessary, the average can be taken over the J components weighted by their degeneracies. For example, $W_{av}(p^2) = \frac{1}{15}[9W(^3P) + 5W(^1D) + W(^1S)]$ provided the J components are not too widely separated.

The VSIE's (given in 1000 cm^{-1}) have been smoothed by subjecting the available data across the transition series to a least-squares fit, for given charge, s, and p character. The fit is quadratic if not more than two VSIE's are missing or omitted, linear otherwise. The fitting parameters A, B, and C in the relationship $VSIE = A(n-3)^2 + B(n-3) + C$ are given, as is the standard deviation of the individual points from the line. Parentheses around a value indicate that it was obtained only by interpolation, with no corresponding data being fed into the least-squares analysis (either because no data were available, or because the originally calculated value was off the line by roughly two standard deviations or more).

I. 3d VSIE's

		q =							
		0 → +1			+1 → +2			+2 → +3	
Atom	n	$3d^n$	$3d^{n-1}4s$	$3d^{n-1}4p$	$3d^{n-1}$	$3d^{n-2}4s$	$3d^{n-2}4p$	$3d^{n-2}$	$3d^{n-3}4s$
Ca	2	(18.7)	(30.2)	(44.3)	(84.2)	(117.2)	(125.7)		
Sc	3	23.2	37.5	49.7	95.0	129.2	138.3	204.7	
Ti	4	27.4	44.6	55.4	105.4	140.9	150.6	217.7	274.1
V	5	31.4	51.4	61.4	(115.2)	(152.4)	(162.7)	230.6	281.4
Cr	6	35.1	57.9	67.7	124.6	163.6	174.4	243.6	288.8
Mn	7	38.6	64.1	74.3	133.5	174.6	(185.8)	256.6	296.1
Fe	8	41.9	70.0	81.2	141.9	185.3	196.9	(269.5)	
Co	9	44.8	75.6	88.4	149.8	195.7	207.8	(282.5)	
Ni	10	47.6	80.9	95.9	157.3	205.8	218.3	(295.4)	
Cu	11		86.0	103.7	164.3	215.7	228.5	(308.4)	
Zn	12					225.4	238.5	321.4	
Standard deviation		1.4	0.8	1.8	0.7	1.1	1.4	3.3	6.8
A		−0.13	−0.15	0.15	−0.24	−0.13	−0.15		
B		4.4	7.2	5.5	10.6	11.9	12.5	13.0	7.4
C		23.2	37.5	49.7	95.0	129.2	138.3		

II. 4s VSIE's

		q =						
		0 → +1			+1 → +2			+2 → +3
Atom	n	$d^{n-1}s$	$d^{n-2}s^2$	$d^{n-2}sp$	$d^{n-2}s$	$d^{n-3}s^2$	$d^{n-3}sp$	$d^{n-3}s$
Ca	2	(43.4)	(50.1)	(55.1)	(96.8)	(109.1)		
Sc	3	46.1	53.8	60.8	102.6	117.4	123.8	179.6
Ti	4	48.6	57.2	66.0	108.3	125.0	130.3	186.6
V	5	51.0	60.4	(70.6)	113.7	(131.9)	(136.7)	193.5
Cr	6	53.2	63.3	74.7	118.8	138.2	(143.2)	200.5
Mn	7	55.3	65.9	78.3	123.8	143.8	(149.7)	207.5
Fe	8	57.3	68.3	81.4	128.5	148.7	156.1	(214.4)
Co	9	59.1	70.5	84.0	133.0	153.0	(162.6)	(221.4)
Ni	10	60.8	72.3	86.0	137.2	(156.7)	169.0	(228.3)
Cu	11	62.3	74.0	87.6	141.2	159.7	175.5	(235.3)
Zn	12		75.3	88.6	145.0	162.0	182.0	242.3
Standard deviation		0.8	0.8	1.4	1.7	4.7	2.8	5.7
A		−0.07	−0.13	−0.26	−0.12	−0.33		
B		2.6	3.55	5.4	5.75	7.9	6.5	7.0
C		46.1	53.8	60.8	102.6	117.4		

III. 4p VSIE's

		q =						
		0 → +1			+1 → +2			+2 → +3
Atom	n	$d^{n-1}p$	$d^{n-2}p^2$	$d^{n-2}sp$	$d^{n-2}p$	$d^{n-3}p^2$	$d^{n-3}sp$	$d^{n-3}p$
Ca	2	(25.3)		(29.1)	(71.8)			
Sc	3	26.1	34.9	31.9	74.9	(89.3)	87.5	140.2
Ti	4	26.9	35.9	34.4	77.8		91.1	144.3
V	5	27.7	(36.8)	36.4	80.6		94.7	(148.4)
Cr	6	28.4	(37.8)	38.1	83.2		(98.2)	152.5
Mn	7	29.2	38.8	39.4	85.7		101.8	(156.6)
Fe	8	29.9	(39.7)	40.3	88.0		105.4	(160.7)
Co	9	30.7	(40.7)	40.8	90.2		(109.0)	(164.8)
Ni	10	31.4	(41.6)	40.9	92.2		(112.6)	(168.9)
Cu	11	32.1	(42.6)	40.6	94.1		116.2	(173.0)
Zn	12		43.5	39.9	95.9		119.8	177.1
Standard deviation		1.1	1.9	2.1	1.7		3.4	3.0
A		−0.01		−0.19	−0.07			
B		0.8	1.0	2.6	3.0		3.6	4.1
C		26.1		31.9	74.9			

Table 8-12 Orbital Ionization Energies[a,b]

Atom	1s	2s	2p	3s	3p	4s	4p
H	110						
He	198						
Li		44					
Be		75					
B		113	67				
C		157	86				
N		206	106				
O		261	128				
F		374	151				
Ne		391	174				
Na				42			
Mg				62			
Al				91	48		
Si				121	63		
P				151	82		
S				167	94		
Cl				204	111		
Ar				236	128		
K						35	
Ca						49	
Zn						76	
Ga						102	48
Ge						126	61
As						142	73
Se						168	87
Br						194	101
Kr						222	115

Atom	$3d^{n-1}4s \rightarrow 3d^{n-2}4s$ 3d	$3d^{n-1}4s \rightarrow 3d^{n-1}$ 4s	$3d^{n-1}4p \rightarrow 3d^{n-1}$ 4p
Sc	38	46	26
Ti	45	49	27
V	51	51	28
Cr	58	53	28
Mn	64	55	29
Fc	70	57	30
Co	76	59	31
Ni	81	61	31
Cu	86	62	32

[a]The one-electron ionization energies of the valence orbitals given, calculated by finding the average energies of both the ground-state and ionized-state configurations (that is, the average energy of all the terms within a particular configuration).

[b]Atom configurations s or s^2p^n; energies in 1000 cm^{-1}.

Table 8-13 Group Overlap Integrals for MnO_4^- [a]

E	$G_E(d,\pi)$	=	0.2601
A_1	$G_{A_1}(s,\sigma_s)$	=	0.6654
	$G_{A_1}(s,\sigma_p)$	=	−0.2191
	$G_{A_1}(\sigma_s,\sigma_p)$	=	−0.0807
T_2	$G_{T_2}(p,\sigma_p)$	=	0.1011
	$G_{T_2}(p,d)$	=	0.0000
	$G_{T_2}(p,\sigma_s)$	=	0.3443
	$G_{T_2}(p,\pi)$	=	−0.2407
	$G_{T_2}(\sigma_p,d)$	=	−0.1724
	$G_{T_2}(\sigma_p,\sigma_s)$	=	0.0303
	$G_{T_2}(\sigma_p,\pi)$	=	0.0464
	$G_{T_2}(d,\sigma_s)$	=	0.2434
	$G_{T_2}(d,\pi)$	=	0.1495
	$G_{T_2}(\sigma_s,\pi)$	=	−0.0412

[a] R(Mn—O) = 1.59 A.

8-16 CALCULATION OF MnO_4^-

The valence orbitals taken for the calculation are 3d, 4s, and 4p for manganese and 2s and 2p for oxygen. Radial functions for Mn are taken from Ref. 3, for O from Ref. 1.

The proper basis functions are given in Table 8-6. There are four secular equations to solve, to find the MO's transforming as a_1(3), e(2), t_2(5), and t_1(1). The first step is to calculate the group overlap integrals, using the expressions in Table 8-9. Two-atom overlaps are either estimated from the available tables[(8)-(11)] or calculated accurately. The calculated group overlap integrals for MnO_4^- are given in Table 8-13.

We obtain the Mn Coulomb integrals from the VSIE data given in

Table 8-11. The d, s, and p VSIE's are strong functions of the charge and orbital configuration of Mn. It is assumed that a VSIE for a particular configuration (e.g., d^n) can be represented quadratically as VSIE = $Aq^2 + Bq + C$, where q is the charge on Mn. The VSIE curves for Mn, calculated from the data in Table 8-11, are given in Table 8-14.

The ligand Coulomb integrals, in this case oxygen 2s and 2p, are given fixed values. Either the neutral atom values from Table 8-12 or the values for an appropriate hydride (e.g., H_2O) are taken. We shall take the 2p Coulomb integral as the I.P. of H_2O, and the 2s value from Table 8-12.

The four secular equations $|H_{ij} - WG_{ij}| = 0$ are solved as follows: For a given cycle, an input electron configuration and charge are assumed for the metal, and the H_{ii} terms are computed. H_{ii} terms for ligand basis functions remain constant throughout the calculation. For each of the MO's calculated in the cycle, a Mulliken population analysis is performed, in which each overlap population is divided equally between the two basis functions involved.[16]

$$\Psi_n = \sum_i C_{ni}\varphi_i \tag{8-58}$$

$$POP_{ni} = \sum_j C_{ni}C_{nj}G_{ij} = C_{ni}^2 + \sum_{j\neq i} C_{ni}C_{nj}G_{ij} \tag{8-59}$$

If the nth MO is occupied by one electron, POP_{ni} represents the

Table 8-14 VSIE Functions for Manganese[a]

VSIE	Starting configuration	A	B	C
d	d^n	14.1	80.8	38.6
d	$d^{n-1}s$	5.5	105.0	64.1
d	$d^{n-1}p$	5.5	106.0	74.3
s	$d^{n-1}s$	7.6	60.9	55.3
s	$d^{n-2}s^2$	7.6	70.3	65.9
s	$d^{n-2}sp$	7.6	63.8	78.3
p	$d^{n-1}p$	7.2	49.3	29.2
p	$d^{n-2}p^2$	7.2	55.2	38.8
p	$d^{n-2}sp$	7.2	55.2	39.4

[a] VSIE = $Aq^2 + Bq + C$; energies in 1000 cm^{-1}.

fraction of the time which the electron spends in basis function i, or the fraction of the electronic charge which resides in basis function i. By adding up the appropriate POP_{ni} terms, the output configuration can be computed for the metal.

In subsequent cycles the input configuration(s) is altered until a self-consistent result is obtained. The result is considered self-consistent if the input and output configurations for the cycle are the same.

To calculate the uncorrected Mn H'_{ii}'s from the VSIE curves, we make use of the following equations:

$$\text{Assumed charge on Mn: } q = 7 - d - s - p$$

$$\text{Assumed configuration on Mn: } d^{(7-s-p-q)}s^{s}p^{p}$$

$$-H'_{dd} = (\text{d VSIE}) = (1 - s - p)(\text{d VSIE: } d^{n}) + s(\text{d VSIE: } d^{n-1}s) + p(\text{d VSIE: } d^{n-1}p) \qquad (8\text{-}60)$$

$$-H'_{ss} = (\text{s VSIE}) = (2 - s - p)(\text{s VSIE: } d^{n-1}s) + (s - 1)(\text{s VSIE: } d^{n-2}s^{2}) + p(\text{s VSIE: } d^{n-2}sp) \qquad (8\text{-}61)$$

$$-H'_{pp} = (\text{p VSIE}) = (2 - s - p)(\text{p VSIE: } d^{n-1}p) + (p - 1)(\text{p VSIE: } d^{n-2}p^{2}) + s(\text{p VSIE: } d^{n-2}sp) \qquad (8\text{-}62)$$

The self-consistent charge and configuration values in this particular calculation turn out to be as follows: $q = 0.6568$, $s = 0.1817$, $p = 0.3436$. For these values the final H'_{ii}'s in 10^3 cm^{-1} are

Mn	d	−121.28
	s	− 93.41
	p	− 58.40
O	s	−260.8
	p	−101.7

Following the treatment in Section 8-15 the H'_{ii}'s are corrected for overlap as follows:

Table 8-15 Molecular Orbitals for MnO_4^-

	Eigenvalues	Eigenvectors				
		s	σ_s	σ_p		
a_1	128.75	1.37	−0.98	0.25		
	−112.11	−0.03	0.13	1.00		
	−276.54	0.12	0.92	−0.01		
		d	π			
e	− 72.98	−0.72	0.90			
	−136.56	0.74	0.51			
		p	σ_p	d	σ_s	π
t_2	− 18.83	1.08	−0.06	0.24	−0.42	0.27
	− 49.29	−0.14	0.67	0.80	−0.30	−0.58
	−105.65	0.02	−0.69	0.13	−0.07	−0.67
	−126.26	−0.17	−0.35	0.63	−0.17	0.49
	−260.45	0.03	−0.02	−0.16	−0.96	0.03
		π				
t_1	− 96.40	1.00				

Input charge: 0.6568
Input configuration: $d^{5.8178} s^{0.1817} p^{0.3436}$
Output charge: 0.6568
Output configuration: $d^{5.8178} s^{0.1817} p^{0.3436}$

Table 8-16 Calculated and Observed Transition Energies in MnO_4^- [a]

Band maxima, cm^{-1}	f	Assignments	Calculated one-electron energies, cm^{-1}
(14,500)	Weak	${}^1A_1 \rightarrow {}^1T_1(t_1 \rightarrow 2e)$	23,400
18,300	0.032	${}^1A_1 \rightarrow {}^1T_2(t_1 \rightarrow 2e)$	23,400
(28,000)[b]	b	${}^1A_1 \rightarrow {}^1T_1(3t_2 \rightarrow 2e)$	32,700
32,200	0.070	${}^1A_1 \rightarrow {}^1T_2(3t_2 \rightarrow 2e)$	32,700
(44,000)[c]	c	${}^1A_1 \rightarrow {}^1T_2(t_1 \rightarrow 4t_2)$	47,100

[a] See Ref. 15.
[b] Weak shoulder.
[c] Shoulder indicating a band with $\epsilon \approx 1500$.

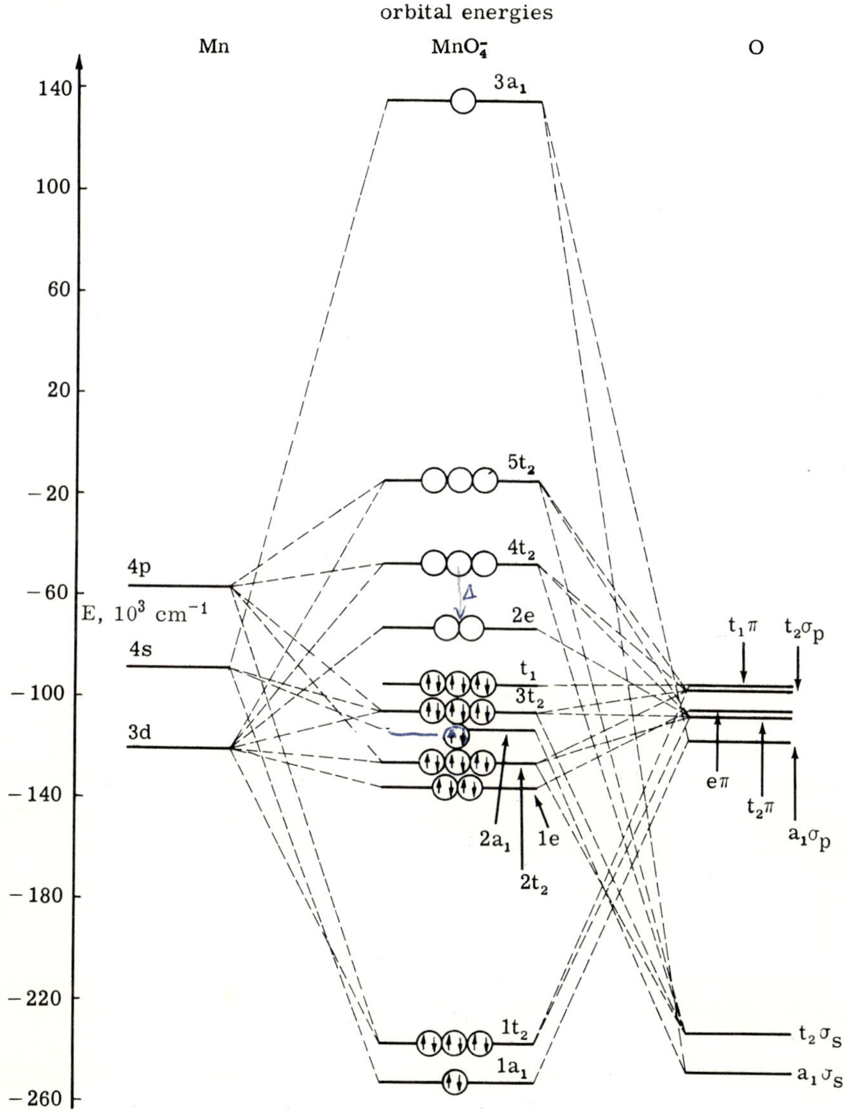

Figure 8-8 Molecular-orbital energy-level diagram for MnO_4^-.

$$H_{ii} = aH'_{ii}$$

For the metal and ligand symmetry orbitals, the a's are calculated to be

		a			a
E	d	1.0000	T_2	p	1.0000
	π	1.0239		σ_p	0.95271
				d	1.0000
A_1	s	1.0000		σ_s	0.98269
	σ_s	1.0485		π	1.0322
	σ_p	1.1193	T_1	π	0.94785

The exchange integrals are estimated from the expression given in Eq. (8-50). The four secular equations are solved to give the final MO's. For example, the final 2×2 E secular equation is as follows:

$$\begin{array}{c} \\ d \\ \pi \end{array} \begin{vmatrix} d & \pi \\ -121.28 - W & -57.77 - .2601W \\ -57.77 - .2601W & -104.13 - W \end{vmatrix} = 0 \qquad (8\text{-}63)$$

The final MO eigenvectors and eigenvalues are assembled in Table 8-15.

A diagram showing the relative MO energies in MnO_4^- is given in Figure 8-8. The ground state is ... $(3t_2)^6 (t_1)^6 = {}^1A_1$. The spectrum of MnO_4^- in aqueous solution shows fairly intense bands at 18,300, 32,200, and 44,000 cm^{-1}. These bands may be assigned[15] as the one-electron transitions $t_1 \rightarrow 2e$, $3t_2 \rightarrow 2e$, and $t_1 \rightarrow 4t_2$, respectively. A comparison of theoretical and experimental values is given in Table 8-16.

8-17 CALCULATION OF CrF_6^{3-}

We present the results of a calculation of the MO's of CrF_6^{3-}, for the benefit of persons wishing to work through the problem. The procedure followed is the same as in the calculation of MnO_4^-.

1. Valence orbitals: Cr 3d, 4s, 4p
   F 2p (we shall not use the F 2s; see the discussion at the end of this section)

2. R(Cr—F) = 1.93 A

3. Group overlap integrals:

$$A_{1g}:\ G_{A_{1g}}(s,\sigma_p) = 0.2673$$

$$E_g:\ G_{E_g}(d,\sigma_p) = 0.2321$$

$$T_{1u}:\ G_{T_{1u}}(p,\sigma_p) = 0.1427$$

$$G_{T_{1u}}(p,\pi_p) = 0.2788$$

$$G_{T_{1u}}(\sigma_p\pi_p, \sigma_p\pi_p) = 0.0312$$

$$T_{2g}:\ G_{T_{2g}}(d,\pi_p) = 0.1728$$

4. Coulomb integrals:

For Cr, from VSIE data in Table 8-11, we obtain VSIE curves given in Table 8-17. For F 2p we obtain VSIE curves from neutral-atom VSIE in Table 8-12.

Table 8-17 VSIE Functions for Cr[a]

VSIE	Starting configuration	A	B	C
d	d^n	14.75	74.75	35.1
d	$d^{n-1}s$	9.75	95.95	57.9
d	$d^{n-1}p$	9.75	96.95	67.7
s	$d^{n-1}s$	8.05	57.55	53.2
s	$d^{n-2}s^2$	8.05	66.85	63.3
s	$d^{n-2}sp$	8.05	60.45	74.7
p	$d^{n-1}p$	7.25	47.55	28.4
p	$d^{n-2}p^2$	7.25	52.85	37.8
p	$d^{n-2}sp$	7.25	52.85	38.1

[a] Energies in 1000 cm^{-1}.

5. Final H'_{ii}'s (in 1000 cm^{-1}); q = 0.7733; s = 0.2128; p = 0.2833

Cr	3d	−122.7
	4s	− 95.6
	4p	− 62.5
F	$2p_\sigma$	−160.4
	$2p_\pi$	−150.4

In the MnO_4^- calculation we took $2p_\sigma = 2p_\pi = 2p$ as a first approximation. However, it is probably better to take $H_{2p_\sigma 2p_\sigma} < H_{2p_\pi 2p_\pi}$, since the σ orbitals are directed toward the metal nucleus. From studies of charge-transfer spectra, it appears that $H_{\sigma\sigma}$ is approximately 10,000 cm^{-1} more stable than $H_{\pi\pi}$. Thus we take $2p_\pi = 2p_\sigma + 10{,}000$ cm^{-1} for F.

6. H_{ii} correction factors (a's):

		a			a
A_{1g}	4s	1.000	T_{1g}	$2p_\pi$	0.9682
	$2p_\sigma$	1.059			
E_g	3d	1.000	T_{2u}	$2p_\pi$	0.9917
	$2p_\sigma$	0.9705			
T_{1u}	4p	1.000	T_{2g}	3d	1.000
	$2p_\sigma$	0.9981		$2p_\pi$	1.030
	$2p_\pi$	1.009			

7. Final form of E_g secular equation:

$$\begin{array}{c} d \\ \sigma \end{array}\begin{vmatrix} (-122.68 - W) & (-65.12 - 0.2321W) \\ (-65.12 - 0.2321W) & (-155.67 - W) \end{vmatrix} = 0$$

Table 8-18. Final results:

Orbital energy	Designation	Degeneracy	Ground-state occupancy
−174.68	$1a_{1g}$	1	2
−169.73	$1e_g$	2	4
−163.48	$1t_{2g}$	3	6
−162.86	$1t_{1u}$	3	6
−150.06	$2t_{1u}$	3	6
−149.15	t_{2u}	3	6
−145.62	t_{1g}	3	6
−105.93	$2t_{2g}$	3	3
− 92.52	$2e_g$	2	0
− 73.12	$2a_{1g}$	1	0
− 43.05	$3t_{1u}$	3	0

	Eigenvalues	Eigenvectors	
		4s	$2p_\sigma$
a_{1g}	−174.68	0.2260	0.9156

	Eigenvalues	Eigenvectors		
		3d	$2p_\sigma$	
e_g	−169.73	0.4387	0.8026	
				$2p_\pi$
t_{1g}	−145.62			1.0000
		4p	$2p_\sigma$	$2p_\pi$
t_{1u}	−162.86	0.0826	0.8664	0.4240
	−150.06	−0.0809	0.4905	−0.8670
		3d		$2p_\pi$
t_{2g}	−163.48	0.3917		0.8549
				$2p_\pi$
t_{2u}	−149.15			1.000

Cr charge = 0.7733
Cr configuration = $d^{4.7306}s^{0.2128}p^{0.2833}$

The calculated value of Δ for CrF_6^{3-} is ca. 13,400 cm^{-1}. The experimental value is 15,200 cm^{-1}. The calculated separation of t_{2u} and $2t_{2g}$ is ca. 43,200 cm^{-1}; the first L → M charge-transfer band in CrF_6^{3-} is probably at even higher energy than the calculated value.

A casual glance at the excellent agreement between theory and experiment in our two examples, MnO_4^- and CrF_6^{3-}, may give one the impression that this type of theory is, in fact, quite good. On close examination, however, it is clear that the ground rules are slightly different in the tetrahedral and octahedral calculations. Specifically, a double σ (ligand) basis set (2s,2p) was used in MnO_4^-, and only a single 2p function in CrF_6^{3-}. Indeed, a calculation with a double σ basis set in CrF_6^{3-} gives a Δ value which is much too large, while a calculation of MnO_4^- with a single σ valence function gives a Δ with the wrong sign. This simply means that the theory is incapable in its present form of quantitatively calculating Δ as a function of geometry. From the calculations published, it is a striking fact that all the successful octahedral or distorted octahedral calculations have employed a single σ function, while the successful tetrahedral calculations have required a double σ set.

Thus, although we feel it is useful to do simple MO calculations to *help* in interpreting and classifying electronic spectra, it is certainly clear that the calculated MO's must be consistent with a goodly amount of experimental information before they can be taken seriously.

REFERENCES

1. E. Clementi, *J. Chem. Phys.*, **40**, 1944 (1964).
2. R. E. Watson, *Phys. Rev.*, **118**, 1036 (1960); **119**, 1934 (1960).
3. (a) J. W. Richardson, W. C. Nieuwpoort, R. R. Powell, and W. F. Edgell, *J. Chem. Phys.*, **36**, 1057 (1962); (b) J. W. Richardson, R. R. Powell, and W. C. Nieuwpoort, *J. Chem. Phys.* **38**, 796 (1963).
4. J. C. Slater, *Phys. Rev.*, **36**, 57 (1930).
5. L. E. Sutton, "Interatomic Distances," Spec. Publ. 11, The Chemical Society, London, 1958.
6. L. Pauling, "The Nature of the Chemical Bond," Cornell University Press, Ithaca, N.Y., 1960.
7. A. F. Wells, "Structural Inorganic Chemistry," Oxford University Press, New York, 1962.
8. R. S. Mulliken, C. A. Rieke, D. Orloff, and H. Orloff, *J. Chem. Phys.*, **17**, 1248 (1949).
9. (a) H. H. Jaffé and G. O. Doak, *J. Chem. Phys.*, **21**, 196 (1953); (b) H. H. Jaffé, *J. Chem. Phys.*, **21**, 258 (1953); (c) J. L. Roberts and H. H. Jaffé, *J. Chem. Phys.*, **27**, 883 (1957).
10. L. Leifer, F. A. Cotton, and J. R. Leto, *J. Chem. Phys.*, **28**, 364, 1253 (1958).
11. D. P. Craig, A. Maccoll, R. S. Nyholm, L. E. Orgel, and L. E. Sutton, *J. Chem. Soc.*, 354 (1954).
12. A. Lofthus, *Mol. Phys.*, **5**, 105 (1962).
13. M. Wolfsberg and L. Helmholz, *J. Chem. Phys.*, **20**, 837 (1952).
14. C. J. Ballhausen and H. B. Gray, *Inorg. Chem.*, **1**, 111 (1962).
15. A. Viste and H. B. Gray, *Inorg. Chem.*, **3**, 1113 (1964).
16. R. S. Mulliken, *J. Chem. Phys.*, **23**, 1833 (1955).
17. C. J. Ballhausen, "Introduction to Ligand Field Theory," McGraw-Hill, New York, 1962, p. 54.

Problems

1. Prove that the 1s orbital of hydrogen is orthogonal to the 2s orbital. Is this also true if Z, the "effective nuclear charge," is different in 1s and 2s?

2. Sketch the periodic system in a world such as ours, but with only two dimensions:

$$\nabla^2 = \frac{1}{r}\frac{\partial}{\partial r}\left(r\frac{\partial}{\partial r}\right) + \frac{1}{r^2}\frac{\partial^2}{\partial \varphi^2}$$

in the plane (r, φ). Set the constant in the separation of the $R(r)$ and $\Phi(\varphi)$ equations equal to k^2.

3. Calculate the energy needed to ionize an electron from the 2s orbital of Be^{3+}, from a 2p orbital of Be^{3+}, and from the 1s orbital of He^+.

4. Write the determinantal wave functions for the ground state of the oxygen atom, assuming the unpaired electrons have α spin. What is the orbital degeneracy of the ground state? What is the spin degeneracy?

5. The elliptical coordinates λ and μ are defined as follows (see the figure):

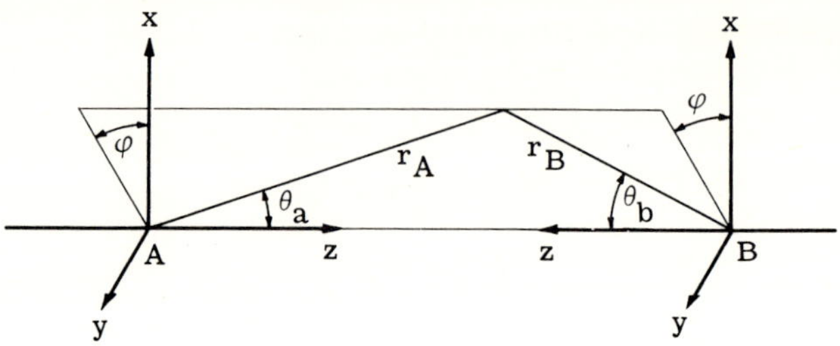

$$\lambda = \frac{r_A + r_B}{R}$$

$$\mu = \frac{r_A - r_B}{R}$$

$$d\tau = \frac{R^3}{8}(\lambda^2 - \mu^2)d\lambda\, d\mu\, d\varphi$$

$$1 \leq \lambda \leq \infty \qquad -1 \leq \mu \leq 1 \qquad 0 \leq \varphi \leq 2\pi$$

R is the distance AB. Further,

$$r_A \cos\theta_a = \frac{R}{2}(1 + \lambda\mu)$$

$$r_B \cos\theta_b = \frac{R}{2}(1 - \lambda\mu)$$

$$r_A \sin\theta_a = r_B \sin\theta_b = \frac{R}{2}[(\lambda^2 - 1)(1 - \mu^2)]^{1/2}$$

Calculate the overlap integral, S(1s,1s), between two 1s orbitals, centered at a distance R from each other. What is the value of S(1s,1s) for $R \to 0$ and ∞?

6. With the normalized $2p_z$ and $2p_x$ orbitals given by $2p_z = R(r) \times \cos\theta$ and $2p_x = R(r)\sin\theta\cos\varphi$, with

$$R(r) = \frac{1}{4\sqrt{2\pi}}\left(\frac{Z}{a_0}\right)^{5/2} re^{-(Z/2a_0)r}$$

calculate the following using elliptical coordinates (see Problem 5):

$$S(p_\sigma) = S(2p_z A, 2p_z B)$$

$$S(p_\pi) = S(2p_x A, 2p_x B) = S\left(2p_y A, 2p_y B\right)$$

The answers should be expressed using the so-called A integrals, defined as

$$A_n(\alpha) = \int_1^{\infty} x^n e^{-\alpha x}\, dx = \frac{n!e^{-\alpha}}{\alpha^{n+1}} \sum_{k=0}^{n} \frac{\alpha^k}{k!}$$

Calculate numerical values for S_π and S_σ, with $Z = 4$ and $R = 2a_0$.

7. In H_2^+ we take

$$\psi_A = \psi_B = 1s = \sqrt{\frac{1}{\pi}}\left(\frac{Z}{a_0}\right)^{3/2} e^{-(Z/a_0)r}$$

and

$$\mathcal{H} = -\frac{\hbar^2}{2m}\nabla^2 - \frac{Ze^2}{r_A} - \frac{Ze^2}{r_B} + \frac{Z^2e^2}{R}$$

Show that $H_{AA} = H_{BB}$, as well as $H_{AB} = H_{BA}$, using elliptical coordinates as in Problem 5. With the overlap integral from Problem 5, obtain the expression for the lowest orbital in H_2^+ as a function of R and Z. Putting $Z = 1$, sketch the curve $W = f(R)$.

8. Construct a molecular-orbital energy-level diagram for HF. Compare and contrast the bonding and the energy levels in HF and LiH.

9. The ground state of C_2 is not known for certain. The $^3\Pi_u$ and $^1\Sigma_g^+$ states are believed to have approximately the same energy. Which orbital configurations are responsible for these states, and under what conditions would they have nearly the same energy?

10. What are the term designations for the ground states of the following molecules: S_2, B_2, Cl_2, Cl_2^+, O_2^-, BF, BN, BO, CH, and NH?

11. A 0.01-M solution of a substance absorbs light between 5500 and 8500 A. In a cell of 1-cm path length, the maximum absorption at 6800 A is 0.6 units, and the absorption at 7500 and 6300 A is 0.3 units. Calculate the ϵ and the f values of the band. Is the transition fully allowed?

12. Calculate the dipole transition integral D for the $^1\Sigma_g^+ \rightarrow {}^1\Sigma_u^+$ transition in H_2 as a function of S and R. For the ground state, $S(1s_A, 1s_B) \approx 0.75$ and $R = 0.74$ A. The transition is observed at some 10^5 cm^{-1}. What is the theoretical f value?

13. Theoretical calculations of the rotatory dispersion curves for molecules require knowledge of the magnetic-dipole transition integral M between two states, ψ_1 and ψ_2. This integral is $M = -i\beta \times \int \psi_1^* \hat{M} \psi_2 \, d\tau$, where β is the Bohr magneton ($= e\hbar/2mc$) and $i = \sqrt{-1}$.

$$\mathbf{M} = \mathbf{i}\left(y\frac{\partial}{\partial z} - z\frac{\partial}{\partial x}\right) + \mathbf{j}\left(z\frac{\partial}{\partial x} - x\frac{\partial}{\partial z}\right) + \mathbf{k}\left(x\frac{\partial}{\partial y} - y\frac{\partial}{\partial x}\right)$$

Calculate the value and direction of **M** for the $^1A_1 \rightarrow {}^1A_2$ transition in formaldehyde.

14. The F_2 molecule shows a continuous absorption band with a maximum at 34,500 cm^{-1}. Discuss the possible assignment and polarization of the transition.

15. When a carbon arc light burns in air, the radical CN is formed. The spectrum of the arc shows two strong absorption bands at about 9000 cm^{-1} and 26,000 cm^{-1}.

a. What is the term designation of the ground state?

b. Make tentative assignments for the two transitions.

c. How would the bands be polarized for your assignments?

16. What is the term designation of the ground state of N_3, assuming the molecule is linear? What is the term designation for a bent N_3 molecule?

17. The N_3^- and NNO molecules are linear. What are the ground-state terms? What are the designations of the excited states which arise on the lowest energy electronic transition? Discuss the intensities and polarizations that might be anticipated for the bands.

18. The molecule HNO is diamagnetic in its ground state. We note that H + N equals 0. What shape does the molecule have? What would be the first electronic transition? If you apply similar ideas to HCN, what do you obtain?

19. Formulate the bonding in NH_3 in terms of delocalized molecular orbitals. The molecule is trigonal-pyramidal (C_{3v} point group). Compare the general molecular-orbital description with a localized "tetrahedral" model for NH_3. Discuss the values of the following bond angles: H—N—H, 107°; H—P—H (in PH_3), 94°; and F—N—F (in NF_3), 103°.

20. Boron trifluoride has a trigonal-planar structure. Formulate the bonding in terms of molecular orbitals for the D_{3h} symmetry. In addition, construct wave functions for three equivalent sp^2 hybrid orbitals, using the $2p_x$, $2p_y$, and 2s boron valence orbitals, which may be used to form three localized bonds with the three fluorines. Compare and contrast the molecular-orbital and the hybrid-orbital descriptions.

21. The VCl_4 molecule has a tetrahedral structure. Construct the various LCAO's that give σ and π bonds between vanadium and the four chlorines. What is the ground state of the molecule? The first electronic transition is found at about 10,000 cm^{-1}. Assign this band.

22. The SF_6 molecule has an octahedral structure. Construct the various σ and π bonding orbitals. What is the ground state of the molecule?

23. Construct wave functions for six equivalent octahedral (d^2sp^3) hybrid orbitals, using $d_{x^2-y^2}$, d_{z^2}, s, p_x, p_y, and p_z valence orbitals.

24. Assume that the Coulomb energies H_{dd} and H_{LL} are the same for an octahedral ML_6 and for a tetrahedral ML_4 complex, and further that no π bonding occurs. Use (2-30) in connection with (8-49) or (8-50) to prove that the tetrahedral splitting $W(t_2^*) - W(e)$ is $\frac{4}{9}$ of the octahedral splitting $W(e_g^*) - W(t_{2g})$.

25. An octahedral complex ion in $(MA_6)^{3+}$ has excited electronic states corresponding to the excitation of an electron from an orbital on the M^{3+} ion into an empty orbital on one of the ligands A, say the jth one (charge transfer excitation). Let us designate the zero-order wave function representing an electron from M excited to an orbital on ligand j as ϕ_j. Because the electron may "tunnel" from one ligand to another, there is an effective interaction between any two excited states ϕ_i, ϕ_j. The matrix elements are as follows:

$$(\phi_j | H' | \phi_j) = 0$$

$$(\phi_j | H' | \phi_i) = V_1 \text{ if i, j are adjacent vertices}$$

$$(\phi_j | H' | \phi_i) = V_2 \text{ if i, j are opposite vertices}$$

In zero-order all the states ϕ_j have excitation energy W_1 above the ground state.

a. Calculate the first-order perturbation energies and the correct zero-order wave functions associated with them.

b. What is the energy of the electric dipole-allowed transition?

Suggested Reading

Ballhausen, C. J., "Introduction to Ligand Field Theory," McGraw-Hill, New York, 1962.

Cartmell, E., and G. W. A. Fowles, "Valency and Molecular Structure," 2nd ed., Butterworths, London, 1961.

Cotton, F. A., "Chemical Applications of Group Theory," Wiley-Interscience, New York, 1963.

Coulson, C. A., "Valence," 2nd ed., Oxford University Press, New York, 1961.

Dodd, R. E., "Chemical Spectroscopy," Elsevier, Amsterdam, 1963.

Eyring, H., J. Walter, and G. E. Kimball, "Quantum Chemistry," Wiley, New York, 1960.

Gray, H. B., "Electrons and Chemical Bonding," Benjamin, New York, 1964.

Herzberg, G., "Spectra of Diatomic Molecules," Van Nostrand, Princeton, N.J., 1950.

Jaffe, H. H., and M. Orchin, "Theory and Applications of Ultraviolet Spectroscopy," Wiley, New York, 1962.

Ramsey, D. A., Electronic Structures of Polyatomic Molecules, in "Determination of Organic Structures by Physical Methods," Academic Press, New York, 1962.

Notes on the Molecular Orbital Treatment of the Hydrogen Molecule.

By Prof. C. A. Coulson and Miss I. Fischer,
Wheatstone Physics Department, King's College, London.

§ 1. Introduction.

The molecular orbital (m.o.) method has been used regularly for many years (Coulson 1947), but relatively little is known about its validity except on semi-empirical grounds. The purpose of these notes is to investigate the fundamentals of the m.o. method in its LCAO form (linear combination of atomic orbitals) somewhat more fully than before. For this study we have chosen the hydrogen molecule. By virtue of its simplicity this molecule allows us to make calculations which would be prohibitively difficult for larger systems. Now both the ground state and several excited states of H_2 have been successfully treated by other authors. Our object is not to achieve a better agreement with experiment than that already obtained, but to examine the possibilities and limitations of the m.o. method in its simplest form, and including certain refinements. Some of our considerations have been published earlier, though in a less completed form (see a paper by Coulson (1937) and a review by Van Vleck and Sherman (1935)), but we believe that the rest, particularly the detailed numerical values, are here made available for the first time.

If we confine ourselves to molecular orbitals composed out of atomic $1s$ functions ψ_a and ψ_b of the atoms A and B, the allowed m.o. are ϕ and χ, where

$$\phi=(\psi_a+\psi_b)/\{2(1+S)\}^{\frac{1}{2}}, \quad (1)$$

$$\chi=(\psi_a-\psi_b)/\{2(1-S)\}^{\frac{1}{2}}, \quad (2)$$

and

$$S=\text{overlap integral}=\int\psi_a(1)\psi_b(1)\,d\tau. \quad (3)$$

If ψ_a and ψ_b are exactly the same as the atomic orbitals for an isolated hydrogen atom (perturbation method of Coulson (1937)), then, in atomic units,

$$\psi_a(1)=\sqrt{\frac{1}{\pi}}e^{-r_{a1}}. \quad (4)$$

But if we include the possibility of a partial screening (variation method of Coulson (1937)), we write

$$\psi_a(1)=\sqrt{\frac{c^3}{\pi}}e^{-cr_{a1}}, \quad (5)$$

where c is a constant which varies with the internuclear distance R and which may be calculated in the standard manner. We shall be concerned with the inter-relations of the various molecular states that arise from different possible occupations of the orbitals ϕ and χ. No details of the manipulative part of our calculations need be given, for they follow completely conventional lines, once the appropriate wave functions for the molecule have been set up.

§ 2. Failure at Large Internuclear Distances in the Ground State.

It is well recognized that at large internuclear distances R the m.o. method fails badly because it takes insufficient account of the electron repulsion. This repulsion ensures that when the molecule dissociates one electron is found on each nucleus. But the conventional m.o. description of the ground state, viz.

$$\Psi=\phi(1)\phi(2)\times\text{spin term}, \qquad (6)$$

divides each electron equally between A and B. It is important to know at what value of R the wave function (6) falls seriously in error. This may be done as follows.

We may imagine that as R increases, one of the two m.o. tends to concentrate more round atom A, and the other around atom B. This could be achieved by replacing the m.o. $\phi(1)$ and $\phi(2)$ in (6) by two different orbitals

$$\psi_a+\lambda\psi_b, \qquad \psi_b+\lambda\psi_a, \qquad (7)$$

where λ is a parameter not necessarily equal to 1. Assigning one electron to each of these orbitals we describe the molecular system by the wave function

$$\{\psi_a(1)+\lambda\psi_b(1)\}\{\psi_b(2)+\lambda\psi_a(2)\}. \qquad (8)$$

Now, the parameter λ has to be determined by a variation method. For those values of R for which λ turns out to be 1, we can say that the ordinary m.o. is a valid type of approximation; but for those values for which λ differs considerably from 1, we interpret the significance of the wave function (8) to be that the repulsion between the electrons, tending to separate them on to different nuclei, is stronger than the additional attraction which arises when both electrons are under the influence of two nuclei.

We have chosen the appropriate value of λ for each R by minimizing the "energy" function $\int\Psi^*H\Psi\,d\tau/\int\Psi^*\Psi\,d\tau$, where H is the Hamiltonian of the hydrogen molecule. For this purpose we used the values of the exponent c in (5) which had previously been calculated for the true m.o. wave function (6) by Coulson (1937). Fig. 1 (curve 1) shows the value of λ plotted against R. Up to R=2·27 a.u. (*i. e.* $R=1{\cdot}6\,R_0$, where R_0 is the equilibrium distance in H_2) λ is exactly equal to 1. But when $R>1{\cdot}6\,R_0$, λ rapidly falls almost to zero, indicating a complete breakdown in the simple m.o. theory.

It is very probable that these conclusions hold for other molecules. In that case our calculation shows very clearly the dangers inherent in too naive an application of m.o. theory to interactions across large

Fig. 1.

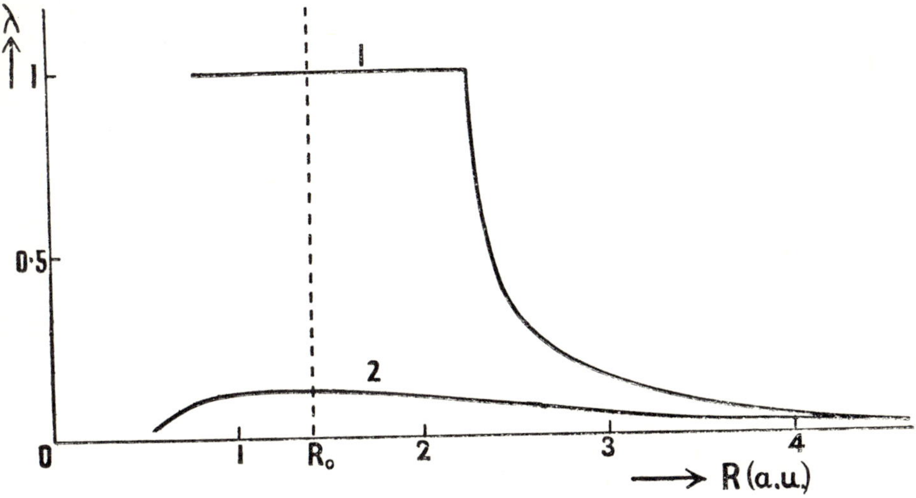

Values of the parameter λ in unsymmetrical m.o.'s (7) giving energy minimum with : curve 1 : an asymmetrical wave function (8) ; curve 2 : a wave function including exchange (9).

distances, as, for example, in trying to follow the complete course of a chemical reaction (Coulson and Dewar 1947).

§ 3. Over-Emphasis of Ionic Terms.

A second criticism of m.o. wave functions such as (6) is that even at the equilibrium distance they overemphasize ionic terms. We may discuss this also with the aid of the unsymmetrical molecular orbitals (7). If we suppose that those are the occupied m.o., then the true wave function, in which antisymmetrization is employed, is, apart from a normalization factor and a spin term :

$$\Psi'=\{\psi_a(1)+\lambda\psi_b(1)\}\{\psi_b(2)+\lambda\psi_a(2)\}+\{\psi_b(1)+\lambda\psi_a(1)\}\{\psi_a(2)+\lambda\psi_b(2)\}. \quad . \quad (9)$$

If we write this in the equivalent form

$$\Psi'=(1+\lambda^2)\{\psi_a(1)\psi_b(2)+\psi_b(1)\psi_a(2)\}+2\lambda\{\psi_a(1)\psi_a(2)+\psi_b(1)\psi_b(2)\}, \quad . \quad (10)$$

we recognize it as the wave function used by Weinbaum (1933), who actually used

$$\Psi'=\{\psi_a(1)\psi_b(2)+\psi_b(1)\psi_a(2)\}+\mu\{\psi_a(1)\psi_a(2)+\psi_b(1)\psi_b(2)\}. \quad . \quad . \quad (11)$$

The expressions in brackets are the simple Heitler–London covalent wave function and the pure ionic wave function; μ shows the degree of mixing between them. It is interesting—and, to us, new—that the conventional covalent–ionic resonance can be described equally well by the introduction of asymmetrical molecular orbitals, as in (9). The significance of this equivalence is that if we wish to refine the m.o. theory, while still retaining the idea of a molecular orbital, one possibility is to abandon the m.o. ϕ in equation (1) in favour of the pair of functions (7). If it turns out that with the wave function (9), the value of λ is not equal to 1, then the best description of the molecule in terms of a pair of m.o. composed linearly of ψ_a and ψ_b, requires the two orbitals to be different.

Values of μ which minimize the energy function for (11) are shown in fig. 2 in terms of the internuclear distance R. The three curves shown relate to different choices of the screening constant c. In curve 1 we used

Fig. 2.

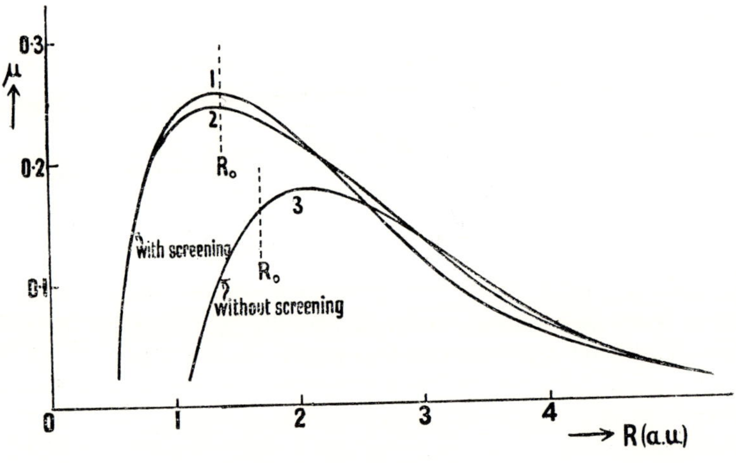

Contribution μ of ionic terms calculated with best screening constant of :—
curve 1 : a molecular orbital wave function ; curve 2 : a Heitler–London wave function ; curve 3 : atomic orbitals.

The dotted lines show the values of the equilibrium distance R_0 calculated with these wave functions.

the values calculated for a pure m.o. wave function by Coulson (1937); in curve 2 we used the values appropriate to a pure Heitler–London wave function; and in curve 3 we used the atomic wave functions (perturbation method) for which $c=1$. Fig. 2 shows that provided some allowance is made for screening, the value of μ is not particularly sensitive to c; but if no allowance whatever is made, values of μ in error by as much as 70 per cent, may occur. As we should expect, the ionic contribution to the ionic–covalent hybrid decreases with increasing R.

A comparison of (10) and (11) allows us to convert our μ-values into λ-values. The conversion appropriate to the second curve in fig. 2 is shown as curve 2 in fig. 1, where it may be contrasted with the curve already drawn on the basis of wave function (8). This shows very clearly how grossly the naive m.o. theory with $\lambda=1$, exaggerates the importance of ionic terms, and how, though in a smaller degree, the asymmetrical function (8) also does so. As we might have anticipated, antisymmetrization, or exchange, has the effect of reducing the two-centre character of the molecular orbitals, *i. e.*, of making them more asymmetrical.

§ 4. Excited States ; Configurational Interaction.

There is a third point of view from which we may question the validity of the naive m.o. treatment. The idea is thoroughly discussed in a qualitative way by Hund (1932), but the only published numerical results that we have found, are incomplete and take no account of any screening (Slater 1930, Mulliken 1932). If we start with the two m.o. ϕ and χ of (1) and (2), we can form four molecular states $\Psi_1 \ldots \Psi_4$. Neglecting spin terms, these are

$$\left.\begin{array}{ll} \Psi_1=\phi(1)\phi(2), & {}^1\Sigma_g \\ \Psi_2=\phi(1)\chi(2)-\chi(1)\phi(2), & {}^3\Sigma_u \\ \Psi_3=\phi(1)\chi(2)+\chi(1)\phi(2), & {}^1\Sigma_u \\ \Psi_4=\chi(1)\chi(2). & {}^1\Sigma_g \end{array}\right\} \quad . \quad . \quad . \quad . \quad . \quad (12)$$

These may be said to be the wave functions associated with the " pure configurations " ϕ^2, $\phi\chi$ and χ^2. If our molecular orbital description of the molecule is sound, these configurations should be practically non-interacting with each other : and also the energy of the singlet function Ψ_3, which is an ionic function and represents the main part of the experimentally found B state, and which corresponds to excitation of one electron in a process $\phi^2 \rightarrow \phi\chi$, should lie between that of ϕ^2, in which there is no excitation, and that of χ^2 in which two electrons are excited. Now Craig (1949) has given reasons for believing that among the π-electrons of ethylene, where states analogous to those of (12) occur, the energy of the configuration χ^2 is either just below, or closely similar to, the energy of $\phi\chi$. This makes it desirable to make similar calculations for H_2, where the approximations in calculating the energy are less severe than for ethylene. We have therefore calculated—and show in fig. 3—the energies appropriate to the four pure configurations $\Psi_1 \ldots \Psi_4$. In these calculations we used the atomic orbitals (4) rather than (5) because that seemed the best compromise between values $c>1$ for the ϕ orbitals and $c<1$ for the χ orbitals (Coulson 1937). Fortunately the main character of these curves is still maintained if, instead, we use the values of c found most suitable for the ϕ orbitals. It will be noticed that at fairly large internuclear distances ($R>3{\cdot}2$ a.u.) the energies of the singlets $\phi\chi$ and χ^2 are in an inverted order. The interpretation of this is that our pure configurations interact with each other. So far as the configurations listed in (12) are concerned, this only involves Ψ_1 and Ψ_4, for these are the only ones with equivalent symmetry. We ought, therefore, to replace Ψ_1 and Ψ_4 by two " mixed configurational " wave functions

$$\left.\begin{aligned}\Psi_5&=\Psi_1+\nu\,\Psi_4,\\ \Psi_8&=\Psi_1+\nu'\Psi_4,\end{aligned}\right\}, \qquad (13)$$

where ν and ν' are two constants such that $\nu\nu'=-1$. Explicit expansion of Ψ_5 shows that it is of exactly the same form as (10), so that we may use

Fig. 3.

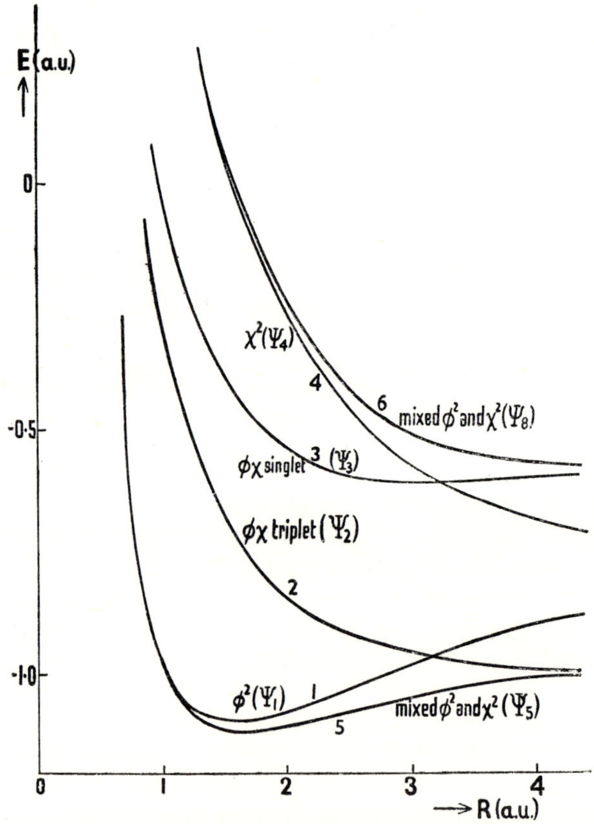

Energy curves for the hydrogen molecule. Curves 1 . . . . 4 correspond to the " pure configurations " Ψ_1 . . . . Ψ_4 of (12), curves 5 and 6 to Ψ_5 and Ψ_8 (13) which account for configurational interaction between Ψ_1 and Ψ_4.

our calculations with this wave function to infer the value of ν corresponding to any chosen R, without further labour. The value $\nu\simeq 1/8$ at $R=R_0$ has already been given by Slater (1930) and used by Mulliken (1932). Our fig. 4 shows that for large distances ν increases, indicating an increasing configurational interaction, until, at $R\rightarrow\infty$, we find $\nu\rightarrow 1$. The figure also shows that the degree of interaction is almost independent of the choice of screening constant.

This configurational interaction lowers the energy of the state Ψ_1 and raises that of Ψ_4, so that, as fig. 3 shows, the unsatisfactory crossing-over of the energies $\phi\chi$ and χ^2 no longer takes place, even at large distances. For reasonably small R the configurational representation in terms of ϕ and χ is valid—though there is always some degree of interaction between configurations—but for R larger than about $1 \cdot 5\ R_0$ this interaction is so large that the ϕ, χ representation becomes quite invalid.

Fig. 4.

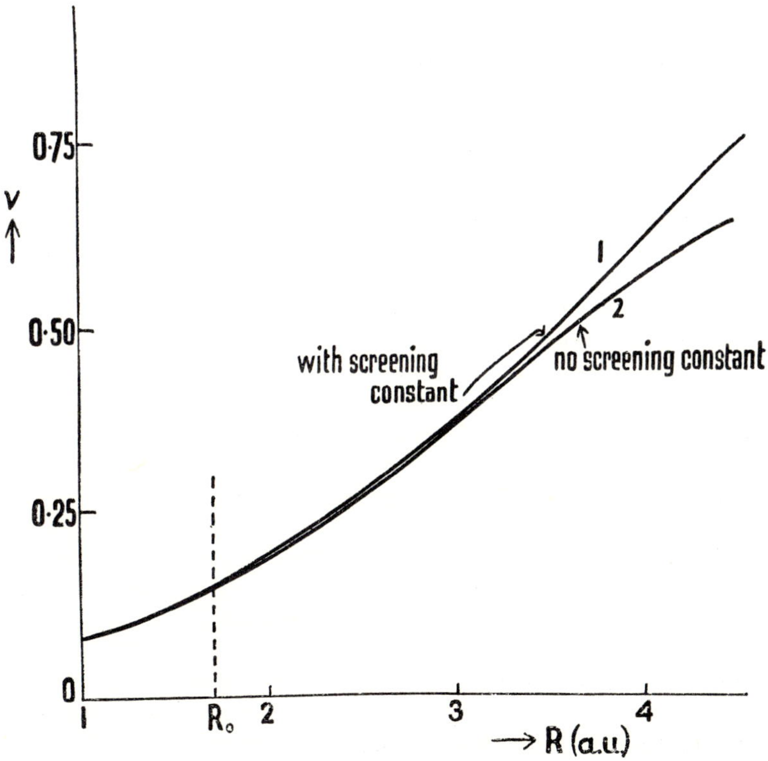

Configurational interaction constant ν calculated with :—curve 1 : best screening constant of molecular orbital wave function ; curve 2 : atomic orbitals.

As already pointed out, Ψ_5 and Ψ_8 are of the same form as (10) and are thus equivalent to the use of asymmetrical molecular orbitals. We may make the parallel between conventional descriptions of the wave functions for H_2 and our new description a little clearer if we label the two asymmetrical m.o.'s (7) A_+ and B_+, and we introduce two related orbitals A_-, B_-, where

$$\left.\begin{aligned} A_{\pm} &= \psi_a \pm \lambda\psi_b, \\ B_{\pm} &= \psi_b \pm \lambda\psi_a. \end{aligned}\right\} \quad \text{. . . . . (14)}$$

It is easily verified that the "mixed configurations" which result from interaction among the group (12) are represented by the four wave functions

$$\left.\begin{array}{ll}\Psi_5=A_+(1)B_+(2)+B_+(1)A_+(2), & {}^1\Sigma_g\\ \Psi_6=\{A_+(1)B_-(2)-B_-(1)A_+(2)\}-\{A_-(1)B_+(2)-B_+(1)A_-(2)\}, & {}^3\Sigma_u\\ \Psi_7=\{A_+(1)B_-(2)+B_-(1)A_+(2)\}-\{A_-(1)B_+(2)+B_+(1)A_-(2)\}, & {}^1\Sigma_u\\ \Psi_8=A_-(1)B_-(2)+B_-(1)A_-(2). & {}^1\Sigma_g\end{array}\right\} \quad (15)$$

The energies appropriate to these four wave functions are given by curves 5, 2, 3 and 6 respectively of fig. 3.

Now the Σ_u states in (15) are identical with those of (12). But the Σ_g states could, if we wished, be regarded as arising from pure configurations A_+B_+ and A_-B_-. Indeed, to this degree of approximation, we can avoid the annoying configurational interaction between Ψ_1 and Ψ_4 if we replace the symmetrical molecular orbitals (1) and (2) by the asymmetrical ones (14), giving the non-interacting configurations Ψ_5 and Ψ_8. If we are to use a configurational representation at all, it must be in terms of these latter functions. In this way we get further light on the discussion in § 3.

In any complete treatment of this molecule, we ought also to include further interaction with other configurations such as those built on the 2*s* or 2*p* atomic orbitals. This would depress all the levels slightly, but it is unlikely that their sequence would be changed.

In conclusion we wish to point out that the configurational interaction to which we have drawn attention is likely to be particularly significant when the energies of the levels are closer together than in H_2. We may therefore anticipate that some of the difficulties already found in interpreting the u.v. spectra of large condensed molecules arise from its neglect. Work in progress in this department suggests that this is indeed the case (Miss J. Jacobs, unpublished calculations).

One of us (I. F.) would like to acknowledge a grant from the Swedish State Council for Scientific Research—(Statens Naturvetenskapliga Forskningsrad)—which has made possible this work.

References.

Coulson, C. A., 1937, *Trans. Faraday Society*, **33**, 1479; 1947, *Quart Reviews*, **1**, 144.
Craig, D. P. In course of publication.
Hund, F., 1932, *Zeit. Physik*, **73**, 1,
Jacobs, Miss J. Unpublished calculations.
Mulliken, R. S., 1932, *Phys. Rev.*, **41**, 49, especially pp. 68–70.
Slater, J. C., 1930, *Phys. Rev.*, **35**, 509, especially pp. 514–5.
Van Vleck, J. H., and Sherman, A., 1935, *Rev. Modern Physics*, **7**, 167
Weinbaum, S., 1933, *J. Chem. Phys.*, **1**, 593.

Directed Valence*

George E. Kimball
Department of Chemistry, Columbia University, New York, New York
(Received October 19, 1939)

The problem of directed valence is treated from a group theory point of view. A method is developed by which the possibility of formation of covalent bonds in any spatial arrangement from a given electron configuration can be tested. The same method also determines the possibilities of double and triple bond formation. Previous results in the field of directed valence are extended to cover all possible configurations from two to eight *s*, *p*, or *d* electrons, and the possibilities of double bond formation in each case. A number of examples are discussed.

Introduction

Problems of directed valence, like most problems of molecular structure, can be attacked by either of two methods: the method of localized electron pairs (Heitler-London) or the method of molecular orbitals (Hund-Mulliken). While it is now realized[1] that these methods are but different starting approximations to the same final solution, each has its advantages in obtaining qualitative results. Theories of directed valence based on the methods of localized pairs have been developed by Slater[2] and Pauling[3] and extended by Hultgren.[4] The method of molecular orbitals has been developed principally by Hund[5] and Mulliken.[6] These methods have been compared extensively by Van Vleck and Sherman.[1]

In the previous papers no attempt has been made to discover *all* the possible stable electron groups which lead to directed valence bonds, nor have the possibilities of double bond formation been completely explored. In the present paper both of these deficiencies in the theory have been removed.

Method

Pauling's method consists of finding linear combinations of *s*, *p*, and *d* orbitals which differ from each other only in direction. Thus he has found combinations which are directed toward the corners of a tetrahedron, others directed toward the corners of a square, and others directed toward the corners of an octohedron. Electrons occupying these new orbitals can then resonate with unpaired electrons occupying orbitals of other atoms lying in the directions of these new orbitals and so form covalent bonds with these atoms. The further these orbitals project in the direction of the surrounding atoms, the stronger should be the resulting bonds.

In order to construct such sets of orbitals, it is most convenient to make use of group theory. Each set of equivalent directed valence orbitals has a characteristic symmetry group. If the operations of this group are performed on the orbitals, a representation, which is usually reducible, is generated. By means of the character table of the group[7] this representation, which we shall call the σ representation, can be reduced to its component irreducible representations. The *s*, *p*, and *d* orbitals of the atom also form representations of the group, and can also be divided into sets which form irreducible representations.[8]

Let us refer to the set of equivalent valence orbitals as the set $\mathfrak{S}$. If the transformation which reduces this set is T, then the set $T\mathfrak{S}$ can be broken up into subsets, each of which forms a basis for one of the irreducible representations of the symmetry group of $\mathfrak{S}$.

$$T\mathfrak{S}=\sum a_i\mathfrak{S}_i, \tag{1}$$

* Presented at the Boston meeting of the American Chemical Society, September 15, 1939. Publication assisted by the Ernest Kempton Adams Fund for Physical Research of Columbia University.

[1] Van Vleck and Sherman, Rev. Mod. Phys. **7**, 167 (1935).
[2] Slater, Phys. Rev. **37**, 841 (1931).
[3] Pauling, J. Am. Chem. Soc. **53**, 1367, 3225 (1931).
[4] Hultgren, Phys. Rev. **40**, 891 (1932).
[5] Hund, Zeits. f. Physik **73**, 1 (1931); **73**, 565 (1931); **74**, 429 (1932).
[6] Mulliken, Phys. Rev. **40**, 55 (1932); **41**, 49 (1932); **41**, 751 (1932); **43**, 279 (1933).
[7] For the group theoretical methods used here see Wigner, *Gruppentheorie* (Vieweg, Braunschweig, 1931); Weyl, *Theory of Groups and Quantum Mechanics* (tr. Robertson) (Methuen, London, 1931); Van der Waerden, *Gruppentheoretische Methoden in der Quantenmechanik* (Springer, Berlin, 1932).
[8] Bethe, Ann. d. Physik (5) **3**, 133 (1929).

where $\mathfrak{S}_i$ belongs to the ith irreducible representation. If we let $\mathfrak{R}$ be the set of available orbitals of the atom, these may always be chosen so that they fall into subsets each of which is a basis for one of the irreducible representations:

$$\mathfrak{R}=\sum b_i\mathfrak{R}_i. \quad (2)$$

If each of the coefficients b_i in (2) is equal to or greater than the corresponding a_i in (1), we may form the new subset $\mathfrak{R}'$, given by

$$\mathfrak{R}'=\sum a_i\mathfrak{R}_i, \quad (3)$$

in which each orbital transforms in exactly the same way as the corresponding orbital in (1). If we now apply the inverse transformation T^{-1} to this set we obtain a set of orbitals $T^{-1}\mathfrak{R}'$ which must have exactly the symmetry properties of the desired set $\mathfrak{S}$.

To illustrate this process, consider the set $\mathfrak{S}=\sigma_1+\sigma_2+\sigma_3$ consisting of three valence orbitals lying in a plane and making equal angles of 120° with each other. If the atom in question has available s, p, and d orbitals, the set $\mathfrak{R}$ consists of the nine orbitals s, p_x, p_y, p_z, d_z, d_{xz}, d_{yz}, d_{xy}, d_{x+y}. The symmetry group of $\mathfrak{S}$ is D_{3h}, and the set $\mathfrak{S}$ may be reduced by the transformation

$$\begin{aligned}\sigma_1'&=1/\sqrt{3}(\sigma_1+\sigma_2+\sigma_3)\\ \sigma_2'&=1/\sqrt{6}(2\sigma_1-\sigma_2-\sigma_3)\\ \sigma_3'&=1/\sqrt{2}(\sigma_2-\sigma_3).\end{aligned} \quad (3)$$

Of these orbitals, σ_1' belongs to the representation[9] A_1' while σ_2' and σ_3' belong to E'. Hence we find for (1)

$$\mathfrak{S}=A_1'+E'. \quad (4)$$

TABLE I. *Character table for trigonal orbitals.*

D_{3h}	E	σ_h	$2C_3$	$2S_3$	$3C_2$	$3\sigma_v$
A_1'	1	1	1	1	1	1
A_2'	1	1	1	1	−1	−1
A_1''	1	−1	1	−1	1	−1
A_2''	1	−1	1	−1	−1	1
E'	2	2	−1	−1	0	0
E''	2	−2	−1	1	0	0
s	1	1	1	1	1	1
p	3	1	0	−2	−1	1
d	5	1	−1	1	1	1
σ	3	3	0	0	1	1
π	6	0	0	0	−2	0

[9] The notation used here is that of Mulliken, Phys. Rev. **43**, 279 (1933).

The set $\mathfrak{R}$ is already reduced, for s and d_z belong to A_1', p_z to A_2'', the pairs p_x, p_y and d_{xy}, $-d_{x+y}$ to E', and the pair d_{xz}, d_{yz} to E''. Hence (2) becomes

$$\mathfrak{R}=2A_1'+A_2''+E'+E''. \quad (5)$$

Since the coefficients in (5) are each not less than the corresponding coefficient in (4), the desired directed orbitals are possible. In fact, since we have a choice between s and d_z for the orbital

TABLE II. *Reduction table for trigonal orbitals.*

D_{3h}	A_1'	A_1''	A_2'	A_2''	E'	E''
s	1	0	0	0	0	0
p	0	0	0	1	1	0
d	1	0	0	0	1	1
σ	1	0	0	0	1	0
π	0	0	1	1	1	1

belonging to A_1', we may construct two different sets of directed orbitals. If we choose for $\mathfrak{R}'$ the set

$$\mathfrak{R}'=s+p_x+p_y \quad (6)$$

and apply the inverse transformation

$$\begin{aligned}\sigma_1&=\frac{1}{\sqrt{3}}\sigma_1'+\frac{2}{\sqrt{6}}\sigma_2',\\ \sigma_2&=\frac{1}{\sqrt{3}}\sigma_1'-\frac{1}{\sqrt{6}}\sigma_2'+\frac{1}{\sqrt{2}}\sigma_3',\\ \sigma_3&=\frac{1}{\sqrt{3}}\sigma_1'-\frac{1}{\sqrt{6}}\sigma_2'-\frac{1}{\sqrt{2}}\sigma_3',\end{aligned} \quad (7)$$

we obtain Pauling's trigonal orbitals

$$\begin{aligned}\sigma_1&=\frac{1}{\sqrt{3}}s+\frac{2}{\sqrt{6}}p_x,\\ \sigma_2&=\frac{1}{\sqrt{3}}s-\frac{1}{\sqrt{6}}p_x+\frac{1}{\sqrt{2}}p_y,\\ \sigma_3&=\frac{1}{\sqrt{3}}s-\frac{1}{\sqrt{6}}p_x-\frac{1}{\sqrt{2}}p_y.\end{aligned} \quad (8)$$

On the other hand, this theory shows that we can everywhere replace s by d_z, or in fact by any linear combination $\alpha s+\beta d_z$ provided $\alpha^2+\beta^2=1$. Similarly p_x and p_y can be replaced by d_{xy} and d_{x+y}.

It is not necessary, however, to carry out the actual determination of T and the reduction of $\mathfrak{S}$ and $\mathfrak{R}$ to decide whether or not a given set of directed orbitals is obtainable from given atomic orbitals, for the coefficients a_i and b_i can be obtained very simply from the character tables of $\mathfrak{S}$ and $\mathfrak{R}$. The transformation matrices of the σ representation can be written down and the trace of each gives the character for the corresponding element of the group. The characters for $\mathfrak{R}$ are easily found by the method of Bethe.[8] We can thus construct a character table for the representations based on $\mathfrak{S}$ and $\mathfrak{R}$. We also can enter in the same table the characters for the irreducible representations, as given for example by Mulliken.[9] Such a character table for the plane trigonal case we have been considering is shown in Table I. By means of the orthogonality theorems for group characters, the coefficients a_i and b_i can easily be found. These are entered in a second table, to which I shall refer as the reduction table. The reduction table for our trigonal case is shown in Table II. Comparing the coefficients a_i (given in the row marked σ) with the coefficients b_i (given in the rows marked s, p, and d) we see immediately that the trigonal orbitals require one orbital which may be either s or d, and two p orbitals (remembering that the representation E' is of degree two) or two d orbitals. Hence the possible valence configurations are sp^2, dp^2, sd^2, and d^3.

It is interesting to note that the method of molecular orbitals leads to identical results, but by a rather different route. In this method we consider first the set of orbitals on the atoms surrounding the central atom. If this set consists of orbitals symmetrical about the line joining each external atom to the central atom, then these external orbitals form a basis for a representation of the symmetry group which is identical with the σ representation. The reduction of this representation then corresponds to the resonance of these external orbitals among themselves. The formation of molecular orbitals then takes place by the interaction between these reduced external orbitals and the orbitals of the central atom. This interaction can only take place, however, between orbitals belonging to the same representation. Hence, to obtain a set of molecular orbitals equal in number to the number of external atoms, it is necessary that each of the reduced external orbitals be matched with an orbital from the set $\mathfrak{R}$ which belongs to the same representation. The condition for this is again that $b_i \geqslant a_i$, so that the same result is reached as before.

The possibilities of double or triple bond formation are most easily discussed in terms of molecular orbitals. The principal type of multiple bond consists of two parts: first, a pair of electrons in an orbital symmetrical about the axis of the bond; and second, one or more pairs of electrons in orbitals which are not symmetrical about the axis. The first pair of electrons form a bond which differs in no way from the ordinary single, or σ, bond. The other pairs are ordinarily in orbitals which are antisymmetric with respect to a plane passed through the axis. They may be regarded as formed by the interaction of two p orbitals, one on each atom, with axes parallel to each other and perpendicular to the axis of the bond. We shall refer to orbitals of this type as π orbitals.

In a polyatomic molecule, consisting of a central atom and a number of external atoms bound to it, bonds of this type may also be formed. As far as the external atoms are concerned, the condition for the formation of π bonds is the presence of p orbitals at right angles to the bond axes. These p orbitals, however, will resonate among themselves to form new orbitals which are bases of irreducible representations of the symmetry group of the molecule. This reduction can be carried out in the same way as the reduction of the σ representation. We first determine the representation generated by the p orbitals of the external atoms. Since there are two such p orbitals per external atom, this representation, which will be referred to as the π representation, will have a degree twice that of the σ representation. The characters of this representation are then computed, and entered in the character table. This has been done for the case of a plane trigonal molecule in Table I. The component irreducible representations are found as before, and entered in the reduction table.

The condition for the formation of a π bond is now that there be an orbital of the central atom belonging to the same representation as one of

TABLE III. *Resolution table for linear bonds.*

$D_{\infty h}$	Σ_g^+	Σ_u^+	Π_g	Π_u	Δ_g	Δ_u
s	1	0	0	0	0	0
p	0	1	0	1	0	0
d	1	0	1	0	1	0
σ	1	1	0	0	0	0
π	0	0	1	1	0	0

TABLE IV. *Resolution table for angular bonds.*

C_{2v}	A_1	A_2	B_1	B_2
s	1	0	0	0
p	1	0	1	1
d	2	1	1	1
σ	1	0	0	1
π	1	1	1	1

the irreducible components of the π representation. Since this molecule is already supposed to be held together by σ bonds, it is not necessary that all of the irreducible components of the π representation be matched by orbitals of the central atom. However, unless at least half of the irreducible components of the π representation are so matched, it will be impossible to localize the π bonds, and the resulting molecule will be of the type ordinarily written with resonating double bonds.

Thus in the plane trigonal case, Table II shows that π bond formation is possible through the p orbital belonging to $A_2''(p_z)$, the d orbitals belonging to $E''(d_{xz},d_{yz})$ or through the two orbitals belonging to E' which are not used in forming the original σ bonds. Since, however, the σ bonds are probably formed by a mixture of both the p and the d E' orbitals, these last two π bonds are probably weaker than the others. In general we shall divide the π bonds into two classes, calling them "strong" if they belong to representations not used in σ bond formation, and "weak" if they belong to those representations already used in σ bond formation.

Because of the "resonating" character of π bonds, it is usually difficult to form a mental picture of them. In this trigonal case which we have been discussing, we may imagine the p_z orbital of the central atom to interact in turn with the p_z orbitals of the external atoms. If the orbitals d_{xz} and d_{yz} are available, these may be combined with p_z, in the same way that p_x and p_y can be combined with s, to form three directed π orbitals. No such simple picture, however, seems to be available for the π bonds formed by d_{xy} and d_{x+y}.

It should be noted that this method does not predict directly the type of bond arrangement formed by any given electron configuration. Instead it merely tells whether or not a given arrangement is possible. In many cases it is found that several arrangements are possible for a single configuration of electrons. In these cases the relative stability of the various arrangements must be decided by other methods, such as Pauling's "strength" criterion, or consideration of the repulsions between nonbonded atoms.

RESULTS

The results of these calculations are most conveniently arranged according to the coordination number of the central atom. For the sake of completeness all of the results, including those previously obtained by Pauling, Hultgren and others, are contained in the following summary.

Coordination number 2

If the central atom forms bonds with two external atoms only two arrangements of the bonds are possible: a linear arrangement (group $D_{\infty h}$) and an angular one (group C_{2v}). The resolution tables for these are given in Tables III and IV. The configurations sp and dp can lead to either arrangement. The linear arrangement, however, is favored by both the repulsive forces and the possibilities for double bond formation and is therefore the stable arrangement for these configurations. The configurations ds, d^2 and p^2 on the other hand must be angular.

In the linear arrangement two p and two d orbitals are available for double bond formation, while in the angular arrangement only two strong π bonds are possible, one of which must be through a d orbital, the other of which may be formed by either a d or a p orbital. The other two possible π bonds are weakened by the fact that their orbitals belong to representations already used by the σ bonds.

The angular nature of the p^2 bonds in such molecules as H_2O and H_2S is well known, and need not be discussed further. As examples of double bond formation in molecules having this

TABLE V. *Resolution table for trigonal pyramid bonds.*

C_{3v}	A_1	A_2	E
s	1	0	0
p	1	0	1
d	1	0	2
σ	1	0	1
π	1	1	2

TABLE VI. *Resolution table for three unsymmetrical bonds in a plane.*

C_{2v}	A_1	B_2	B_1	B_2
s	1	0	0	0
p	1	0	1	1
d	2	1	1	1
σ	2	0	1	0
π	1	1	2	2

primary valence structure the nitroso compounds are typical. The prototype of these compounds is the nitrite ion NO_2^-. In this ion each oxygen atom is joined to the nitrogen by a σ bond involving one of the p orbitals of the nitrogen. In addition to these bonds, however, there is a π bond which belongs to the representation B_1 and cannot be localized. Its component orbitals are the p_x orbital of the nitrogen atom and the normalized sum of the p_x orbitals of the two oxygen atoms. The remaining parts of the oxygen p orbitals are occupied by unshared pairs of electrons. Because of the distributed character of the π bond no single valence bond picture can be drawn for this molecule, but only the resonating pair

In the true nitroso compounds RNO, and also in the nitrosyl halides ClNO, etc., the situation is very similar. The lack of complete symmetry does not change any of the essential features of the structure of these molecules. In these cases, however, the difference in electro-negativity between oxygen and the atom or group R may cause ionic structures to play an important role. Pauling has suggested for example that in ClNO the important structures are

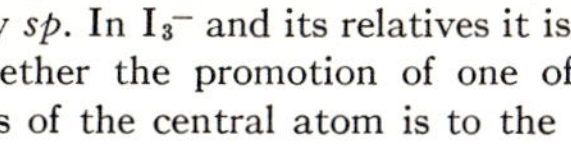

The bond formation in both these structures is in accord with the present theory.

Examples of the linear configuration are found in the ions of the $Ag(NH_3)_2^+$ type and those of the I_3^- type. In $Ag(NH_3)_2^+$ the configuration is certainly sp. In I_3^- and its relatives it is not certain whether the promotion of one of the $5p$ electrons of the central atom is to the $5d$ or $6s$ orbital, but in either case the resulting configuration (ps or pd) should give the observed linear arrangement.

The effect of double bond formation is shown clearly by CO_2 and SO_2. In CO_2 the valence configuration of the carbon is sp^3. The primary structure of σ bonds is sp, which requires the molecule to be linear. The other two p orbitals form the two π bonds. In SO_2, on the other hand, the primary bonds are certainly formed by two p orbitals, thus producing angular arrangement. The customary way of writing the structure of this molecule,

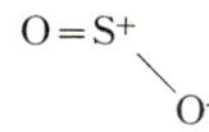

(analogous to NO_2^-) is quite possible, since the π bond can be formed by the third p orbital of the sulfur, but the structure

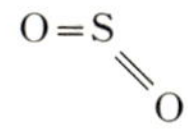

is also possible and probably as important as the first. In this structure one of the d orbitals of the sulfur acts as an acceptor for one of the electron pairs of the oxygen. It should be noted that the formation of two double bonds by an atom does not require the linear arrangement.

Coordination number 3

For this number of σ bonds three arrangements are important: the plane trigonal arrangement (group D_{3h}) already discussed, the trigonal pyramid (group C_{3v}), and an unsymmetrical plane arrangement with two of the bond angles equal, but not equal to the third (group C_{2v}). The resolution tables for the last two are given in Tables V and VI. As has already been shown, the plane arrangement is stable for the configurations sp^2, sd^2, dp^2, and d^3. (The pyramidal configuration is also possible, but less stable because of repulsive forces. Such cases will simply be omitted in the future.) The configurations p^3 and d^2p are therefore the only ones leading to a pyramidal structure.[10] The configuration dsp leads to the unsymmetrical plane arrangement.

The possibilities of double bond formation in the plane trigonal arrangement have already been discussed. In the pyramidal arrangement no strong double bonds are possible, although weak π bonds are possible with d orbitals. Thus the sulfite ion, for example, is restricted to the structure

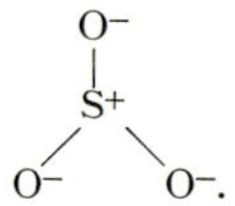

It is interesting to contrast this structure with that of SO_3. The most easily removed pair of electrons in $SO_3^{=}$ is the pair of $3s$ electrons in the sulfur atom. If these are removed, however, one of the $3p$ pairs forming a σ bond falls to the $3s$ level, thus making the valence configuration sp^2 and the molecule plane. The vacated $3p$ orbital is then filled by an unshared pair from the oxygen atoms to give a π bond. Two more unshared pairs of electrons from the oxygen atoms may interact with the $3d$ orbitals thus forming two more π bonds, which are somewhat weaker than the first. The SO_3 molecule has therefore three comparatively strong double bonds, while the $SO_3^{=}$ ion has only weak double bonds, if any at all. It is this lack of stabilizing double bonds which gives $SO_3^{=}$ its relative instability.

The remaining configuration, dsp, should give rise to three bonds in a plane forming two right angles. No examples of molecules of this configuration are known. This is hardly surprising, however, in view of the instability of this arrangement compared to the others. If molecules of this arrangement are to be found at all they would be complex ions of the transition elements. For example, if the ion $Ni(NH_3)_3^{++}$ existed it would be of this structure.

Coordination number 4

With a coordination number of four there are three arrangements which need consideration. The reduction tables for these are given in Tables VII, VIII, IX and X. Table VII is for the regular tetrahedral arrangement of the bonds. It is easily seen that the configurations sp^3 and d^3s lead to this arrangement. For double bond formation there remain two d orbitals which can form strong π bonds and two other d orbitals (if the σ bonds are sp^3) or two p orbitals (if the σ bonds are d^3s) which can form weaker π bonds. These possibilities of double bond formation are in accord with Pauling's suggestion that the structure of ions of the type XO_4 should be written with double bonds. Thus for the sulfate ion the primary σ structure arises from the configuration sp^3, but two of the unshared pairs of the O^- ions can be donated to the vacant d

TABLE VII. *Reduction table for tetrahedral bonds.*

T_d	A_1	A_2	E	T_1	T_2
s	1	0	0	0	0
p	0	0	0	0	1
d	0	0	1	0	1
σ	1	0	0	0	1
π	0	0	1	1	1

TABLE VIII. *Reduction table for tetragonal plane bonds.*

D_{4h}	A_{1g}	A_{1u}	A_{2g}	A_{2u}	B_{1g}	B_{1u}	B_{2g}	B_{2u}	E_g	E_u
s	1	0	0	0	0	0	0	0	0	0
p	0	0	0	1	0	0	0	0	0	1
d	1	0	0	0	1	0	1	0	1	0
σ	1	0	0	0	0	0	1	0	0	1
π	0	0	1	1	1	1	0	0	1	1

TABLE IX. *Reduction table for tetragonal pyramidal bonds.*

C_{4v}	A_1	A_2	B_1	B_2	E
s	1	0	0	0	0
p	1	0	0	0	1
d	1	0	1	1	1
σ	1	0	0	1	1
π	1	1	1	1	2

TABLE X. *Reduction table for irregular tetrahedral bonds.*

C_{3v}	A_1	A_2	E
s	1	0	0
p	1	0	1
d	1	0	2
σ	2	0	1
π	1	1	3

[10] The bond angles of 90° in this structure are not predicted by the reduction table alone, but can be easily found by the process of forming the actual orbitals as shown in the second section.

orbitals of the sulfur atom, thus giving the structure

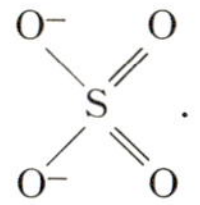

It must, however, be doubtful that two further π bonds are formed to give the structure

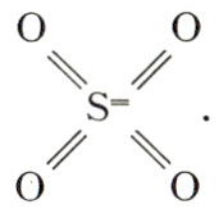

The situation in such molecules as $SiCl_4$ is similar.

From Table VIII it is seen that tetragonal planar (square) bonds are possible with configurations dsp^2 and d^2p^2. Beside the primary bonds, four strong π bonds are possible, using one p and three d orbitals. In $Ni(CN)_4^{=}$, for example, the primary valence configuration of the Ni is dsp^2, so that the structure

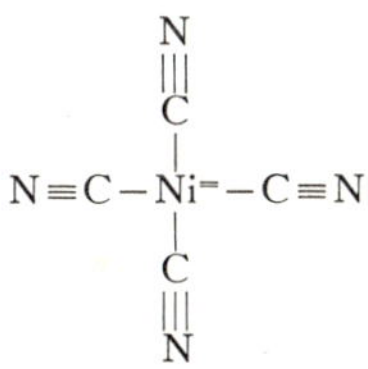

is one possible structure for this ion. By a shift of one of the π pairs forming the C–N triple bond to the nitrogen, however, the carbon atom is left with an empty orbital, which can accept an electron pair donated by the nickel atom. In this way three π bonds can be formed between the nickel and carbon. At the same time one of the triple bond π pairs can shift in the opposite direction and be donated by the carbon to the empty p orbital of the nickel. These shifts would lead to the structure

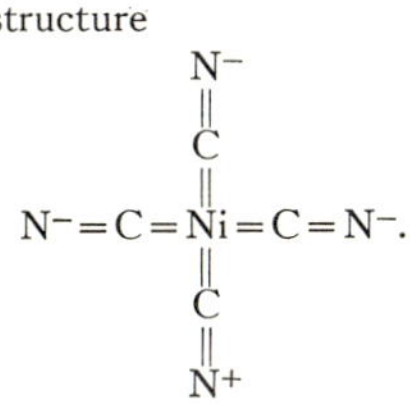

This structure affords an interesting contrast with that of $Ni(CO)_4$, in which one more pair of electrons must be accommodated. This pair occupies the d orbital of the nickel, thus pushing the valence configuration up to sp^3 and making the molecule tetrahedral. There are then only two π bonds possible, and the structure is

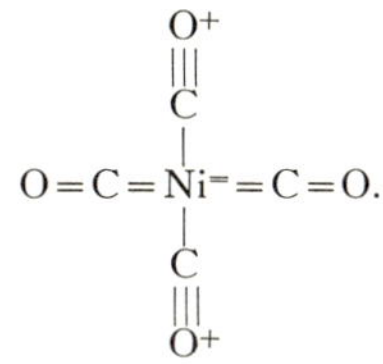

The configuration d^2p^2 is found in such ions as ICl_4^-, which are known to be plane.

From Table IX it appears that the configurations dp^3 and d^3p can lead to a tetragonal pyramid structure, while from Table X it appears that the "irregular tetrahedron" is also possible for these same configurations and also for the configuration d^2sp. By "irregular tetrahedron" is meant a structure in which three of the bonds are directed to the corners of an equilateral triangle and the fourth along the line perpendicular to the triangle at its center. The central atom is not necessarily located at the center of the triangle, but it is probably somewhat above it in the direction of the fourth bond. The choice between these two structures in the cases of the configurations dp^3 and d^3p is very close, with the advantage somewhat on the side of the irregular tetrahedron. The configuration d^4 can only have the pyramidal structure.

Examples of these configurations are rare. In the tetrahalides of the sulfur family we have the configuration p^3d produced by the promotion of one of the p electrons of the central atom to a d orbital. Unfortunately the spatial arrangements of these molecules are unknown. The other two configurations seem to be unstable. Such ions as $Fe(NH_3)_4^{++}$ and $Fe(CN)_4^{=}$ would have the configuration d^2sp, but do not seem to exist, which is perhaps some indication of this instability.

Coordination number 5

For this coordination number four bond arrangements are possible. In Table XI is given the

TABLE XI. *Reduction table for trigonal bipyramidal bonds.*

D_{3h}	A_1'	A_1''	A_2'	A_2''	E'	E''
s	1	0	0	0	0	0
p	0	0	0	1	1	0
d	1	0	0	0	1	1
σ	2	0	0	1	1	0
π	0	1	0	1	2	2

reduction table for a trigonal bipyramid, which is seen to be stable for the configurations dsp^3 and d^3sp. In Table XII the bonds are directed from the center to the corners of a square pyramid. This is the arrangement expected for the configurations d^2sp^2, d^4s, d^2p^3, and d^4p. Table XIII is for the case of five bonds in a plane, directed toward the corners of a regular pentagon, the arrangement for the configuration d^3p^2, and Table XIV is for the case of five bonds directed along the slant edges of a pentagonal pyramid, the arrangement for the configuration d^5.

In PCl_5 and other molecules of this type, the pentavalent state is formed by the promotion of an s electron to the vacant d shell. The valence configuration is therefore dsp^3 and the bipyramidal structure is to be expected, in agreement with the results of electron diffraction studies. Table XI shows that two of the remaining d orbitals are capable of forming strong π bonds; the other two, two weak π bonds by accepting pairs of electrons. Neglecting weak bonds, the structure of PCl_5 is best written

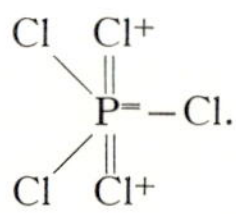

In the molecule $Fe(CO)_5$ the valence configuration is again dsp^3, and the bipyramidal structure is to be expected. Here π bonds can be formed by donation of pairs of d electrons from the iron through the carbon to the oxygens, as in $Ni(CO)_4$, giving the structure

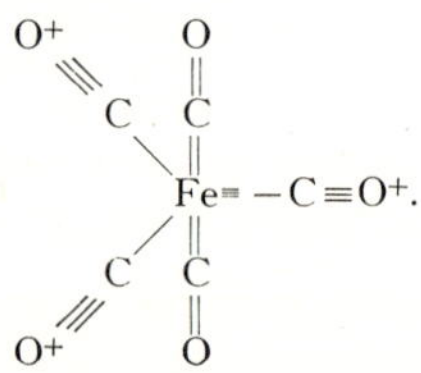

TABLE XII. *Reduction table for five tetragonal pyramidal bonds.*

C_{4v}	A_1	A_2	B_1	B_2	E
s	1	0	0	0	0
p	1	0	0	0	1
d	1	0	1	1	1
σ	2	0	0	1	1
π	1	1	1	1	3

TABLE XIII. *Reduction table for pentagonal plane bonds.*

D_{5h}	A_1'	A_1''	A_2'	A_2''	E_1'	E_1''	E_2'	E_2''
s	1	0	0	0	0	0	0	0
p	0	1	0	0	1	0	0	0
d	1	0	0	0	0	1	1	0
σ	1	0	0	0	1	0	1	0
π	0	1	1	0	1	1	1	1

TABLE XIV. *Reduction table for pentagonal pyramidal bonds.*

C_{5v}	A_1	A_2	E_1	E_2
s	1	0	0	0
p	1	0	1	0
d	1	0	1	0
σ	1	0	1	1
π	1	1	2	2

In IF_5, however, the valence configuration is p^3d^2, and the molecule should have the square pyramid structure. Here again two strong π bonds can be formed by donations of pairs of electrons from the F to the I, so that the structure may be

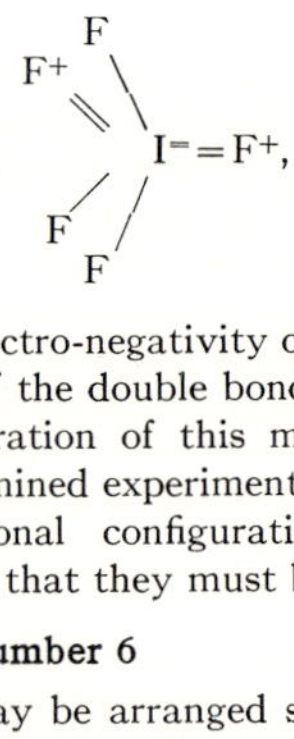

but the high electro-negativity of fluorine makes the existence of the double bonds doubtful. The spatial configuration of this molecule has not yet been determined experimentally.

The pentagonal configurations crowd the atoms so much that they must be unstable.

Coordination number 6

Six bonds may be arranged symmetrically in space in three ways: to the corners of a regular octahedron, to those of a trigonal prism, or to those of a trigonal antiprism (an octahedron stretched or compressed along one of the tri-

TABLE XV. *Reduction table for octahedral bonds.*

O_h	A_{1g}	A_{1u}	A_{2g}	A_{2u}	E_g	E_u	T_{1g}	T_{1u}	T_{2g}	T_{2u}
s	1	0	0	0	0	0	0	0	0	0
p	0	0	0	0	0	0	0	1	0	0
d	0	0	0	0	1	0	0	0	1	0
σ	1	0	0	0	1	0	0	1	0	0
π	0	0	0	0	0	0	1	1	1	1

TABLE XVI. *Reduction table for trigonal prismatic bonds.*

D_{3h}	A_1'	A_1''	A_2'	A_2''	E'	E''
s	1	0	0	0	0	0
p	0	0	0	1	1	0
d	1	0	0	0	1	1
σ	1	0	0	1	1	1
π	1	1	1	1	2	2

TABLE XVII. *Reduction table for trigonal antiprismatic bonds.*

D_{3d}	A_{1g}	A_{1u}	A_{2g}	A_{2u}	E_g	E_u
s	1	0	0	0	0	0
p	0	0	0	1	0	1
d	1	0	0	0	2	0
σ	1	0	0	1	1	1
π	1	1	1	1	2	2

gonal axes). The reduction tables for these possibilities are given in Tables XV, XVI, and XVII.

Table XV shows that octahedral bonds are formed by the configuration d^2sp^3 and no other. The commonness of this arrangement is therefore due to the fact that the configuration d^2sp^3 is the usual configuration of six valence electrons, rather than any particular virtue of the octahedral arrangement. This configuration arises both in the 6-coordinated ions of the transition elements, and in the molecules of the SF_6 type in which one s and one p electron are promoted to the next higher d level. Table XV also shows that all three remaining d orbitals are capable of forming strong π bonds. In the ferrocyanide ion $Fe(CN)_6^{-4}$, these d orbitals can donate their pairs of electrons to the nitrogen atoms, giving Pauling's structure

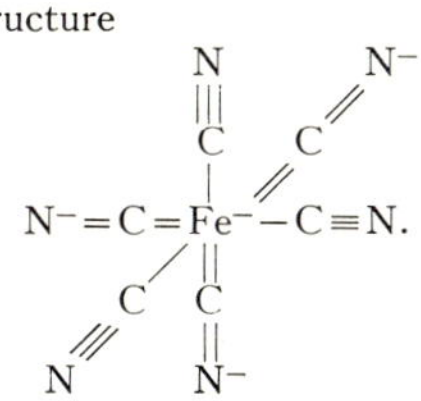

If SF_6 the remaining d orbitals are empty, and can accept pairs of electron from the fluorine atoms, giving the structure

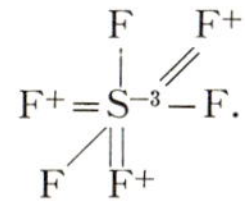

It is this structure which accounts for the great resistance of SF_6 to hydrolysis. It has been suggested by Sidgwick[11] that the hydrolysis of halides takes place either by their accepting a pair of electrons from the oxygen of a water molecule, with subsequent decomposition, or by donation of an unshared pair on the central atom to a hydrogen atom, with loss of hypohalous acid. With the usual single bonded structure of SF_6, the first mechanism is possible, and one would expect the hydrolysis to take place easily. With the double bonded structure, however, all of the electrons in the outer shell of the sulfur are taking part in bond formation, so that one should expect the same inertness as that displayed by CCl_4. It is interesting to note that SeF_6 is like SF_6, but TeF_6 is easily hydrolyzed. This must be due to the possibility of the Te atom's accepting a pair of oxygen electrons in the vacant $4f$ orbital.

In MoS_2 and WS_2 in the crystalline form the valence configuration is d^4sp and the arrangement is prismatic, as Hultgren[4] has pointed out. Table XVI indicates that the two empty p orbitals of the metal atom cannot form strong π bonds to the sulfur atoms. The configuration d^5p should also lead to the prismatic arrangement, but no examples are known.

The cases of the ions $SeBr_6^{-2}$ and $SbBr_6^{-3}$ are interesting in that the configuration in each of these should be p^3d^3. This configuration should lead to the antiprismatic arrangement, i.e., the octahedral symmetry should not be perfect. If, however, the unshared pair of s electrons is promoted to a d orbital, and one of the valence pairs slips into its place, the configuration will be the octahedral d^2sp^3. The deciding factor here may be the possibility of double bond formation offered by the octahedral but not by the antiprismatic arrangement.

[11] Sidgwick, *The Electronic Theory of Valency* (Oxford, 1927), p. 157.

The configurations d^3sp^2, d^5s and d^4p^2 do not permit any arrangement in which all of the bonds are equivalent. While the configuration d^3sp^2 might form bonds of a mixed type, e.g., the tetrahedral d^3s+the angular p^2, the bonds will be weak, and the configuration unstable. Molecules which might be expected to have this arrangement, e.g., $Fe(CN)_6^{-3}$, rearrange the electrons to obtain the more stable d^2sp^3 configuration. In $Fe(CN)_6^{-3}$ for example, the odd electron, which normally should occupy a $4p$ orbital, instead goes to a $3d$ orbital making the third $4p$ orbital available for the formation of octahedral bonds.

Coordination number 7

The coordination number 7 is extremely rare. The fact that it appears only in the heavier atoms, such as Zr, Cb, I, and Ta, leads one to suspect that for stability f electrons are necessary, although it is possible that the determining factor is ion size. Two arrangements of seven bonds have been observed:[12] the ZrF_7^{-3} structure, which may be obtained from the octahedron by adding an atom at the center of one face; and the TaF_7^{-2} structure, which may be obtained from the trigonal prism by adding an atom at the center of one of the square faces. The reduction tables for these two arrangements are given in Tables XVIII and XIX.

From Table XVIII we see that if no f orbitals are involved, the ZrF_7^{-3} type bonds can arise from the configurations d^3sp^3 and d^5sp. The configuration expected for ZrF_7^{-3} is d^5sp, so that it is not necessary to appeal to the f orbitals to form these bonds. This, however, does not preclude the possibility that f orbitals are used to strengthen the bonds.

Table XIX shows that the TaF_7^{-2} structure is possible with the configurations d^3sp^3, d^4sp^2, d^5sp, d^4p^3 and d^5p^2. The ion TaF_7^{-2} itself is isoelectronic with ZrF_7^{-3}, so that it is difficult to understand why it should prefer its structure to that of ZrF_7^{-3}. It is quite possible that the bonds in all the 7-coordinated molecules are so ionic that the

TABLE XVIII. *Reduction table for ZrF_7^{-3} type bonds.*

C_{3v}	A_1	A_2	E
s	1	0	0
p	1	0	1
d	1	0	2
f	2	1	2
σ	3	0	2
π	2	2	5

TABLE XIX. *Reduction table for TaF_7^{-2} type bonds.*

C_{2v}	A_1	A_2	B_1	B_2
s	1	0	0	0
p	1	0	1	1
d	2	1	1	1
f	2	1	2	2
σ	3	1	2	1
π	3	3	4	4

TABLE XX. *Reduction table for cubic bonds.*

O_h	A_{1g}	A_{1u}	A_{2g}	A_{2u}	E_g	E_u	T_{1g}	T_{1u}	T_{2g}	T_{2u}
s	1	0	0	0	0	0	0	0	0	0
p	0	0	0	0	0	0	0	1	0	0
d	0	0	0	0	1	0	0	0	1	0
f	0	0	0	1	0	0	0	1	0	1
σ	1	0	0	1	0	0	0	1	1	0
π	0	0	0	0	0	0	1	1	1	1

TABLE XXI. *Reduction table for tetragonal antiprismatic bonds.*

D_{4d}	A_{1g}	A_{1u}	A_{2g}	A_{2u}	E_1	E_2	E_3
s	1	0	0	0	0	0	0
p	0	0	0	1	1	0	0
d	1	0	0	0	0	1	1
σ	1	0	0	1	1	1	1
π	1	1	1	1	2	2	2

TABLE XXII. *Reduction table for dodecahedral bonds.*

V_d	A_1	A_2	B_1	B_2	E
s	1	0	0	0	0
p	0	0	0	1	0
d	1	0	1	1	1
σ	2	0	0	2	2
π	2	2	2	2	4

TABLE XXIII. *Reduction table for face-centered prismatic bonds.*

C_{2v}	A_1	A_2	B_1	B_2
s	1	0	0	0
p	1	0	1	1
d	2	1	1	1
σ	3	0	2	1
π	4	4	4	4

[12] Hampson and Pauling, J. Am. Chem. Soc. **60**, 2702 (1938); Hoard, J. Am. Chem. Soc. **61**, 1252 (1939).

TABLE XXIV. *Summary of stable bond arrangements and multiple bond possibilities.*

Coordination Number	Configuration	Arrangement	Strong π Orbitals	Weak π Orbitals	Table
2	sp	linear	p^2d^2	—	III
	dp	linear	p^2d^2	—	III
	p^2	angular	$d(pd)$	$d(sd)$	IV
	ds	angular	$d(pd)$	$p(pd)$	IV
	d^2	angular	$d(pd)$	$p(spd)$	IV
3	sp^2	trigonal plane	pd^2	d^2	II
	dp^2	trigonal plane	pd^2	d^2	II
	d^2s	trigonal plane	pd^2	p^2	II
	d^3	trigonal plane	pd^2	p^2	II
	dsp	unsymmetrical plane	pd^2	$(pd)d$	VI
	p^3	trigonal pyramid	—	$(sd)d^4$	V
	d^2p	trigonal pyramid	—	$(sd)p^2d^2$	V
4	sp^3	tetrahedral	d^2	d^3	VII
	d^3s	tetrahedral	d^2	p^3	VII
	dsp^2	tetragonal plane	d^3p	—	VIII
	d^2p^2	tetragonal plane	d^3p	—	VIII
	d^2sp	irregular tetrahedron	—	d	X
	dp^3	irregular tetrahedron	—	s	X
	d^3p	irregular tetrahedron	—	s	X
	d^4	tetragonal pyramid	d	$(sp)p$	IX
5	dsp^3	bipyramid	d^2	d^2	XI
	d^3sp	bipyramid	d^2	p^2	XI
	d^2sp^2	tetragonal pyramid	d	pd^2	XII
	d^4s	tetragonal pyramid	d	p^3	XII
	d^2p^3	tetragonal pyramid	d	sd^2	XII
	d^4p	tetragonal pyramid	d	sp^2	XII
	d^3p^2	pentagonal plane	pd^2	—	XIII
	d^5	pentagonal pyramid	—	$(sp)p^2$	XIV
6	d^2sp^3	octahedron	d^3	—	XV
	d^4sp	trigonal prism	—	p^2d	XVI
	d^5p	trigonal prism	—	p^2s	XVI
	d^3p^3	trigonal antiprism	—	sd	XVII
	d^3sp^2	mixed			
	d^5s	mixed			
	d^4p^2	mixed			
7	d^3sp^3	ZrF_7^{-3}	—	d^2	XVIII
	d^5sp	ZrF_7^{-3}	—	p^2	XVIII
	d^4sp^2	TaF_7^{-2}	—	dp	XIX
	d^4p^3	TaF_7^{-2}	—	ds	XIX
	d^5p^2	TaF_7^{-2}	—	ps	XIX
8	d^4sp^3	dodecahedron	d	—	XXII
	d^5p^3	antiprism	—	s	XXI
	d^5sp^2	face-centered prism	p	—	XXIII

directed nature of covalent bonds has little importance in determining the atomic arrangement.

Coordination number 8

Until very recently the spatial arrangement of no molecule involving a coordinate number of 8 had been determined experimentally. It had been supposed that the arrangements which were most probable were the cubic and the tetragonal antiprismatic. Hoard and Nordsieck[13] have now determined the arrangement of the ion $Mo(CN)_8^{-4}$ and have found that the structure is neither cubic nor antiprismatic, but instead that of a dodecahedron with triangular faces and symmetry $V_d(D_2^d)$. The reduction tables for these three cases are given in Tables XX, XXI, and XXII. Table XXIII is the reduction table for an arrangement which is derivable from a trigonal prism by placing two atoms at the centers of two of the rectangular faces. It is seen that the cubic requires the use of f orbitals, and even then arises only from the configurations d^3fsp^3 and d^3f^4s. The cubic arrangement should therefore not be ordinarily found. The antiprismatic arrangement is stable for the configurations d^4sp^3 and d^5p^3, and the dodecahedral for only d^4sp^3. The observed arrangement for $Mo(CN)_8^{-4}$, which has the configuration d^4sp^3, indicates that the dodecahedral arrangement has greater stability than the antiprismatic. The structure of such ions as TaF_8^{-3} and the molecule OsF_8 which should have the configuration d^5sp^2 should be the face-centered prismatic structure of Table XXIII.

Table XXIV summarizes these results. In it are given the most stable bond arrangements for each configuration of two to eight electrons in s, p or d orbitals. The orbitals available for strong π bonds and weak π bonds are also given. In those cases where there is a choice of two or more orbitals when only one can be chosen, the orbitals are enclosed in parentheses.

[13] Hoard and Nordsieck, J. Am. Chem. Soc. **61**, 2853 (1939).

Formulas and Numerical Tables for Overlap Integrals*

R. S. Mulliken, C. A. Rieke,† D. Orloff, and H. Orloff††
Departments of Physics and Mathematics, University of Chicago, Chicago, Illinois

(Received April 27, 1949)

Explicit formulas and numerical tables for the overlap integral S between AO's (atomic orbitals) of two overlapping atoms a and b are given. These cover all the most important combinations of AO pairs involving ns, $np\sigma$, and $np\pi$ AO's. They are based on approximate AO's of the Slater type, each containing two parameters μ [equal to $Z/(n-\delta)$], and $n-\delta$, where $n-\delta$ is an effective principal quantum number. The S formulas are given as functions of two parameters p and t, where $p=\frac{1}{2}(\mu_a+\mu_b)R/a_{\rm H}$, R being the interatomic distance, and $t=(\mu_a-\mu_b)/(\mu_a+\mu_b)$. Master tables of computed values of S are given over wide ranges of p and t values corresponding to actual molecules, and also including the case $p=0$ (intra-atomic overlap integrals). In addition, tables of computed S values are given for several cases involving 2-quantum s, p hybrid AO's.

Hybrid S values for any desired type of hybrid can be obtained very easily from the tables as simple linear combinations of non-hybrid S values. It is shown how S values corresponding to orthogonalized Slater AO's and approximate S values for SCF (self-consistent-field) AO's can also be obtained as linear combinations of the Slater-AO S values. S values for carbon-carbon $2p\sigma$- and $2p\pi$-bonds using SCF carbon AO's have been computed (see Table in Section Vb); they correspond to stronger overlap than for Slater AO's. Non-localized MO group-orbital S values are also discussed, and are illustrated by an application to H_2O. The use of the tables to obtain dipole moments for electronic transitions in certain cases is also mentioned. The use of the tables to obtain S values for various specific atom-pairs and bond-types, and resulting conclusions, will be discussed in another paper.

I. INTRODUCTION

As is well known, the quantity known as the overlap integral is of considerable importance in the theory of molecular structure. Although the existing literature[1] contains a number of formulas and some numerical values for overlap integrals, it was thought worth while to carry through the much more systematic and comprehensive study whose results are presented below.

The overlap integral S for a pair of overlapping AO's (atomic orbitals) χ_a and χ_b of a pair of atoms a and b is defined for any value of the internuclear distance R by

$$S(\chi_a, \chi_b; R) = \int \chi_a^* \chi_b dv. \tag{1}$$

In the present paper we give formulas for S for all *AO pairs* involving ns, $np\sigma$, and $np\pi$ AO's for $n=1, 2, 3,$ and 5 using Slater AO's. We also give numerical tables for the most important of these cases, applicable to a wide variety of *atom pairs* over a wide range of R. In addition, we show how the tables, though based on Slater AO's, may be used to obtain S values for central-field AO's, as well as for hybrid AO's. Explicit tables of hybrid S values are also given for $n=2$.

In a subsequent paper, tables and figures for a variety of selected atom-pairs will be presented, together with some comments on their significance for the theory of chemical binding.

II. THE CHOICE OF ATOMIC ORBITAL FORMS

In order to evaluate overlap integrals, it is necessary first to specify the forms of the AO's. As a background, and also to obtain certain needed formulas, we first give a brief review of useful approximate AO forms. We are primarily concerned with *central-field* orbitals, classified under the familiar designations ns, np, nd, nf, $\cdots$ corresponding to $l=0, 1, 2, 3, \cdots$, with $n \geqslant l+1$. However, since the application of our computations is to atom-pairs in which each atom is under the influence of the cylindrically-symmetrical field of its partner, we need a sub-classification according to values of the diatomic quantum number λ ($\lambda=|m_l|$). This gives the AO types ns, $np\sigma$, $nd\sigma$, $\cdots (\lambda=0)$; $np\pi$, $nd\pi$, $\cdots (\lambda=1)$; and so on.

Let us consider an electron belonging to either of a pair of atoms a and b, using spherical polar coordinates about either center, as in Fig. 1. Intentionally, we take

* This work was assisted by the ONR under Task Order IX of Contract N6ori-20 with the University of Chicago.

† Present address: Physics Department, Purdue University, Lafayette, Indiana.

†† The work was begun before the war with computations on a considerable variety of individual atom-pairs (see abstract by R. S. Mulliken and C. A. Rieke, Rev. Mod. Phys. **14**, 259 (1942)). Recently, with the indispensable assistance of Mr. Tracy J. Kinyon in making the numerical computations, it was extended to the comprehensive effort reported here.

[1] (a) W. Heitler and F. London, Zeits. f. Physik **44**, 455 (1927), for homopolar $S(1s, 1s)$; (b) J. H. Bartlett, Jr., Phys. Rev. **37**, 507 (1931), for homopolar $S(2p\sigma, 2p\sigma)$ and $S(2p\pi, 2p\pi)$ and a table of values of the integrals A (see Eq. (15) below); (c) J. H. Bartlett, Jr. and W. H. Furry, Phys. Rev. **38**, 1615 (1931); **39** 210 (1932), especially table of values of homopolar $S(2s, 2s)$, $S(2s, 2p\sigma)$, $S(2p\sigma, 2p\sigma)$, $S(2p\pi, 2p\pi)$: Table VII, p. 222; (d) N. Rosen, Phys. Rev. **38**, 255 (1931) for formulas and extensive tables of values of the integrals A and B (see Eqs. (15), (16) below); (e) Kotani, Amemiya, and Simose, Proc. Phys. Math. Soc. Japan **20**, 1 (1938); **22**, 1 (1940). These authors give equations for S for all AO combinations with $n=1$ and 2, and (1940, p. 13) numerical tables for $S(1s, 2s)$ and $S(1s, 2p\sigma)$ for various t values. They also (1938, pp. 24–30, corrections 1940, pp. 17–18) give very complete and useful numerical tables for the integrals A and B. (f) C. A. Coulson, Proc. Camb. Phil. Soc. **38**, 210 (1941). Coulson gives very extensive formulas but no numerical tables, for integrals which differ from S only by multiplicative factors. These correspond to S for all combinations of ns, $np\sigma$, and $np\pi$ AO's with $n=1-3$ (and a few more), both for $t\neq 0$ (in our notation) and for $t=0$. (g) Several of the S formulas and a few numerical values are also given elsewhere.

the positive direction of the z axis for each atom to be directed toward the other atom, since this ensures a positive sign for S in all ordinary cases. Every central-field AO of atom a is of the form

$$\chi_a{}^{nl\lambda}=R_{nl}(r_a)Y_{l,\lambda}(\theta_a,\phi). \tag{2}$$

In normalized form, the Y's for the cases considered in this paper are:

$$\left.\begin{aligned} Y_{ns}&=(1/4\pi)^{\frac{1}{2}};\\ Y_{np\sigma}&=(3/4\pi)^{\frac{1}{2}}\cos\theta_a;\\ Y_{np\pi}&=(3/4\pi)^{\frac{1}{2}}\sin\theta_a\{\cos\phi \text{ or } \sin\phi\}.\end{aligned}\right\} \tag{3}$$

Corresponding expressions hold for atom b.

When there is only one electron (hydrogenic AO's), R_{nl} takes the well-known form

$$R_{nl}(r)=\left(\sum_{k=l}^{n-1} c_{nlk}r^k\right)e^{-Zr/na_{\mathrm{H}}}, \tag{4}$$

where Z is the nuclear charge, and a_{H} (0.529A) is the 1-quantum Bohr radius. The coefficients in Eq. (4) are those of the associated Laguerre polynomials, and the R_{nl}'s for each l form an orthogonal set.

When there is more than one electron, the best R_{nl}'s are of the SCF (self-consistent-field) type obtainable by the method of Hartree and Fock. These are not given by analytical expressions, and are usually presented in the form of numerical tables. Moreover, these SCF R_{nl}'s as tabulated are not always all orthogonal; however, it is always possible to find an equivalent set of equally good SCF R_{nl}'s which *are* orthogonal.[2] For practical computations, as Slater has shown,[3] the SCF R_{nl}'s may be approximated passably well by a finite series similar to that of Eq. (4) but with a different exponential factor in each term:

$$R_{nl}(r)=\sum_{k=l}^{n-1} d_k r^k e^{-\mu_k r/a_{\mathrm{H}}}, \tag{5}$$

where the d_k's and μ_k's depend on n, l, and on the particular atom and electronic state. A considerably better series is obtained if the exponential factor in the term for which $k=n-1$ is replaced by a sum of two or three exponential terms, as follows:[3]

$$R_{nl}(r)=\sum_{\substack{k=l\\(\neq n-1)}}^{n-2} d_k r^k e^{-\mu_k r/a_{\mathrm{H}}}+r^{n-1}\sum_i d_i e^{-\mu_i r/a_{\mathrm{H}}}. \tag{6}$$

For atoms containing $1s$, $2s$, and $2p$ electrons, the best approximate R_{nl}'s of the type of Eq. (5), so far as we know, are those obtained by Morse, Young, and Haurwitz, and Duncanson and Coulson.[4] These were obtained for several states of several atoms by minimizing the total energy of a complete antisymmetrized wave function built of AO's having R_{nl}'s of the type given by Eq. (5), subject also to the condition that the $2s$ AO is kept orthogonal to the $1s$. These AO's were also chosen in such a way that the virial theorem for the mean total kinetic and potential energies is satisfied.

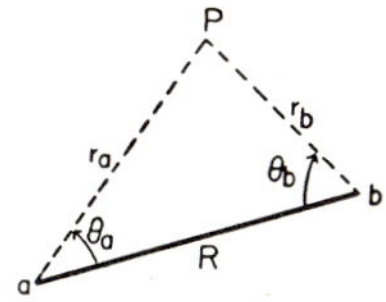

FIG. 1. Polar coordinates for an electron at P in an AO of atom a or b. In addition (not shown), $\phi=\phi_a=\phi_b$.

The well-known Slater AO's[5] are obtained by approximating the series of Eq. (5) by the single term $N_{nl}r^{n-1}e^{-\mu r/a_{\mathrm{H}}}$ (or by a modification of this, $N_{nl}r^{n-1-\delta}e^{-\mu r/a_{\mathrm{H}}}$). The values of μ and δ are specified by Slater for any AO of any atom by a simple recipe. This gives μ-values surprisingly close,[4b] for 2-quantum AO's, to those obtained for the r^{n-1} term by minimizing the total energy. Unlike most of the forms given by Eqs. (2)–(5), Slater's AO's have no radial nodes. This, however, does not much impair their ability to represent the *outer parts* of AO's, hence they should be especially useful for the computation of overlap integrals. Following is a summary of Slater's recipe as applied to s and p AO's in normalized form:

$$\left.\begin{aligned} &R_{nl}(r)=N_{nl}r^{n-1-\delta}e^{-\mu r/a_{\mathrm{H}}};\\ &\qquad\delta=0 \text{ for } n=1 \text{ to } 3;\ 0.3 \text{ for } 4s,\ 4p;\\ &\qquad\qquad 1 \text{ for } 5s,\ 5p;\\ &\mu_{ns}=\mu_{np}=Z_{a,n}/(n-\delta),\\ &1/N_{nl}^2=\int_0^\infty r^{2n-2\delta}e^{-2\mu_{nl}r/a_{\mathrm{H}}}dr,\end{aligned}\right\} \tag{7}$$

with Z values dependent on the atom a and on n. For convenience of later reference, Slater μ-values for the *valence-shell* ns and np AO's of several atoms and ions are given in Table I. Inner-electron AO's if present are assigned larger Z values.[3]

As a consequence of their nodeless character, Slater AO's of equal l and different n (e.g., $1s$ and $2s$) are very far from being orthogonal, but they can easily be orthogonalized if desired. This process restores the missing nodes, giving AO's of the form of Eq. (5), and surprisingly close to the best possible of this form.[6] As an example, if χ_{1s} and χ_{2s} refer to Slater AO's, and $\chi_{2s}{}^{\mathrm{or}}$

[2] See, e.g., C. C. J. Roothaan, forthcoming paper.

[3] J. C. Slater, Phys. Rev. **42**, 33 (1932).

[4] (a) Morse, Young, and Haurwitz, Phys. Rev. **48**, 948 (1935); (b) extended and partially corrected by W. E. Duncanson and C. A. Coulson, Proc. Roy. Soc. Edinburgh **62A**, 37 (1944).

[5] J. C. Slater, Phys. Rev. **36**, 57 (1930); C. Zener, Phys. Rev. **36**, 51 (1930).

[6] W. E. Moffitt and C. A. Coulson, Phil. Mag. [7], **38**, 634 (1937). These authors also give instructive contour diagrams of carbon atom *2te*, *2tr*, and *2di* hybrid AO's, using orthogonalized Slater *2s* AO's.

TABLE I. Slater μ-values[a] for valence-shell ns, np AO's.

H	1.00	C^-	1.45	O^-	2.10	Na	0.733	S	1.817
Li	0.65	C	1.625	O	2.275	Mg	0.95	Cl	2.033
Be	0.975	C^+	1.80	O^+	2.45	Al	1.167	Br	2.054
B^-	1.125	N	1.95	F	2.60	Si	1.383	I	1.90
B	1.30	N^+	2.125			P	1.60		

[a] For $1s$ inner-shell AO's, $\mu = Z - 0.3$ in all cases, where Z is the actual nuclear charge.

refers to the orthogonalized Slater $2s$ AO, we have

$$\left.\begin{aligned} \chi_{2s}^{or} &= (\chi_{2s} - Q\chi_{1s})/(1-Q^2)^{\frac{1}{2}} \\ Q &\equiv S(1s_a, 2s_a) \equiv \int \chi_{2s_a}\chi_{1s_a} d\tau. \end{aligned}\right\} \quad (8)$$

Although Q is often fairly large (for example, 0.220 if atom a is carbon), the process of orthogonalization has little effect on the computed values of overlap integrals (see Section V).

In addition to overlap integrals for pure central-field AO's, those for hybrid AO's are obviously of great interest for molecular problems. Hybrid AO's, although they can be constructed as linear combinations of central-field AO's, are not themselves central-field AO's. Hybrid overlap integrals can be obtained as simple linear combinations of Slater-AO S's (see Section V), so that no laborious new computations are required. For purposes of computing S's, approximate hybrid AO's built from Slater AO's without orthogonalization should according to our previous reasoning give satisfactory results.

The hybrid AO's which can be formed from s and p AO's are all of σ-type, that is, have cylindrical symmetry. Let χ be a normalized hybrid σ AO of the form

$$\chi = \alpha\chi_{ns} \pm (1-\alpha^2)^{\frac{1}{2}}\chi_{np\sigma}, \quad (9)$$

where $0 \leqslant \alpha \leqslant 1$, and χ_{ns} and $\chi_{np\sigma}$ are as defined by Eqs. (2)–(3) and Fig. 1, with R_{nl} given by Eqs. (7) for the case of hybrid Slater AO's. In the following, the abbreviations *te*, *tr*, and *di* (tetrahedral, trigonal, and digonal) will be used for hybrids with $\alpha^2 = \frac{1}{4}$, $\frac{1}{3}$, and $\frac{1}{2}$ respectively and with the plus sign in Eq. (9). A further, complementary, set of hybrid AO's *te'*, *tr'*, and *di'* is obtained by using the minus sign in Eq. (9). The hybrids *te*, *tr*, *di* of an atom give *large overlap* (large S) with corresponding or similar AO's of a second atom, while *te'*, *tr'*, and *di'* give *small overlap*.

III. NOMENCLATURE AND PARAMETERS FOR OVERLAP INTEGRALS

In discussing and tabulating overlap integrals, a careful choice of nomenclature will prove to be important. In order to characterize an overlap integral fully, we need to specify (a) the two AO's involved; (b) the two atoms involved, and which AO belongs to each atom; (c) the value of R. Two suitable types of general symbols for this purpose, also a symbol for an internal overlap or non-orthogonality integral, are indicated by the following examples. Simplified symbols can be used in special cases.

(a) Atom-pair at any distance R:
  general AO's and atoms: $S(n_a x, n_b y; R)$
  specific AO's, general atoms:
    $S(1s_a, 2p\sigma_b; R)$ or $S(1s, 2p\sigma; R)$
  specific AO's and atoms:
    $S(2p\pi_C, 3p\pi_{Si}; 1.75A)$
(b) Atom-pair involved in specified bond at equilibrium distance:
    $S(2s, 2p\sigma; C=C)$; $S(2te, 3p\sigma; C-Cl)$;
    $S(1s, 2s; H-Li)$; $S(2p\pi, 2p\pi, O^+\equiv C)$
(c) Single atom a (internal overlap integral):
    $S(1s_a, 2s_a)$.
                                                   (10)

In the atom-pair symbols above, it is intended to be understood that the *first AO symbol always goes with the first atom, the second with the second*. However, it is not essential here to follow any fixed rule as to which AO (or atom) symbol shall be written first. Nevertheless, in another type of notation which will be required below, it *does* become essential to adopt certain fixed conventions in this matter, and it will diminish the possibilities of error if we follow the same conventions also in using the types of notation given above. These conventions are: the AO of smaller n is to be written first, or if both n's are equal, the AO belonging to the atom of larger Z is to be written first. These rules were followed in the examples given above.

We turn now to the matter of coordinates and parameters to be used in the evaluation of the S's. The computation is best effected after transforming from polar coordinates of the two atoms (see Fig. 1) to spheroidal coordinates ξ, η, ϕ given by:

$$\xi = (r_a + r_b)/R; \quad \eta = (r_a - r_b)/R; \quad \phi = \phi_a = \phi_b. \quad (11)$$

The coordinate ξ ranges from 1 to ∞, η from -1 to $+1$.

For any given AO pair, we shall wish to obtain S values for various pairs of atoms, each for various interatomic distances R. To accomplish this, the best procedure is first to set up for each AO pair a single master formula expressed in terms of suitable parameters depending on the μ's of the two AO's and on R, and from this to compute and tabulate numerical values of S as a function of the chosen parameters. For this purpose, the two parameters p and t defined as follows are found to be appropriate for the case of Slater AO's (see Eqs. (7) and Table I).

$$\left.\begin{aligned} p &\equiv \tfrac{1}{2}(\mu_a + \mu_b)R/a_H, \\ t &\equiv (\mu_a - \mu_b)/(\mu_a + \mu_b). \end{aligned}\right\} \quad (12)$$

To find S for a given Slater-AO pair, atom pair and R value, one then first computes p and t for the particular case, then looks up the value of S in the master table for the given AO pair. Methods of obtaining S values for other types of AO's will be described in Section V.

Since in connection with the master tables we replace atom designations by parameters p and t, it is now necessary to introduce a type of S symbol different from those given in (10), and indicated by the following examples:

(a) Atom pair characterized by parameters p, t:
General form: $S(n_a x, n_b y; p, t)$;
Specific form: $S(2s_a, 3p\sigma_b; p, t)$,
or more briefly, but with the same meaning,
$S(2s, 3p\sigma; p, t)$.
(b) Single atom (internal overlap integral):
General form: $S(n_a x, n_a' y)$
Specific form: $S(1s, 2s; \mathrm{O}, t)$. (13)

The signs of the coordinate η in Eqs. (11) and of the parameter t in Eqs. (12) obviously depend on the assignment of the labels a and b to the two atoms in any atom pair. In order to make the master formulas and tables unambiguous with respect to the signs of η and t, it is therefore necessary to adopt suitable conventions to fix this assignment. The following two (necessarily arbitrary) conventions will be used in this paper:

(1) *In case the two AO's have unequal n, the atom whose AO has smaller n will always be identified with atom a, and its AO will always be written first in the symbol for S*, for example $S(2s_a, 3p\sigma_b; p, t)$ or more briefly $S(2s, 3p\sigma; p, t)$; but *never* $S(3p\sigma_a, 2s_b; p, t)$ or $S(3p\sigma, 2s; p, t)$. As another example, we may write $S(2p\sigma, 3s)$, but never $S(3s, 2p\sigma)$. Further examples are $S(2s, 3s)$ and $S(2p\sigma, 3p\sigma)$. In general formulations, symbols such as $S(n_a s, n_b p\sigma)$ or $S(n_a p\sigma, n_b s)$ will be used, always with the implication that $n_a < n_b$. It will be noticed that for the present case of two AO's with unequal n, both positive and negative t values are possible, depending on the sign of $\mu_a - \mu_b$. For example (see Table I),

$$t(2p\sigma_{\mathrm{F}}, 3s_{\mathrm{Mg}}) = (2.60-0.95)/(2.60+0.95) > 0,$$

but

$$t(2p\sigma_{\mathrm{Be}}, 3s_{P}) = (1.30-1.60)/(1.30+1.60) < 0.$$

Also, but only by accident, $t=0$ is possible. The relative practical importance of the positive and negative ranges of t varies according to what AO pair-type one considers.

(2) *In case the two AO's have equal n, the atom with larger μ will always be identified with atom a, and its AO will always be written first in the symbol for S; if the μ's are equal, either atom may be called a.* For example, $S(ns_a, np\sigma_b)$ or briefly $S(ns, np\sigma)$; $S(np\sigma_a, ns_b)$ or briefly $S(np\sigma, ns)$; and $S(2p\pi_a, 2p\pi_b)$ or briefly $S(2p\pi, 2p\pi)$, all imply $Z_a \geqslant Z_b$. Hence, according to Eqs. (12), convention (2) here restricts t to the range $t \geqslant 0$. The integrals $S(ns_a, np\sigma_b)$ and $S(np\sigma_a, ns_b)$—see Eqs. (14) below, taking $n_a = n_b$—require a little special comment because of their somewhat confusing similarity.[7] For $t>0$, the two integrals are different and have different values, but for $t=0$ they become identical. For instance, $S(2p\sigma_{\mathrm{O}}, 2s_{\mathrm{C}})$ and $S(2s_{\mathrm{O}}, 2p\sigma_{\mathrm{C}})$, which are examples of the two cases written in accordance with convention (2), have different values; but $S(2p\sigma_{\mathrm{C}}, 2s_{\mathrm{C}})$ and $S(2s_{\mathrm{C}}, 2p\sigma_{\mathrm{C}})$ are identical.

IV. MASTER FORMULAS AND COMPUTATION METHODS FOR MASTER TABLES OF SLATER-AO OVERLAP INTEGRALS

With the foregoing conventions, the S master formulas fall into five basic types. Using Eqs. (1)-(3), (7), (11)-(12), letting

$$m = n - \delta,$$

and integrating over ϕ, these five types take the forms:

$$\left.\begin{aligned}
S(n_a s, n_b s; p, t) &= \tfrac{1}{2} N_a N_b (\tfrac{1}{2}R)^{m_a+m_b+1} \int_1^\infty \int_{-1}^1 (\xi+\eta)^{m_a-1}(\xi-\eta)^{m_b-1}(\xi^2-\eta^2) e^{-p(\xi+\eta t)} d\eta d\xi,\\
S(n_a s, n_b p\sigma; p, t) &= (\tfrac{1}{2}\sqrt{3}) N_a N_b (\tfrac{1}{2}R)^{m_a+m_b+1} \int_1^\infty \int_{-1}^1 (\xi+\eta)^{m_a-1}(\xi-\eta)^{m_b-2}(-\xi\eta+1)(\xi^2-\eta^2) e^{-p(\xi+\eta t)} d\eta d\xi,\\
S(n_a p\sigma, n_b s; p, t) &= (\tfrac{1}{2}\sqrt{3}) N_a N_b (\tfrac{1}{2}R)^{m_a+m_b+1} \int_1^\infty \int_{-1}^1 (\xi+\eta)^{m_a-2}(\xi-\eta)^{m_b-1}(-\xi\eta-1)(\xi^2-\eta^2) e^{-p(\xi+\eta t)} d\eta d\xi,\\
S(n_a p\sigma, n_b p\sigma; p, t) &= (3/2) N_a N_b (\tfrac{1}{2}R)^{m_a+m_b+1} \int_1^\infty \int_{-1}^1 (\xi+\eta)^{m_a-2}(\xi-\eta)^{m_b-2}(\xi^2\eta^2-1)(\xi^2-\eta^2) e^{-p(\xi+\eta t)} d\eta d\xi,\\
S(n_a p\pi, n_b p\pi; p, t) &= (3/4) N_a N_b (\tfrac{1}{2}R)^{m_a+m_b+1} \int_1^\infty \int_{-1}^1 (\xi+\eta)^{m_a-2}(\xi-\eta)^{m_b-2}(\xi^2-1)(1-\eta^2)(\xi^2-\eta^2) e^{-p(\xi+\eta t)} d\eta d\xi.
\end{aligned}\right\} \quad (14)$$

[7] Without convention (2), $S(ns_a, np\sigma_b)$ would be defined for positive, zero, and negative values of t ($Z_a > Z_b$, $Z_a = Z_b$, $Z_a < Z_b$; see Eqs. (12)). The same would be true of $S(np\sigma_a, ns_b)$. The two integrals would then differ only in the manner in which the labels a and b are associated with the AO forms ns and $np\sigma$. Now as will be seen from Eqs. (11), (12), an exchange of the labels a and b merely changes the definitions of η and t in such a way that the positive domains of η and t for the one labeling (say η', t') become the corresponding negative domains for the other (say η'', t''). Consequently, if in Eqs. (14) for $S(ns, np\sigma)$ and $S(np\sigma, ns)$ we write η', t' in the one equation and η'', t'' in the other, and then make the substitution $\eta'' = -\eta'$, $t'' = -t'$, the two equations

The integrals over ξ and η in Eqs. (14) may be evaluated by making use of the following mathematical relations:[1]

$$A_k(p) \equiv \int_1^\infty \xi^k e^{-p\xi} d\xi = e^{-p} \sum_{\mu=1}^{k+1} [k!/p^\mu (k-\mu+1)!] \tag{15}$$

$$\left.\begin{aligned} B_k(pt) &\equiv \int_{-1}^{1} \eta^k e^{-pt\eta} d\eta \\ &= -e^{-pt} \sum_{\mu=1}^{k+1} [k!/(pt)^\mu (k-\mu+1)!] - e^{pt} \sum_{\mu=1}^{k+1} [(-1)^{k-\mu} k!/(pt)^\mu (k-\mu+1)!]; \end{aligned}\right\} \tag{16}$$

but $B_k(0) = 2/(k+1)$ for k even; $=0$ for k odd. The factors N in Eqs. (14) (see N_{nl} in Eqs. (7)) may be evaluated by means of:

$$C(k, q) \equiv \int_0^\infty \lambda^k e^{-q\lambda} d\lambda = k!/q^{k+1}. \tag{17}$$

The numerical computation of the overlap integrals is comparatively simple if the A_k's and B_k's are first computed for appropriate values of the parameters p and t; here the simplest procedure (rather than using Eqs. (15), (16) directly) is to obtain the higher k A's and B's from those with lower k by using certain *recursion formulas*.[1d, 1e] We first note that any particular S involves the integral of a polynomial in ξ and η. Then it is plain that each value of S is the product of a factor times a linear combination of A_k's and B_k's all computed at the same values of p and t.

For the computation of the master tables given below, the necessary A_k's and B_k's were first computed[8] for the desired values of p and t, and then combined in accordance with the following Eqs. (18)–(50) to give the S values.[9] These formulas were obtained by the use of Eqs. (15)–(17) in connection with Eqs. (14) and (7). It will be noted that a separate formula is given in each case for $t=0$, for reasons that will be understood on looking at Eqs. (16). Certain additional special formulas will now be described.

Although in general the computation of the A's and B's followed by the use of Eqs. (18)–(50) is the most expeditious procedure, an alternative set of formulas very considerably simplifies the computation in the important special case of two identical atoms ($n_a = n_b$, $t=0$), if one is interested *only* in this case. This case includes not only integrals of the type $S(nx, nx; p, 0)$, but also such integrals as $S(ns, np\sigma; p, 0)$ or $S(np\sigma, ns; p, 0)$; these are particularly important in the computation of hybrid-AO S values (see Section V). The simplified formulas were obtained by substituting for the A's in the $t=0$ formulas in Eqs. (18)–(50), using Eq. (15). Although we did not use them in computing the master tables, they are given below as Eqs. (51)–(63) for the convenience of readers who may wish to make their own computations for additional p values.[9]

One further set of special formulas is required, for the internal overlap integrals $S(n_a x, n_b y; 0, t)$,—see (10) and (13). This type of integral occurs if one orthogonalizes Slater AO's, and also if one wishes to compute SCF S's from Slater S's (see Section V). The necessary formulas are given below as Eqs. (64)–(73).

MASTER FORMULAS FOR SLATER-AO OVERLAP INTEGRALS

$$\left.\begin{aligned} &t=0: \\ &\quad S(1s, 1s) = (6)^{-1} p^3 [3A_2 - A_0] \\ &t>0: \\ &\quad S(1s, 1s) = (4)^{-1} p^3 (1-t^2)^{+3/2} [A_2 B_0 - A_0 B_2] \end{aligned}\right\} \tag{18}$$

become identical except for the sign of t in the exponential, and completely identical for $t=0$. Hence $S(ns, np\sigma)$ for positive t would be equal to $S(np\sigma, ns)$ for negative t, and *vice versa*, if negative t here were not excluded by convention (2); and for $t=0$, $S(ns, np\sigma) \equiv S(np\sigma, ns)$.

[8] Most of this labor could have been avoided by the use of the A and B tables of references 1d or especially 1e. Actually, however, independent computations were made. Afterward, checks were made of some of our A and B values against those in the tables of reference 1e, and excellent agreement was found.

[9] Coulson (reference 1f) also gives formulas corresponding to the majority of the S integrals listed here. However, his formulas for $t \neq 0$ are in terms of μ_a, μ_b, and R, in a form not adapted to the present computation and tabulation. His formulas for $t=0$ are identical with our Eqs. (51)–(63) except that they are not normalized like ours to give S values directly.

$t=0$:
$$S(1s,\,2s)=(12)^{-1}(3)^{-\frac{1}{2}}p^4[3A_3-A_1]$$
$t\neq 0$: (19)
$$S(1s,\,2s)=(8)^{-1}(3)^{-\frac{1}{2}}p^4(1+t)^{3/2}(1-t)^{5/2}[A_3B_0-A_2B_1-A_1B_2+A_0B_3]$$

$t=0$:
$$S(1s,\,2p\sigma)=(12)^{-1}p^4[3A_2-A_0]$$
$t\neq 0$: (20)
$$S(1s,\,2p\sigma)=(8)^{-1}p^4(1+t)^{3/2}(1-t)^{5/2}[-A_3B_1+A_2B_0+A_1B_3-A_0B_2]$$

$t=0$:
$$S(1s,\,3s)=(60)^{-1}(10)^{-\frac{1}{2}}p^5[5A_4-A_0]$$
$t\neq 0$: (21)
$$S(1s,\,3s)=(24)^{-1}(10)^{-\frac{1}{2}}p^5(1+t)^{3/2}(1-t)^{7/2}[A_4B_0-2A_3B_1+2A_1B_3-A_0B_4]$$

$t=0$:
$$S(1s,\,3p\sigma)=(15)^{-1}(30)^{-\frac{1}{2}}p^5[5A_3-2A_1]$$
$t\neq 0$: (22)
$$S(1s,\,3p\sigma)=(8)^{-1}(30)^{-\frac{1}{2}}p^5(1+t)^{3/2}(1-t)^{7/2}[A_3(B_0+B_2)-A_1(B_2+B_4)-B_1(A_2+A_4)+B_3(A_0+A_2)]$$

$t=0$:
$$S(1s,\,5s)=(720)^{-1}(35)^{-\frac{1}{2}}p^6[15A_5+10A_3-9A_1]$$
$t\neq 0$: (23)
$$S(1s,\,5s)=(96)^{-1}(35)^{-\frac{1}{2}}p^6(1+t)^{3/2}(1-t)^{9/2}[A_5B_0-3A_4B_1+2A_3B_2+2A_2B_3-3A_1B_4+A_0B_5]$$

$t=0$:
$$S(1s,\,5p\sigma)=(240)^{-1}(105)^{-\frac{1}{2}}p^6[25A_4-6A_2-3A_0]$$
$t\neq 0$: (24)
$$S(1s,\,5p\sigma)=(32)^{-1}(105)^{-\frac{1}{2}}p^6(1+t)^{3/2}(1-t)^{9/2}[A_4(B_0+2B_2)+A_1(2B_3+B_5)-B_1(2A_3+A_5)-B_4(A_0+2A_2)]$$

$t=0$:
$$S(2s,\,2s)=(360)^{-1}p^5[15A_4-10A_2+3A_0]$$
$t>0$: (25)
$$S(2s,\,2s)=(48)^{-1}p^5(1-t^2)^{5/2}[A_4B_0-2A_2B_2+A_0B_4]$$

$t=0$:
$$S(2s,\,2p\sigma)=(60)^{-1}(3)^{-\frac{1}{2}}p^5[5A_3-A_1]$$
$t\neq 0$: (26)
$$S(2s,\,2p\sigma)=(16)^{-1}(3)^{-\frac{1}{2}}p^5(1-t^2)^{5/2}[A_3(B_0-B_2)+A_1(B_4-B_2)+B_1(A_2-A_4)+B_3(A_2-A_0)]$$
$S(2p\sigma,\,2s)$: same as $S(2s,\,2p\sigma)$ except each $B_k(t)$ replaced by $B_k(-t)$

$t=0$:
$$S(2s,\,3s)=(360)^{-1}(30)^{-\frac{1}{2}}p^6[15A_5-10A_3+3A_1]$$
$t\neq 0$: (27)
$$S(2s,\,3s)=(48)^{-1}(30)^{-\frac{1}{2}}p^6(1+t)^{5/2}(1-t)^{7/2}[A_5B_0-A_4B_1-2A_3B_2+2A_2B_3+A_1B_4-A_0B_5]$$

$t=0$:
$$S(2s,\,3p\sigma)=(360)^{-1}(10)^{-\frac{1}{2}}p^6[15A_4-10A_2+3A_0]$$
$t\neq 0$: (28)
$$S(2s,\,3p\sigma)=(48)^{-1}(10)^{-\frac{1}{2}}p^6(1+t)^{5/2}(1-t)^{7/2}[-A_5B_1+A_4B_0+2A_3B_3-2A_2B_2-A_1B_5+A_0B_4]$$

$t=0$:
$$S(2s,\,5s)=(10{,}080)^{-1}(105)^{-\frac{1}{2}}p^7[105A_6-35A_4-21A_2+15A_0]$$
$t\neq 0$: (29)
$$S(2s,\,5s)=(192)^{-1}(105)^{-\frac{1}{2}}p^7(1+t)^{5/2}(1-t)^{9/2}[A_6B_0-2A_5B_1-A_4B_2+4A_3B_3-A_2B_4-2A_1B_5+A_0B_6]$$

$t=0$:
$$S(2s,\,5p\sigma)=(2520)^{-1}(35)^{-\frac{1}{2}}p^7[35A_5-28A_3+9A_1]$$
$t\neq 0$: (30)
$$S(2s,\,5p\sigma)=(192)^{-1}(35)^{-\frac{1}{2}}p^7(1+t)^{5/2}(1-t)^{9/2}[-A_6B_1+A_5(B_0+B_2)+A_4(2B_3-B_1)-2A_3(B_2+B_4)$$
$$+A_2(2B_3-B_5)+A_1(B_4+B_6)-A_0B_5]$$

$t=0$:
$$S(2p\sigma,2p\sigma)=(120)^{-1}p^5[5A_4-18A_2+5A_0]$$
$t>0$:
$$S(2p\sigma,2p\sigma)=(16)^{-1}p^5(1-t^2)^{5/2}[B_2(A_0+A_4)-A_2(B_0+B_4)] \quad (31)$$

$t=0$:
$$S(2p\sigma,3s)=(360)^{-1}(10)^{-\frac{1}{2}}p^6[5A_4+6A_2-3A_0]$$
$t\neq 0$:
$$S(2p\sigma,3s)=(48)^{-1}(10)^{-\frac{1}{2}}p^6(1+t)^{5/2}(1-t)^{7/2}[A_4(B_0-2B_2)+A_1(2B_3-B_5)+B_1(A_5-2A_3)+B_4(2A_2-A_0)] \quad (32)$$

$t=0$:
$$S(2p\sigma,3p\sigma)=(120)^{-1}(30)^{-\frac{1}{2}}p^6[5A_5-18A_3+5A_1]$$
$t\neq 0$:
$$S(2p\sigma,3p\sigma)=(16)^{-1}(30)^{-\frac{1}{2}}p^6(1+t)^{5/2}(1-t)^{7/2}[A_2(B_1+B_5)-A_3(B_0+B_4)-B_3(A_0+A_4)+B_2(A_1+A_5)] \quad (33)$$

$t=0$:
$$S(2p\sigma,5s)=(630)^{-1}(35)^{-\frac{1}{2}}p^7[7A_3-3A_1]$$
$t\neq 0$:
$$S(2p\sigma,5s)=(192)^{-1}(35)^{-\frac{1}{2}}p^7(1+t)^{5/2}(1-t)^{9/2}[A_6B_1+A_5(B_0-3B_2)+A_4(2B_3-3B_1)+2A_3(B_2+B_4)+A_2(2B_3-3B_5)+A_1(B_6-3B_4)+A_0B_5] \quad (34)$$

$t=0$:
$$S(2p\sigma,5p\sigma)=(3360)^{-1}(105)^{-\frac{1}{2}}p^7[35A_6-105A_4-15A_2+21A_0]$$
$t\neq 0$:
$$S(2p\sigma,5p\sigma)=(64)^{-1}(105)^{-\frac{1}{2}}p^7(1+t)^{5/2}(1-t)^{9/2}[A_6B_2-A_4B_0+2A_3(B_1+B_5)-2B_3(A_1+A_5)+B_4A_0-B_6A_2] \quad (35)$$

$t=0$:
$$S(3s,3s)=(25{,}200)^{-1}p^7[35A_6-35A_4+21A_2-5A_0]$$
$t>0$:
$$S(3s,3s)=(1440)^{-1}p^7(1-t^2)^{7/2}[A_6B_0-3A_4B_2+3A_2B_4-A_0B_6] \quad (36)$$

$t=0$:
$$S(3s,3p\sigma)=(12{,}600)^{-1}(3)^{-\frac{1}{2}}p^7[35A_5-14A_3+3A_1]$$
$t\neq 0$:
$$S(3s,3p\sigma)=(480)^{-1}(3)^{-\frac{1}{2}}p^7(1-t^2)^{7/2}[-A_6B_1+A_5(B_0-B_2)+A_4(B_1+2B_3)+2A_3(B_4-B_2)-A_2(2B_3+B_5)+A_1(B_4-B_6)+A_0B_5] \quad (37)$$

$S(3p\sigma,3s)$: same as $S(3s,3p\sigma)$ except each $B_k(t)$ replaced by $B_k(-t)$.

$t=0$:
$$S(3s,5s)=(50{,}400)^{-1}(14)^{-\frac{1}{2}}p^8[35A_7-35A_5+21A_3-5A_1]$$
$t\neq 0$:
$$S(3s,5s)=(2880)^{-1}(14)^{-\frac{1}{2}}p^8(1+t)^{7/2}(1-t)^{9/2}[A_7B_0-A_6B_1-3A_5B_2+3A_4B_3+3A_3B_4-3A_2B_5-A_1B_6+A_0B_7] \quad (38)$$

$t=0$:
$$S(3s,5p\sigma)=(16{,}800)^{-1}(42)^{-\frac{1}{2}}p^8[35A_6-35A_4+21A_2-5A_0]$$
$t\neq 0$:
$$S(3s,5p\sigma)=(960)^{-1}(42)^{-\frac{1}{2}}p^8(1+t)^{7/2}(1-t)^{9/2}[-A_7B_1+A_6B_0+3A_5B_3-3A_4B_2-3A_3B_5+3A_2B_4+A_1B_7-A_0B_6] \quad (39)$$

$t=0$:
$$S(3p\sigma,3p\sigma)=(25{,}200)^{-1}p^7[35A_6-147A_4+85A_2-21A_0]$$
$t>0$:
$$S(3p\sigma,3p\sigma)=(480)^{-1}p^7(1-t^2)^{7/2}[A_6B_2-A_4(B_0+2B_4)+A_2(B_6+2B_2)-A_0B_4] \quad (40)$$

$t=0$:
$$S(3p\sigma,5s)=(50{,}400)^{-1}(42)^{-\frac{1}{2}}p^8[35A_6+49A_4-51A_2+15A_0]$$
$t\neq 0$: (41)
$$S(3p\sigma,5s)=(960)^{-1}(42)^{-\frac{1}{2}}p^8(1+t)^{7/2}(1-t)^{9/2}[A_7B_1+A_6(B_0-2B_2)-A_5(B_3+2B_1)+A_4(4B_4-B_2)+A_3(4B_3-B_5)-A_2(2B_6+B_4)+A_1(B_7-2B_5)+A_0B_6]$$

$t=0$:
$$S(3p\sigma,5p\sigma)=(50{,}400)^{-1}(14)^{-\frac{1}{2}}p^8[35A_7-147A_5+85A_3-21A_1]$$

(42)

$t\neq 0$:

$$S(3p\sigma, 5p\sigma)=(960)^{-1}(14)^{-\frac{1}{2}}p^{8}(1+t)^{7/2}(1-t)^{9/2}[A_7B_2-A_6B_3-A_5(B_0+2B_4)+A_4(B_1+2B_5)+A_3(2B_2+B_6)-A_2(2B_3+B_7)-A_1B_4+A_0B_5]$$

(43)

$t=0$:

$$S(5s, 5s)=(12{,}700{,}800)^{-1}p^{9}[315A_8-420A_6+378A_4-180A_2+35A_0]$$

$t>0$:

$$S(5s, 5s)=(80{,}640)^{-1}p^{9}(1-t^2)^{9/2}[A_8B_0-4A_6B_2+6A_4B_4-4A_2B_6+A_0B_8]$$

(44)

$t=0$:

$$S(5s, 5p\sigma)=(2{,}116{,}800)^{-1}(3)^{-\frac{1}{2}}p^{9}[105A_7-63A_5+27A_3-5A_1]$$

$t\neq 0$:

$$S(5s, 5p\sigma)=(26{,}880)^{-1}(3)^{-\frac{1}{2}}p^{9}(1-t^2)^{9/2}[-A_8B_1+A_7(B_0-B_2)+A_6(B_1+3B_3)+3A_5(B_4-B_2)-3A_4(B_3+B_5)+3A_3(B_4-B_6)+A_2(3B_5+B_7)+A_1(B_8-B_6)-A_0B_7]$$

$S(5p\sigma, 5s)$: same as $S(5s, 5p\sigma)$ except each $B_k(t)$ replaced by $B_k(-t)$.

(45)

$t=0$:

$$S(5p\sigma, 5p\sigma)=(4{,}233{,}600)^{-1}p^{9}[105A_8-504A_6+450A_4-224A_2+45A_0]$$

$t>0$:

$$S(5p\sigma, 5p\sigma)=(26{,}880)^{-1}p^{9}(1-t^2)^{9/2}[A_8B_2-A_6(B_0+3B_4)+3A_4(B_2+B_6)-A_2(B_8+3B_4)+A_0B_6]$$

(46)

$t=0$:

$$S(2p\pi, 2p\pi)=(120)^{-1}p^{5}[5A_4-6A_2+A_0]$$

$t>0$:

$$S(2p\pi, 2p\pi)=(32)^{-1}p^{5}(1-t^2)^{5/2}[A_4(B_0-B_2)+A_2(B_4-B_0)+A_0(B_2-B_4)]$$

(47)

$t=0$:

$$S(2p\pi, 3p\pi)=(120)^{-1}(30)^{-\frac{1}{2}}p^{6}[5A_5-6A_3+A_1]$$

$t\neq 0$:

$$S(2p\pi, 3p\pi)=(32)^{-1}(30)^{-\frac{1}{2}}p^{6}(1+t)^{5/2}(1-t)^{7/2}[A_5(B_0-B_2)+A_4(B_3-B_1)+A_3(B_4-B_0)+A_2(B_1-B_5)+A_1(B_2-B_4)+A_0(B_5-B_3)]$$

(48)

$t=0$:

$$S(2p\pi, 5p\pi)=(3360)^{-1}(105)^{-\frac{1}{2}}p^{7}[35A_6-35A_4-3A_2+3A_0]$$

$t\neq 0$:

$$S(2p\pi, 5p\pi)=(128)^{-1}(105)^{-\frac{1}{2}}p^{7}(1+t)^{5/2}(1-t)^{9/2}[A_6(B_0-B_2)+2A_5(B_3-B_1)+A_4(B_2-B_0)+2A_3(B_1-B_5)+A_2(B_6-B_4)+2A_1(B_5-B_3)+A_0(B_4-B_6)]$$

(49)

$t=0$:

$$S(3p\pi, 3p\pi)=(25{,}200)^{-1}p^{7}[35A_6-49A_4+17A_2-3A_0]$$

$t>0$:

$$S(3p\pi, 3p\pi)=(960)^{-1}p^{7}(1-t^2)^{7/2}[A_6(B_0-B_2)+A_4(2B_4-B_0-B_2)+A_2(2B_2-B_4-B_6)+A_0(B_6-B_4)]$$

(49a)

$t=0$:

$$S(3p\pi, 5p\pi)=(50{,}400)^{-1}(14)^{-\frac{1}{2}}p^{8}[35A_7-49A_5+17A_3-3A_1]$$

$t\neq 0$:

$$S(3p\pi, 5p\pi)=(1920)^{-1}(14)^{-\frac{1}{2}}p^{8}(1+t)^{7/2}(1-t)^{9/2}[A_7(B_0-B_2)+A_6(B_3-B_1)+A_5(2B_4-B_2-B_0)+A_4(B_1+B_3-2B_5)+A_3(2B_2-B_4-B_6)+A_2(B_5-2B_3+B_7)+A_1(B_6-B_4)+A_0(B_5-B_7)]$$

(50)

$t=0$:

$$S(5p\pi, 5p\pi)=(4{,}233{,}600)^{-1}p^{9}[105A_8-168A_6+90A_4-32A_2+5A_0]$$

$t>0$:

$$S(5p\pi, 5p\pi)=(53{,}760)^{-1}p^{9}(1-t^2)^{9/2}[A_8(B_0-B_2)+A_6(3B_4-2B_2-B_0)+3A_4(B_2-B_6)+A_2(B_8+2B_6-3B_4)+A_0(B_6-B_8)]$$

SPECIAL FORMULAS FOR SLATER-AO OVERLAP INTEGRALS FOR $t=0$

$$S(1s, 1s)=e^{-p}[1+p+(1/3)p^2] \quad (51)$$

$$S(2s, 2s)=e^{-p}[1+p+(4/9)p^2+(1/9)p^3+(1/45)p^4] \quad (52)$$

$$S(3s, 3s)=e^{-p}[1+p+(7/15)p^2+(2/15)p^3+(2/75)p^4+(1/225)p^5+(1/1575)p^6] \quad (53)$$

$$S(5s,5s)=e^{-p}[1+p+(10/21)p^2+(1/7)p^3+(16/525)p^4+(8/1575)p^5+(8/11{,}025)p^6+(1/11{,}025)p^7+(1/99{,}225)p^8] \quad (54)$$

$$S(2p\sigma,2p\sigma)=e^{-p}[-1-p-(1/5)p^2+(2/15)p^3+(1/15)p^4] \quad (55)$$

$$S(3p\sigma,3p\sigma)=e^{-p}[-1-p-(9/25)p^2-(2/75)p^3+(34/1575)p^4+(13/1575)p^5+(1/525)p^6] \quad (56)$$

$$S(5p\sigma,5p\sigma)=e^{-p}[-1-p-(29/70)p^2-(17/210)p^3-(1/735)p^4+(1/294)p^5+(67/66{,}150)p^6+(13/66{,}150)p^7+(1/33{,}075)p^8] \quad (57)$$

$$S(2s,2p\sigma)=S(2p\sigma,2s)=(\sqrt{3}/6)e^{-p}[p+p^2+(7/15)p^3+(2/15)p^4] \quad (58)$$

$$S(3s,3p\sigma)=S(3p\sigma,3s)=(\sqrt{3}/9)e^{-p}[p+p^2+(12/25)p^3+(11/75)p^4+(17/525)p^5+(1/175)p^6] \quad (59)$$

$$S(5s,5p\sigma)=S(5p\sigma,5s)=(\sqrt{3}/12)e^{-p}[p+p^2+(17/35)p^3+(16/105)p^4+(128/3675)p^5+(23/3675)p^6+(31/33{,}075)p^7+(4/33{,}075)p^8] \quad (60)$$

$$S(2p\pi,2p\pi)=e^{-p}[1+p+(2/5)p^2+(1/15)p^3] \quad (61)$$

$$S(3p\pi,3p\pi)=e^{-p}[1+p+(34/75)p^2+(3/25)p^3+(31/1575)p^4+(1/525)p^5] \quad (62)$$

$$S(5p\pi,5p\pi)=e^{-p}[1+p+(33/70)p^2+(29/210)p^3+(41/1470)p^4+(1/245)p^5+(29/66{,}150)p^6+(1/33{,}075)p^7] \quad (63)$$

SPECIAL FORMULAS FOR SLATER-AO OVERLAP INTEGRALS FOR $p=0$

$$S(1s,1s)=(1-t^2)^{3/2} \quad (64)$$

$$S(1s,2s)=[(3/4)(1+t)^3(1-t)^5]^{\frac{1}{2}} \quad (65)$$

$$S(1s,3s)=[(2/5)(1+t)^3(1-t)^7]^{\frac{1}{2}} \quad (66)$$

$$S(1s,5s)=[(5/28)(1+t)^3(1-t)^9]^{\frac{1}{2}} \quad (67)$$

$$S(2s,2s)=-S(2p\sigma,2p\sigma)=S(2p\pi,2p\pi)=(1-t^2)^{5/2} \quad (68)$$

$$S(2s,3s)=-S(2p\sigma,3p\sigma)=S(2p\pi,3p\pi)=[(5/6)(1+t)^5(1-t)^7]^{\frac{1}{2}} \quad (69)$$

$$S(2s,5s)=-S(2p\sigma,5p\sigma)=S(2p\pi,5p\pi)=[(15/28)(1+t)^5(1-t)^9]^{\frac{1}{2}} \quad (70)$$

$$S(3s,3s)=-S(3p\sigma,3p\sigma)=S(3p\pi,3p\pi)=(1-t^2)^{7/2} \quad (71)$$

$$S(3s,5s)=-S(3p\sigma,5p\sigma)=S(3p\pi,5p\pi)=[(7/8)(1+t)^7(1-t)^9]^{\frac{1}{2}} \quad (72)$$

$$S(5s,5s)=-S(5p\sigma,5p\sigma)=S(5p\pi,5p\pi)=(1-t^2)^{9/2} \quad (73)$$

V. USE OF THE TABLES TO OBTAIN NON-SLATER OVERLAP INTEGRALS

Although the master formulas above and the master tables below are for Slater-AO overlap integrals, they can also be used in a relatively simple way to obtain S's for orthogonalized Slater AO's, for SCF AO's, for hybrid AO's, and in other ways. This follows from the definition of S in Eq. (1) and the fact that other types of AO's can be written as linear combinations of Slater AO's.

Va. Orthogonalized-Slater-AO Overlap Integrals

As a simple example for the case of orthogonalized Slater AO's, we have for $S^{or}(2s,2s)$ from Eqs. (1) and (8), using the atom-pair notation of (10):

$$S^{or}(2s_a,2s_b;R)=[S(2s_a,2s_b;R)-2QS(1s_a,2s_b;R)+Q^2S(1s_a,1s_b;R)]/(1-Q^2), \quad (74)$$

where $Q\equiv S(1s_a,2s_a)$. Using the parameter notation of (13), Eq. (74) becomes

$$S^{or}(2s,2s;p,t)=[S(2s,2s;p,t)-2QS(1s,2s;p,t)+Q^2S(1s,1s;p,t)]/(1-Q^2), \quad (75)$$

with $Q=S(1s,2s;0,t)$. It will be noted that the computation of S's corresponding to orthogonalized AO's involves the knowledge of internal Slater S's, like $S(1s_a,2s_a)$, and of other Slater S's involving inner-shell AO's, like $S(1s_a,2s_b)$ and $S(1s_a,1s_b)$ in Eq. (74). These, however, differ from like-designated outer-shell S's only in the particular ranges of p and t values needed. For internal S's we have $p=0$, but $t\neq0$,—see Eqs. (64)–(73).

In the master tables, an effort has been made to include such ranges of p and t that orthogonalized-AO S's can be computed if desired. However, we believe that these are not likely to be important for the range

TABLE II. Comparison of Slater-AO with orthogonalized Slater-AO S's.

Integral	Slater AO's	Orthogonalized Slater AO's
$S(1s_H, 2s_C; 1.06A)$	0.575	0.572
$S(1s_H, 2s_{Li}; 1.60A)$	0.423	0.415
$S(2s_C, 2s_C; 1.20A)$	0.507	0.503
$S(2s_C, 2p\sigma_C; 1.20A)$	0.470	0.480

of R values (medium and large R) that occur for molecular problems. This conclusion is based on the sample cases in Table II, where it will be noted that the two sets of S values differ little, in spite of the rather considerable magnitude (0.220) of the integral $S(1s_C, 2s_C)$,—Q of Eq. (74),—which is involved in three of the cases. A wider variety of cases might of course sometimes disclose larger differences, but it seems unlikely that these would ever become important.

Vb. SCF-AO Overlap Integrals

We have made no extensive study of SCF-AO S's, but it is obvious from Eqs. (1) and (5) or (6) that these are approximately expressible as linear combinations of Slater S's similar to Eq. (75), although usually containing a number of terms. As an example, and for its own intrinsic interest, we have, however, computed $S^{SCF}(2p\pi_C, 2p\pi_C; R)$ and $S^{SCF}(2p\sigma_C, 2p\sigma_C; R)$. This was done after fitting the following normalized formula to the SCF 2p AO as given by Jucys in tabular numerical form[10] for the state $1s)^2 2s)^2 2p)^2$, ^1D:

$$\chi^{SCF}(2p_C) \approx 0.260\chi(2p; 2.694) + 0.518\chi(2p; 1.416) + 0.309\chi(2p; 0.898). \quad (76)$$

The formula applies equally to $2p\sigma$ or $2p\pi$. Each of the symbols $\chi(2p; \mu)$ denotes a 2p (σ or π) normalized Slater AO with R_{2p} of the form $N_\mu re^{-\mu r/a_H}$ (see Eqs. (7)). A three-term formula (see Eq. (6)) was found necessary here to obtain a reasonably good fit to the numerical values of Jucys.

When Eq. (76), used for either $2p\sigma$ or $2p\pi$, is combined with Eq. (1), one obtains an expression for $S^{SCF}(2p\sigma_C, 2p\sigma_C; R)$ or $S^{SCF}(2p\pi_C, 2p\pi_C; R)$ as a linear combination of six terms involving Slater-AO S's. For a given R, all of these have different p values; three of them have $t=0$, the rest, other t values >0. Referring to the master tables for the necessary Slater S's, we obtained the $S^{SCF}(R)$ values given in Table III. The Table also includes comparisons with the corresponding simple Slater-AO quantities.

The foregoing example shows that good SCF AO's may give quite different S values than Slater AO's,

TABLE III. Comparison of SCF and Slater S's for carbon-carbon 2p bonds.

R (A)	p (for Slater AO)	$S(2p\sigma_C, 2p\sigma_C; R)$ SCF	Slater	$S(2p\pi_C, 2p\pi_C; R)$ SCF	Slater
0.00	0.00	−1.00	−1.00	1.00	1.00
0.50	1.54	−0.53	−0.46	0.82	0.80
1.00	3.07	0.02	0.18	0.53	0.45
1.20	3.68	0.14	0.29	0.43	0.34
1.35	4.15	0.19	0.33	0.36	0.265
1.39	4.27	0.21	0.33	0.34	0.25
1.54	4.73	0.24	0.33	0.29	0.19
2.00	6.14	0.25	0.24	0.16	0.08
2.41	7.40	0.21	0.14	0.10	0.035
2.78	8.55	0.16	0.08	0.06	0.016
4.00	12.3	0.05	0.01	0.01	0.001

especially at large R values.[11] When one examines the forms of SCF AO's as given by Hartree and other authors in numerical tables for such atoms as carbon and oxygen, it is seen that they differ characteristically from those of Slater and hydrogenic AO's in their more gradual decrease at large R values, especially for p AO's.[11a] This typical difference makes understandable the results set forth in Table III. For the computation of S values (and of other interatomic integrals used in molecular calculations) these results evidently raise much more serious questions than does the matter of orthogonalization. In particular, they suggest that non-neighbor interactions in polyatomic molecules may be relatively large. It might then seem that we ought to proceed to compute SCF S values for suitable parameter-ranges for the more important AO pairs. However, second thought brings to mind the fact that the best AO's for *atoms in molecules* must be very considerably different than in free atoms, even though we are not very well informed as to the exact nature of these differences and as to their variety in different molecular states. Still, for the case of AO's used in molecular states with stable binding, we would probably

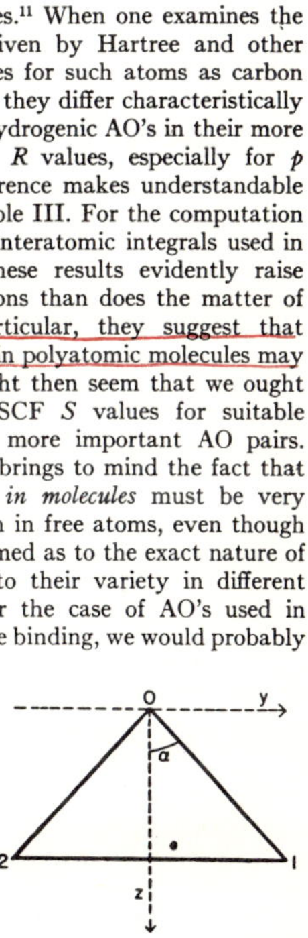

FIG. 2. Axes and notation for H_2O.

[10] A. Jucys, Proc. Roy. Soc. **173A**, 59 (1939); accurate SCF method including exchange. Jucys' 2p AO's are appreciably different for the states ^3P, ^1D, and ^1S, of which we have selected the ^1D, where the AO form approximates the average of those for the three cases. We should have preferred the presumably somewhat different SCF 2p AO for $2s2p^3$, but Jucys did not compute this.

[11] Our conclusions given above as to the smallness of the effect on S of orthogonalization (which introduces inner nodes into Slater's radially nodeless AO's) suggests that the inner nodes in SCF AO's might be dropped without much error for the computation of S values. That is, it might be adequate to retain only the outermost loop as Slater does; *provided*, however, this is represented by a sum of two or more exponential terms as in Eq. (76), instead of by a single Slater function.

[11a] *Note added in proof.* In subsequent work in this laboratory, Dr. H. Shull has found that in the case of 2s AO's there is little difference between SCF and Slater AO's except for small R values where the SCF AO has an inner loop. Hence S values computed using Slater ns AO's are nearly the same as using SCF ns AO's.

expect the free-atom AO's to be so modified as to correspond to *increased* rather than decreased overlap; and this increase should be relative to SCF AO's, and therefore all the more so relative to Slater AO's. Obviously the matter deserves further study. Obviously also, the exact values of Slater S's should not be taken too seriously.

Vc. Hybrid-AO Overlap Integrals

The computation of S values for hybrid AO's is a rather satisfying matter, since it is very easy to obtain these from Slater S's, and since the results are very striking. As discussed at the end of Section II, only σ-hybrid AO's can be formed from s and p AO's. The notation (see (10), (13)) and conventions of Section II for Slater-AO S's can be used without change for hybrid-Slater-AO S's.

In general, for a σ-bond between two atoms a and b, formed by AO's with respective hybridization coefficients α_a and α_b (see Eq. (9)), Eqs. (1) and (9) give

$$\begin{aligned}S^{\text{hy}}(n_ax, n_by; p, t) &= \alpha_a\alpha_b S(n_as, n_bs; p, t)\\ &+\sigma_b\alpha_a(1-\alpha_b^2)^{\frac{1}{2}}S(n_as, n_bp\sigma; p, t)\\ &+\sigma_a\alpha_b(1-\alpha_a^2)^{\frac{1}{2}}S(n_ap\sigma, n_bs; p, t)\\ &+\sigma_a\sigma_b(1-\alpha_a^2)^{\frac{1}{2}}(1-\alpha_b^2)^{\frac{1}{2}}S(n_ap\sigma, n_bp\sigma; p, t), \quad (77)\end{aligned}$$

where σ_a and σ_b denote the signs before $(1-\alpha^2)^{\frac{1}{2}}$ in Eq. (9), and the parameter-pair p, t have the same values in all the integrals. For a homopolar bond this reduces to

$$\begin{aligned}S^{\text{hy}}(nx, nx; p, 0) &= \alpha^2 S(ns, ns; p, 0)\\ &\pm 2\alpha(1-\alpha^2)^{\frac{1}{2}}S(ns, np\sigma; p, 0)\\ &+(1-\alpha^2)S(np\sigma, np\sigma; p, 0). \quad (78)\end{aligned}$$

For a bond between a $1s$ hydrogen atom AO and a hybrid σ-AO of another atom b, it becomes

$$\begin{aligned}S^{\text{hy}}(1s_{\text{H}}, n_by; p, t) &= \alpha_b S(1s_{\text{H}}, ns_b; p, t)\\ &\pm(1-\alpha_b^2)^{\frac{1}{2}}S(1s_{\text{H}}, np\sigma_b; p, t). \quad (79)\end{aligned}$$

In order to obtain numerical values of S for any given hybrid case, it is necessary only to specify α_a and α_b, then to look up the appropriate Slater-AO S values in the master tables, and finally to form linear combinations in accordance with Eqs. (77)–(79). If desired, S values corresponding to hybrid orthogonalized Slater AO's,[6] or to hybrid SCF AO's, can similarly be obtained by using Eqs. (77)–(79) in connection with orthogonalized or SCF pure-AO S values.

Since S values for any desired pair of Slater-AO hybrids can readily be obtained using Eqs. (77)–(79) and the master tables, no extensive hybrid tables will be given here. Nevertheless, two sets of Slater-AO hybrid S tables (Tables XXIV–XXVIII) have been computed explicitly in order to aid the reader in appreciating quantitatively the rather remarkable effects of hybridization on S values. Further examples of hybrid S values for specific molecules, and a discussion, will be given in a following paper.[11b]

[11b] For a preliminary report, see J. Chem. Phys. **17**, 510 (1949).

Vd. Group Overlap Integrals

A further simple application of the present tables is to the computation of overlap integrals in which one or both atoms are replaced by groups of atoms. S's of this sort frequently occur when one is working with non-localized MO's (molecular orbitals) in LCAO approximation. The non-localized MO structure of H_2O furnishes a convenient example. Neglecting s, p hybridization, the electron configuration may be written:

$$(1s_{\text{O}})^2(2s_{\text{O}})^2(a2py_{\text{O}}+b[1s_1-1s_2])^2(c2pz_{\text{O}} +d[1s_1+1s_2])^2(2px_{\text{O}})^2. \quad (80)$$

Here the two bonding MO's have been written out in LCAO form (see Fig. 2 for notation). In connection with these MO's, the overlap integrals

$$\left.\begin{aligned}S(y_{\text{O}}, s-s) &\equiv \int 2py_{\text{O}}[(1s_1-1s_2)/(2-2S_{1s_11s_2})^{\frac{1}{2}}]d\tau\\ S(z_{\text{O}}, s+s) &\equiv \int 2pz_{\text{O}}[(1s_1+1s_2)/(2+2S_{1s_11s_2})^{\frac{1}{2}}]d\tau\end{aligned}\right\} \quad (81)$$

are of interest. The internal H—H group normalization factors $(2\pm 2S_{1s,1s})^{\frac{1}{2}}$ are obtained using the $S(1s, 1s)$ master table. The remaining integrals are easily evaluated by writing $2py_{\text{O}}$ or $2pz_{\text{O}}$ as a linear combination of a $2p\sigma$ and a $2p\pi$ function, relative to the axis of *either* the O—H$_1$ or the O—H$_2$ bond. Thus

$$\begin{aligned}2py_{\text{O}} &= (\sin\alpha)(2p\sigma_1)+(\cos\alpha)(2p\pi_1)\\ &= -(\sin\alpha)(2p\sigma_2)-(\cos\alpha)(2p\pi_2),\\ 2pz_{\text{O}} &= (\cos\alpha)(2p\sigma_1)+(\sin\alpha)(2p\pi_1)\\ &= (\cos\alpha)(2p\sigma_2)+(\sin\alpha)(2p\pi_2).\end{aligned}$$

On substituting into the expressions for $S(y_{\text{O}}, s-s)$ and $S(z_{\text{O}}, s+s)$ above, and noting that $\int 2p\pi_i 1s_i d\tau = 0$, we obtain

$$\left.\begin{aligned}S(y_{\text{O}}, s-s) &= \sqrt{2}\sin\alpha[1-S(1s_{\text{H}}, 1s_{\text{H}}; R_{12})]^{-\frac{1}{2}}\\ &\quad\times S(1s, 2p\sigma; \text{H—O}),\\ S(z_{\text{O}}, s+s) &= \sqrt{2}\cos\alpha[1+S(1s_{\text{H}}, 1s_{\text{H}}; R_{12})]^{-\frac{1}{2}}\\ &\quad\times S(1s, 2p\sigma; \text{H—O}).\end{aligned}\right\} \quad (82)$$

Here the notation of (10), (13) has been used in a convenient way for the S's.

Equations (82) illustrate how the computation or even the mere formulation of group overlap integrals affords added insight into the bonding characteristics of non-localized MO's. If 2α in H_2O were 90°, Eqs. (82) would give

$$S(y_{\text{O}}, s-s) \approx S(z_{\text{O}}, s+s) \approx S(1s, 2p\sigma; \text{H—O}),$$

the same integral that would occur using localized MO's or using the valence-bond AO method. Actually, the experimental $2\alpha = 105°$ makes $S(y_{\text{O}}, s-s)$ distinctly larger and $S(z_{\text{O}}, s+s)$ distinctly smaller than this; the factors $(1\mp S_{1s,1s})^{\frac{1}{2}}$ enhance this effect considerably. To the extent that S values can be taken as measures of bonding strengths (see following paper), this indicates that the $y_O+(1s-1s)$ MO is more strongly bonding in H_2O than the $z_O+(1s+1s)$; also that the binding (i.e., ionization) energies of the two MO's are in the same order.

The discussion could be further elaborated by a consideration of s, p hybridization, which is undoubtedly of some importance here. For reasons of symmetry, this can affect only the $[z_O+(1s+1s)]$ MO in the non-localized MO treatment; the result must be that the $2s_O$ AO in (80) acquires some MO bonding properties and the bonding character of $[z_O+(1s+1s)]$ is further weakened; but with an over-all gain in bonding. $S(z_O, s+s)$ in Eqs. (81) now is replaced by a linear combination including $S(2s_O, s+s)$, the final S being easily computed as a function of the extent of hybridization if this were known.

Group overlap integrals can be used to obtain insight into bond strengths in many molecular problems, for example in the study of hyperconjugation. This subject will not be developed here, however, since the object of the present paper is to give methods rather than to discuss applications.

Ve. Transition Moment Integrals

Certain transition moment integrals for electronic transitions in homopolar diatomic molecules can be obtained very simply from overlap integrals,[12] for example:

$$Q_{\pi\sigma'} \equiv \int \chi_{np\pi_z} X \chi_{np\sigma'} d\tau = \tfrac{1}{2} RS(np\pi, np\pi; p, 0). \quad (83)$$

ACKNOWLEDGMENT

We are very greatly indebted to Mr. Tracy J. Kinyon for his cooperation in the preparation of the tables. Mr. Kinyon carried through the arduous task of making all the extensive numerical computations for the tables in Sections VI and VII.

VI. MASTER TABLES FOR SLATER-AO OVERLAP INTEGRALS**

TABLE IV.

	$S(1s, 1s)$								
p	$t=0.0$	$t=0.1$	$t=0.2$	$t=0.3$	$t=0.4$	$t=0.5$	$t=0.6$	$t=0.7$	$t=0.8$
0.0	1.000	0.985	0.941	0.868	0.770	0.650	0.512	0.364	0.216
0.5	0.960	0.946	0.905	0.837	0.744	0.630	0.499	0.357	0.213
1.0	0.858	0.847	0.812	0.756	0.678	0.580	0.465	0.337	0.205
1.2	0.807	0.797	0.766	0.715	0.644	0.554	0.447	0.327	0.200
1.3	0.780								
1.4	0.753	0.744	0.717	0.671	0.608	0.526	0.428	0.316	0.196
1.5	0.725								
1.6	0.697	0.689	0.666	0.626	0.570	0.498	0.409	0.305	0.191
1.7	0.669								
1.8	0.641	0.635	0.615	0.581	0.533	0.469	0.388	0.293	0.185
1.9	0.614								
2.0	0.586	0.581	0.565	0.536	0.495	0.439	0.368	0.281	0.180
2.1	0.560								
2.2	0.533	0.529	0.515	0.493	0.458	0.410	0.348	0.269	0.175
2.3	0.508								
2.4	0.483	0.479	0.469	0.451	0.423	0.382	0.328	0.256	0.169
2.5	0.458								
2.6	0.435	0.432	0.425	0.411	0.388	0.355	0.308	0.244	0.164
2.7	0.412								
2.8	0.390	0.388	0.383	0.373	0.356	0.329	0.289	0.233	0.158
2.9	0.369								
3.0	0.349	0.348	0.344	0.338	0.325	0.304	0.271	0.221	0.153
3.2	0.310	0.310	0.309	0.305	0.297	0.281	0.254	0.210	0.148
3.4	0.275	0.276	0.276	0.274	0.270	0.259	0.237	0.200	0.143
3.5	0.259								
3.6	0.244	0.244	0.246	0.247	0.245	0.238	0.221	0.189	0.138
3.8	0.215	0.216	0.218	0.221	0.222	0.219	0.206	0.180	0.133
4.0	0.189	0.190	0.194	0.198	0.201	0.201	0.192	0.170	0.128
4.2	0.166	0.167	0.171	0.176	0.182	0.184	0.179	0.161	0.123
4.4	0.146	0.147	0.151	0.157	0.164	0.168	0.166	0.152	0.119
4.5	0.136								
4.6	0.127	0.129	0.133	0.140	0.148	0.154	0.155	0.144	0.114
5.0	0.097	0.098	0.103	0.110	0.120	0.129	0.134	0.129	0.106
5.5	0.068	0.069	0.074	0.082	0.091	0.102	0.111	0.112	0.097
6.0	0.047	0.049	0.053	0.060	0.070	0.081	0.092	0.097	0.088
6.5	0.032	0.034	0.037	0.044	0.053	0.064	0.076	0.084	0.080
7.0	0.022	0.023	0.026	0.032	0.040	0.050	0.063	0.073	0.072
7.5	0.015	0.016	0.018	0.023	0.030	0.040	0.052	0.063	0.066
8.0	0.010	0.011	0.013	0.017	0.023	0.031	0.043	0.054	0.059
9.0	0.005	0.005	0.006	0.009	0.013	0.019	0.029	0.040	0.049
10.0	0.002	0.002	0.003	0.004	0.007	0.012	0.019	0.030	0.040
12.0	0.000	0.000	0.001	0.001	0.002	0.004	0.009	0.017	0.027
15.0	0.000	0.000	0.000	0.000	0.000	0.001	0.003	0.007	0.015
20.0	0.000	0.000	0.000	0.000	0.000	0.000	0.000	0.002	0.005

TABLE V.

	$S(1s, 2s)$													
p	$t=-0.5$	$t=-0.4$	$t=-0.3$	$t=-0.2$	$t=-0.1$	$t=0.0$	$t=0.1$	$t=0.2$	$t=0.3$	$t=0.4$	$t=0.5$	$t=0.6$	$t=0.7$	$t=0.8$	$t=0.9$
0.0	0.844	0.933	0.977	0.978	0.938	0.866	0.768	0.652	0.526	0.400	0.281	0.177	0.095	0.037	0.007
0.5	0.829	0.916	0.959	0.960	0.923	0.854	0.760	0.647	0.525	0.401	0.284	0.180	0.097	0.039	0.007
1.0	0.787	0.866	0.906	0.907	0.875	0.814	0.730	0.628	0.516	0.400	0.288	0.186	0.102	0.042	0.008
1.5	0.722	0.790	0.823	0.825	0.799	0.749	0.679	0.593	0.496	0.393	0.290	0.192	0.109	0.046	0.009
2.0	0.644	0.697	0.722	0.723	0.702	0.664	0.610	0.542	0.463	0.376	0.285	0.195	0.114	0.050	0.011
2.2	0.611	0.657	0.679	0.679	0.661	0.627	0.579	0.519	0.447	0.367	0.282	0.196	0.116	0.051	0.011
2.4	0.577	0.617	0.635	0.635	0.619	0.589	0.547	0.494	0.430	0.357	0.277	0.195	0.117	0.053	0.012
2.6	0.543	0.578	0.592	0.591	0.576	0.550	0.514	0.468	0.411	0.345	0.272	0.194	0.119	0.054	0.012
2.8	0.510	0.539	0.550	0.547	0.534	0.512	0.481	0.441	0.392	0.333	0.265	0.192	0.119	0.055	0.013
3.0	0.478	0.501	0.508	0.505	0.493	0.474	0.448	0.414	0.372	0.320	0.258	0.190	0.120	0.057	0.013
3.2	0.446	0.464	0.468	0.464	0.453	0.437	0.416	0.388	0.352	0.306	0.251	0.187	0.120	0.058	0.014
3.4	0.416	0.428	0.430	0.425	0.415	0.402	0.384	0.361	0.331	0.292	0.243	0.184	0.120	0.058	0.014
3.6	0.386	0.395	0.394	0.388	0.379	0.368	0.354	0.336	0.311	0.278	0.234	0.180	0.119	0.059	0.014

[12] R. S. Mulliken, J. Chem. Phys. 8, 238 (1940), footnote 14. (In this footnote, Eq. (12) should read Eq. (15).)

** See Section III, especially Eqs. (12), (13), for definitions and conventions.

TABLE V.—(*Con't*)

p	$S(1s, 2s)$ $t=-0.5$	$t=-0.4$	$t=-0.3$	$t=-0.2$	$t=-0.1$	$t=0.0$	$t=0.1$	$t=0.2$	$t=0.3$	$t=0.4$	$t=0.5$	$t=0.6$	$t=0.7$	$t=0.8$	$t=0.9$
3.8						0.335	0.325	0.311	0.291	0.263	0.225	0.176	0.119	0.060	0.015
4.0	0.332	0.333	0.327	0.320	0.312	0.305	0.297	0.287	0.272	0.249	0.216	0.172	0.118	0.060	0.015
4.2						0.276	0.271	0.264	0.253	0.235	0.207	0.167	0.116	0.061	0.016
4.4						0.250	0.246	0.242	0.235	0.221	0.198	0.162	0.115	0.061	0.016
4.5	0.272	0.265	0.256	0.247	0.241										
4.6						0.225	0.223	0.221	0.217	0.208	0.189	0.157	0.113	0.061	0.016
5.0	0.221	0.209	0.197	0.188	0.183	0.181	0.182	0.184	0.185	0.182	0.170	0.147	0.109	0.062	0.017
5.5						0.136	0.139	0.144	0.149	0.152	0.149	0.134	0.104	0.061	0.018
6.0	0.143	0.127	0.114	0.105	0.101	0.101	0.105	0.111	0.119	0.126	0.128	0.120	0.098	0.060	0.018
6.5						0.074	0.078	0.085	0.094	0.104	0.110	0.108	0.092	0.059	0.019
7.0	0.091	0.075	0.063	0.056	0.053	0.054	0.058	0.065	0.074	0.085	0.093	0.096	0.086	0.058	0.019
7.5						0.039	0.042	0.048	0.058	0.068	0.079	0.085	0.079	0.056	0.020
8.0	0.057	0.044	0.034	0.029	0.027	0.027	0.030	0.036	0.044	0.055	0.066	0.075	0.073	0.054	0.020
9.0						0.014	0.016	0.020	0.026	0.035	0.046	0.057	0.061	0.050	0.020
10.0						0.007	0.008	0.010	0.015	0.022	0.032	0.043	0.051	0.046	0.020
12.0						0.001	0.002	0.003	0.005	0.008	0.014	0.023	0.034	0.037	0.020
15.0						0.000	0.000	0.000	0.001	0.002	0.004	0.009	0.017	0.025	0.019
20.0						0.000	0.000	0.000	0.000	0.000	0.000	0.002	0.005	0.013	0.015

TABLE VI.

p	$S(1s, 2p\sigma)$ $t=-0.5$	$t=-0.4$	$t=-0.3$	$t=-0.2$	$t=-0.1$	$t=0.0$	$t=0.1$	$t=0.2$	$t=0.3$	$t=0.4$	$t=0.5$	$t=0.6$	$t=0.7$	$t=0.8$
0.0	0.000	0.000	0.000	0.000	0.000	0.000	0.000	0.000	0.000	0.000	0.000	0.000	0.000	0.000
0.5	0.117	0.155	0.189	0.216	0.234	0.240	0.235	0.218	0.191	0.157	0.119	0.080	0.046	0.022
1.0	0.208	0.276	0.336	0.385	0.417	0.429	0.421	0.394	0.348	0.288	0.220	0.151	0.087	0.038
1.5	0.264	0.348	0.423	0.484	0.525	0.544	0.538	0.507	0.454	0.381	0.296	0.206	0.121	0.054
2.0	0.287	0.375	0.454	0.518	0.564	0.586	0.585	0.558	0.506	0.433	0.344	0.245	0.148	0.067
2.2	0.289	0.375	0.453	0.517	0.563	0.587	0.587	0.563	0.515	0.444	0.355	0.256	0.156	0.071
2.4	0.287	0.371	0.447	0.509	0.554	0.579	0.582	0.561	0.517	0.449	0.363	0.264	0.164	0.076
2.6	0.282	0.363	0.435	0.495	0.540	0.565	0.570	0.553	0.513	0.450	0.368	0.271	0.170	0.079
2.8	0.275	0.352	0.420	0.477	0.520	0.546	0.553	0.540	0.505	0.447	0.369	0.275	0.175	0.083
3.0	0.266	0.338	0.402	0.456	0.497	0.523	0.532	0.522	0.492	0.441	0.368	0.278	0.179	0.086
3.2	0.255	0.323	0.382	0.432	0.471	0.497	0.508	0.502	0.477	0.431	0.364	0.279	0.182	0.089
3.4	0.244	0.306	0.361	0.407	0.443	0.468	0.481	0.479	0.459	0.419	0.359	0.278	0.184	0.091
3.6	0.232	0.289	0.338	0.380	0.414	0.439	0.453	0.454	0.439	0.406	0.351	0.276	0.186	0.093
3.8						0.409	0.424	0.428	0.418	0.390	0.342	0.273	0.186	0.095
4.0	0.207	0.253	0.293	0.327	0.356	0.379	0.394	0.401	0.396	0.374	0.332	0.269	0.186	0.097
4.2						0.349	0.365	0.374	0.373	0.357	0.321	0.264	0.186	0.098
4.4						0.320	0.337	0.348	0.350	0.339	0.310	0.258	0.185	0.099
4.5	0.176	0.210	0.239	0.264	0.286									
4.6						0.293	0.309	0.322	0.328	0.321	0.298	0.252	0.183	0.100
5.0	0.147	0.171	0.190	0.208	0.225	0.241	0.258	0.273	0.284	0.286	0.272	0.238	0.179	0.101
5.5						0.186	0.202	0.219	0.234	0.243	0.241	0.219	0.172	0.102
6.0	0.099	0.108	0.115	0.122	0.130	0.141	0.155	0.172	0.189	0.204	0.210	0.199	0.163	0.101
6.5						0.105	0.118	0.133	0.151	0.169	0.181	0.179	0.154	0.100
7.0	0.064	0.066	0.066	0.068	0.071	0.078	0.088	0.102	0.120	0.139	0.155	0.160	0.144	0.098
7.5						0.057	0.065	0.077	0.094	0.113	0.132	0.142	0.134	0.095
8.0	0.041	0.039	0.037	0.036	0.037	0.041	0.048	0.058	0.073	0.092	0.111	0.126	0.124	0.092
9.0						0.021	0.025	0.032	0.043	0.059	0.078	0.097	0.104	0.086
10.0						0.010	0.013	0.017	0.025	0.037	0.054	0.073	0.087	0.078
12.0						0.002	0.003	0.005	0.008	0.014	0.025	0.040	0.058	0.063
15.0						0.000	0.000	0.001	0.001	0.003	0.007	0.015	0.030	0.044
20.0						0.000	0.000	0.000	0.000	0.000	0.001	0.003	0.009	0.022

TABLE VII.

p	$S(1s, 3s)$ $t=-0.4$	$t=-0.3$	$t=-0.2$	$t=-0.1$	$t=0.0$	$t=0.1$	$t=0.2$	$t=0.3$	$t=0.4$
0.0	0.954	0.928	0.857	0.754	0.632	0.505	0.381	0.269	0.175
0.5	0.945	0.920	0.851	0.751	0.632	0.507	0.384	0.273	0.179
1.0	0.915	0.893	0.832	0.739	0.628	0.510	0.393	0.284	0.190
1.5	0.864	0.847	0.792	0.713	0.615	0.508	0.399	0.295	0.203
2.0	0.794	0.780	0.736	0.670	0.588	0.495	0.399	0.304	0.215
2.5	0.712	0.701	0.665	0.612	0.546	0.470	0.389	0.306	0.224
3.0	0.623	0.612	0.584	0.544	0.494	0.436	0.370	0.299	0.227
3.2	0.587	0.575	0.551	0.515	0.471	0.419	0.360	0.295	0.226
3.4	0.551	0.539	0.516	0.486	0.447	0.402	0.347	0.289	0.225
3.6	0.516	0.504	0.483	0.456	0.423	0.383	0.334	0.283	0.223
3.8	0.481	0.469	0.450	0.426	0.398	0.364	0.320	0.275	0.220
4.0	0.448	0.435	0.418	0.397	0.373	0.344	0.309	0.267	0.217
4.2	0.416	0.403	0.386	0.369	0.348	0.324	0.295	0.257	0.212
4.4	0.385	0.371	0.356	0.341	0.324	0.305	0.280	0.248	0.207
4.6	0.355	0.341	0.327	0.314	0.301	0.285	0.265	0.238	0.202
4.8	0.327	0.312	0.300	0.288	0.279	0.266	0.249	0.227	0.195
5.0	0.301	0.287	0.274	0.265	0.256	0.248	0.235	0.217	0.190
5.5	0.242	0.227	0.216	0.210	0.206	0.204	0.200	0.190	0.173
6.0	0.193	0.178	0.168	0.164	0.163	0.165	0.166	0.164	0.155
6.5	0.152	0.137	0.129	0.126	0.127	0.132	0.137	0.140	0.137
7.0	0.119	0.105	0.098	0.096	0.098	0.104	0.111	0.117	0.120
7.5	0.092	0.080	0.073	0.072	0.074	0.080	0.089	0.098	0.104
8.0	0.071	0.060	0.054	0.053	0.056	0.062	0.070	0.080	0.089
9.0	0.042	0.033	0.029	0.028	0.030	0.035	0.043	0.053	0.064
10.0	0.024	0.018	0.015	0.015	0.016	0.020	0.025	0.034	0.044

Table VIII.

$S(1s, 3p\sigma)$

p	$t=-0.4$	$t=-0.3$	$t=-0.2$	$t=-0.1$	$t=0.0$	$t=0.1$	$t=0.2$	$t=0.3$	$t=0.4$
0.0	0.000	0.000	0.000	0.000	0.000	0.000	0.000	0.000	0.000
0.5	0.162	0.183	0.194	0.192	0.179	0.158	0.130	0.100	0.071
1.0	0.302	0.343	0.363	0.362	0.340	0.302	0.251	0.195	0.139
1.5	0.404	0.461	0.490	0.491	0.466	0.418	0.353	0.278	0.202
2.0	0.464	0.529	0.565	0.571	0.547	0.497	0.427	0.343	0.255
2.5	0.483	0.550	0.590	0.601	0.582	0.538	0.470	0.387	0.294
3.0	0.470	0.534	0.574	0.589	0.578	0.543	0.485	0.408	0.320
3.2	0.458	0.520	0.559	0.575	0.567	0.536	0.483	0.411	0.326
3.4	0.444	0.502	0.541	0.558	0.553	0.526	0.478	0.412	0.330
3.6	0.427	0.482	0.519	0.536	0.535	0.512	0.470	0.409	0.332
3.8	0.408	0.460	0.495	0.513	0.514	0.496	0.459	0.404	0.332
4.0	0.388	0.436	0.469	0.487	0.490	0.477	0.446	0.397	0.330
4.2	0.367	0.411	0.442	0.460	0.466	0.457	0.431	0.388	0.327

$S(1s, 3p\sigma)$

p	$t=-0.4$	$t=-0.3$	$t=-0.2$	$t=-0.1$	$t=0.0$	$t=0.1$	$t=0.2$	$t=0.3$	$t=0.4$
4.4	0.346	0.386	0.415	0.433	0.440	0.435	0.415	0.377	0.323
4.6	0.325	0.360	0.387	0.405	0.414	0.412	0.397	0.366	0.317
4.8	0.303	0.335	0.360	0.377	0.387	0.388	0.378	0.352	0.309
5.0	0.283	0.311	0.333	0.350	0.361	0.365	0.359	0.339	0.302
5.5	0.234	0.254	0.270	0.285	0.297	0.307	0.310	0.302	0.279
6.0	0.190	0.203	0.215	0.227	0.240	0.253	0.262	0.264	0.253
6.5	0.153	0.160	0.168	0.178	0.190	0.204	0.218	0.227	0.225
7.0	0.121	0.125	0.129	0.137	0.148	0.162	0.178	0.192	0.198
7.5	0.096	0.096	0.098	0.104	0.114	0.127	0.144	0.161	0.173
8.0	0.074	0.073	0.074	0.078	0.086	0.099	0.115	0.133	0.149
9.0	0.044	0.041	0.040	0.042	0.048	0.057	0.071	0.088	0.107
10.0	0.026	0.023	0.021	0.022	0.026	0.032	0.042	0.057	0.075

Table IX.

$S(1s, 5s)$

p	$t=-0.4$	$t=-0.3$	$t=-0.2$	$t=-0.1$	$t=0.0$	$t=0.1$	$t=0.2$	$t=0.3$	$t=0.4$
0.0	0.893	0.806	0.687	0.554	0.423	0.303	0.204	0.126	0.070
0.5	0.888	0.803	0.687	0.556	0.426	0.308	0.208	0.129	0.073
1.0	0.873	0.795	0.685	0.560	0.435	0.313	0.219	0.139	0.080
1.5	0.846	0.776	0.676	0.562	0.444	0.333	0.234	0.153	0.091
2.0	0.805	0.745	0.658	0.557	0.450	0.346	0.251	0.169	0.104
2.5	0.749	0.699	0.627	0.540	0.448	0.354	0.265	0.185	0.118
3.0	0.682	0.642	0.583	0.513	0.435	0.354	0.273	0.198	0.131
3.2	0.653	0.616	0.562	0.498	0.427	0.351	0.275	0.202	0.135
3.4	0.624	0.589	0.540	0.482	0.417	0.347	0.275	0.205	0.140
3.6	0.593	0.561	0.517	0.465	0.406	0.342	0.275	0.207	0.143
3.8	0.562	0.533	0.493	0.447	0.394	0.335	0.273	0.209	0.147
4.0	0.531	0.504	0.468	0.427	0.380	0.327	0.270	0.209	0.149

$S(1s, 5s)$

p	$t=-0.4$	$t=-0.3$	$t=-0.2$	$t=-0.1$	$t=0.0$	$t=0.1$	$t=0.2$	$t=0.3$	$t=0.4$
4.5	0.454	0.432	0.405	0.376	0.341	0.303	0.257	0.208	0.154
5.0	0.384	0.363	0.343	0.323	0.300	0.273	0.240	0.200	0.154
5.5	0.319	0.300	0.285	0.272	0.258	0.242	0.219	0.190	0.152
6.0	0.261	0.244	0.232	0.224	0.218	0.209	0.196	0.176	0.147
6.5	0.213	0.197	0.187	0.183	0.181	0.179	0.173	0.161	0.140
7.0	0.170	0.155	0.147	0.145	0.147	0.149	0.149	0.144	0.130
7.5	0.135	0.121	0.115	0.114	0.117	0.123	0.127	0.127	0.120
8.0	0.106	0.094	0.089	0.089	0.093	0.100	0.107	0.111	0.109
9.0	0.065	0.055	0.051	0.052	0.056	0.063	0.073	0.082	0.087
10.0	0.039	0.031	0.028	0.029	0.032	0.039	0.048	0.058	0.067

Table X.

$S(1s, 5p\sigma)$

p	$t=-0.4$	$t=-0.3$	$t=-0.2$	$t=-0.1$	$t=0.0$	$t=0.1$	$t=0.2$	$t=0.3$	$t=0.4$
0.0	0.000	0.000	0.000	0.000	0.000	0.000	0.000	0.000	0.000
0.5	0.153	0.161	0.157	0.143	0.121	0.096	0.071	0.047	0.029
1.0	0.293	0.310	0.304	0.278	0.238	0.190	0.141	0.096	0.059
1.5	0.409	0.435	0.429	0.397	0.344	0.279	0.210	0.146	0.091
2.0	0.493	0.526	0.524	0.490	0.431	0.355	0.273	0.194	0.124
2.5	0.540	0.580	0.581	0.550	0.493	0.414	0.326	0.238	0.157
3.0	0.553	0.596	0.603	0.579	0.526	0.453	0.366	0.274	0.186
3.2	0.550	0.593	0.603	0.581	0.532	0.462	0.377	0.286	0.197
3.4	0.543	0.586	0.597	0.578	0.534	0.467	0.385	0.296	0.207
3.6	0.532	0.574	0.587	0.572	0.531	0.470	0.392	0.304	0.215
3.8	0.518	0.559	0.573	0.561	0.526	0.469	0.395	0.311	0.223
4.0	0.501	0.541	0.556	0.548	0.516	0.465	0.396	0.316	0.230

$S(1s, 5p\sigma)$

p	$t=-0.4$	$t=-0.3$	$t=-0.2$	$t=-0.1$	$t=0.0$	$t=0.1$	$t=0.2$	$t=0.3$	$t=0.4$
4.5	0.453	0.488	0.504	0.502	0.482	0.445	0.390	0.321	0.242
5.0	0.396	0.426	0.442	0.445	0.436	0.412	0.372	0.316	0.247
5.5	0.339	0.363	0.377	0.384	0.383	0.371	0.345	0.303	0.246
6.0	0.285	0.302	0.315	0.324	0.328	0.326	0.313	0.285	0.240
6.5	0.237	0.248	0.258	0.268	0.277	0.282	0.279	0.263	0.230
7.0	0.192	0.199	0.206	0.216	0.227	0.237	0.242	0.237	0.216
7.5	0.155	0.158	0.163	0.172	0.184	0.197	0.208	0.211	0.200
8.0	0.123	0.124	0.127	0.135	0.147	0.161	0.175	0.185	0.182
9.0	0.076	0.074	0.074	0.080	0.090	0.104	0.121	0.137	0.147
10.0	0.046	0.042	0.042	0.045	0.052	0.064	0.079	0.098	0.114

Table XI.

$S(2s, 2s)$

p	$t=0.0$	$t=0.1$	$t=0.2$	$t=0.3$	$t=0.4$	$t=0.5$	$t=0.6$
0.0	1.000	0.975	0.903	0.790	0.647	0.487	0.328
0.5	0.986	0.962	0.892	0.782	0.642	0.486	0.328
1.0	0.948	0.926	0.862	0.760	0.629	0.481	0.329
1.5	0.890	0.871	0.815	0.725	0.608	0.472	0.329
2.0	0.815	0.799	0.753	0.678	0.578	0.458	0.327
2.5	0.729	0.717	0.679	0.622	0.540	0.438	0.322
3.0	0.637	0.629	0.603	0.559	0.496	0.413	0.312
3.2	0.600	0.593	0.571	0.533	0.477	0.402	0.308
3.4	0.563	0.557	0.539	0.507	0.457	0.389	0.302
3.6	0.527	0.522	0.507	0.480	0.438	0.377	0.296
3.8	0.491	0.487	0.476	0.454	0.417	0.364	0.290
4.0	0.456	0.454	0.445	0.427	0.397	0.350	0.283
4.2	0.423	0.421	0.415	0.402	0.377	0.336	0.276
4.4	0.390	0.389	0.386	0.376	0.357	0.323	0.268
4.6	0.360	0.359	0.358	0.352	0.337	0.309	0.260
4.8	0.330	0.331	0.331	0.328	0.318	0.295	0.252
5.0	0.302	0.307	0.305	0.305	0.299	0.281	0.244
5.5	0.240	0.242	0.247	0.252	0.255	0.248	0.223

$S(2s, 2s)$

p	$t=0.0$	$t=0.1$	$t=0.2$	$t=0.3$	$t=0.4$	$t=0.5$	$t=0.6$
6.0	0.188	0.190	0.197	0.206	0.215	0.216	0.203
6.5	0.145	0.148	0.155	0.167	0.179	0.187	0.182
7.0	0.111	0.113	0.121	0.133	0.148	0.160	0.163
7.5	0.083	0.086	0.093	0.106	0.121	0.136	0.145
8.0	0.062	0.064	0.071	0.083	0.099	0.116	0.128
8.5	0.046	0.048	0.054	0.065	0.080	0.097	0.112
9.0	0.034	0.035	0.041	0.050	0.064	0.082	0.098
9.5	0.024	0.026	0.030	0.039	0.051	0.068	0.086
10.0	0.018	0.019	0.023	0.030	0.041	0.057	0.075
10.5	0.013	0.014	0.017	0.023	0.032	0.047	0.065
11.0	0.009	0.010	0.012	0.017	0.026	0.039	0.056
11.5	0.006	0.007	0.009	0.013	0.020	0.032	0.048
12.0	0.004	0.005	0.007	0.010	0.016	0.026	0.041
12.5	0.003	0.004	0.005	0.007	0.012	0.021	0.036
13.0	0.002	0.002	0.003	0.006	0.010	0.017	0.030
13.5	0.002	0.002	0.002	0.004	0.008	0.014	0.026
14.0	0.001	0.001	0.002	0.003	0.006	0.012	0.022

TABLE XII.

p	S(2pσ, 2s) t=0.6	t=0.5	t=0.4	t=0.3	t=0.2	t=0.1	S(2s, 2pσ) t=0.0	t=0.1	t=0.2	t=0.3	t=0.4	t=0.5	t=0.6
0.0	0.000	0.000	0.000	0.000	0.000	0.000	0.000	0.000	0.000	0.000	0.000	0.000	0.000
0.5	−0.008	0.002	0.020	0.047	0.079	0.112	0.143	0.167	0.180	0.180	0.166	0.139	0.103
1.0	−0.010	0.011	0.047	0.098	0.158	0.220	0.276	0.319	0.343	0.342	0.315	0.265	0.198
1.5	−0.005	0.028	0.081	0.151	0.231	0.313	0.386	0.443	0.473	0.472	0.437	0.369	0.278
2.0	0.006	0.049	0.114	0.197	0.289	0.381	0.464	0.526	0.561	0.560	0.521	0.446	0.340
2.5	0.019	0.070	0.143	0.218	0.327	0.420	0.504	0.567	0.603	0.602	0.569	0.493	0.383
3.0	0.033	0.089	0.163	0.251	0.342	0.431	0.509	0.570	0.606	0.612	0.582	0.513	0.407
3.2	0.037	0.096	0.169	0.254	0.342	0.427	0.503	0.562	0.598	0.606	0.580	0.515	0.412
3.4	0.042	0.099	0.173	0.255	0.339	0.421	0.492	0.549	0.586	0.596	0.573	0.513	0.415
3.6	0.046	0.103	0.175	0.254	0.335	0.411	0.479	0.534	0.570	0.582	0.563	0.509	0.416
3.8	0.049	0.106	0.176	0.252	0.327	0.399	0.463	0.515	0.551	0.565	0.550	0.502	0.415
4.0	0.052	0.109	0.176	0.247	0.318	0.385	0.444	0.494	0.529	0.545	0.535	0.492	0.412
4.2	0.055	0.110	0.174	0.241	0.307	0.369	0.425	0.471	0.505	0.523	0.518	0.481	0.407
4.4	0.057	0.111	0.172	0.234	0.295	0.351	0.402	0.446	0.480	0.500	0.499	0.469	0.401
4.6	0.059	0.111	0.168	0.226	0.281	0.333	0.380	0.421	0.455	0.476	0.479	0.454	0.394
4.8	0.060	0.110	0.164	0.217	0.267	0.314	0.357	0.396	0.428	0.451	0.458	0.439	0.386
5.0	0.061	0.109	0.159	0.208	0.253	0.295	0.334	0.370	0.402	0.426	0.436	0.423	0.377
5.5	0.062	0.104	0.145	0.182	0.216	0.248	0.278	0.308	0.337	0.364	0.382	0.382	0.351
6.0	0.061	0.096	0.129	0.157	0.181	0.203	0.226	0.250	0.277	0.305	0.328	0.339	0.324
6.5	0.058	0.088	0.112	0.132	0.148	0.163	0.180	0.200	0.224	0.251	0.278	0.297	0.295
7.0	0.055	0.079	0.096	0.109	0.119	0.129	0.141	0.157	0.178	0.205	0.233	0.258	0.266
7.5	0.051	0.070	0.082	0.089	0.095	0.101	0.109	0.122	0.140	0.165	0.193	0.222	0.238
8.0	0.047	0.061	0.068	0.072	0.074	0.077	0.083	0.093	0.109	0.131	0.159	0.189	0.212
8.5	0.043	0.053	0.057	0.057	0.057	0.059	0.063	0.070	0.084	0.103	0.130	0.160	0.187
9.0	0.039	0.045	0.047	0.045	0.044	0.044	0.046	0.053	0.064	0.081	0.105	0.135	0.165
9.5	0.035	0.038	0.038	0.036	0.033	0.033	0.034	0.039	0.048	0.063	0.085	0.113	0.144
10.0	0.031	0.033	0.031	0.028	0.025	0.024	0.025	0.029	0.036	0.048	0.068	0.095	0.126
10.5							0.018	0.021	0.027	0.037	0.054	0.079	0.109
11.0	0.024	0.023	0.020	0.016	0.014	0.013	0.013	0.015	0.020	0.028	0.043	0.065	0.095
11.5							0.009	0.011	0.015	0.022	0.034	0.054	0.082
12.0	0.018	0.016	0.012	0.010	0.008	0.007	0.007	0.008	0.011	0.016	0.027	0.044	0.070
12.5							0.005	0.006	0.008	0.012	0.021	0.036	0.060
13.0							0.003	0.004	0.006	0.009	0.016	0.030	0.052
13.5							0.002	0.003	0.004	0.007	0.013	0.024	0.044
14.0							0.002	0.002	0.003	0.005	0.010	0.020	0.038

TABLE XIII.

p	S(2s, 3s) t=−0.6	t=−0.5	t=−0.4	t=−0.3	t=−0.2	t=−0.1	t=0.0	t=0.1	t=0.2	t=0.3	t=0.4	t=0.5	t=0.6
0.0	0.479	0.667	0.826	0.937	0.989	0.979	0.913	0.801	0.659	0.505	0.354	0.222	0.120
0.5	0.477	0.663	0.820	0.929	0.979	0.970	0.905	0.796	0.657	0.505	0.356	0.225	0.122
1.0	0.473	0.651	0.801	0.904	0.952	0.943	0.884	0.782	0.650	0.505	0.361	0.231	0.127
1.5	0.466	0.634	0.771	0.865	0.909	0.902	0.849	0.758	0.638	0.504	0.367	0.240	0.136
2.0	0.457	0.611	0.733	0.815	0.853	0.847	0.803	0.725	0.620	0.499	0.372	0.250	0.146
2.5	0.445	0.582	0.688	0.756	0.789	0.785	0.747	0.682	0.594	0.488	0.373	0.258	0.156
3.0	0.430	0.549	0.637	0.692	0.716	0.712	0.684	0.632	0.560	0.472	0.371	0.265	0.165
3.5	0.411	0.512	0.582	0.623	0.639	0.636	0.614	0.575	0.519	0.448	0.362	0.268	0.173
4.0	0.390	0.472	0.524	0.552	0.562	0.557	0.541	0.514	0.473	0.418	0.349	0.267	0.178
4.5	0.365	0.429	0.466	0.483	0.485	0.481	0.469	0.451	0.423	0.384	0.329	0.261	0.181
5.0	0.339	0.387	0.408	0.415	0.413	0.407	0.395	0.389	0.372	0.347	0.307	0.252	0.182
5.5	0.312	0.344	0.354	0.352	0.346	0.339	0.334	0.329	0.322	0.308	0.282	0.239	0.180
6.0	0.285	0.304	0.303	0.295	0.285	0.278	0.275	0.275	0.275	0.270	0.255	0.225	0.175
6.5	0.259	0.266	0.258	0.245	0.233	0.226	0.225	0.227	0.232	0.234	0.229	0.209	0.170
7.0	0.232	0.230	0.216	0.200	0.187	0.180	0.179	0.184	0.191	0.199	0.201	0.191	0.162
7.5	0.207	0.198	0.180	0.162	0.149	0.142	0.142	0.147	0.157	0.168	0.176	0.173	0.153
8.0	0.184	0.169	0.148	0.130	0.117	0.111	0.111	0.117	0.127	0.140	0.152	0.156	0.144
9.0	0.143	0.121	0.099	0.082	0.071	0.065	0.066	0.071	0.081	0.095	0.111	0.123	0.124
10.0	0.110	0.085	0.065	0.050	0.041	0.037	0.037	0.042	0.050	0.062	0.079	0.095	0.104

TABLE XIV.

p	$S(2s, 3p\sigma)$ $t=-0.6$	$t=-0.5$	$t=-0.4$	$t=-0.3$	$t=-0.2$	$t=-0.1$	$t=0.0$	$t=0.1$	$t=0.2$	$t=0.3$	$t=0.4$	$t=0.5$	$t=0.6$
0.0	0.000	0.000	0.000	0.000	0.000	0.000	0.000	0.000	0.000	0.000	0.000	0.000	0.000
0.5	0.001	0.020	0.048	0.081	0.113	0.140	0.156	0.160	0.150	0.130	0.101	0.070	0.041
1.0	0.004	0.042	0.096	0.158	0.219	0.269	0.300	0.307	0.290	0.251	0.198	0.138	0.082
1.5	0.013	0.067	0.143	0.229	0.312	0.380	0.422	0.433	0.410	0.358	0.285	0.201	0.122
2.0	0.027	0.095	0.187	0.289	0.386	0.465	0.515	0.529	0.505	0.445	0.359	0.258	0.159
2.5	0.044	0.122	0.224	0.334	0.438	0.522	0.576	0.593	0.570	0.509	0.416	0.306	0.193
3.0	0.060	0.146	0.252	0.363	0.466	0.550	0.605	0.624	0.605	0.548	0.457	0.343	0.222
3.5	0.076	0.164	0.267	0.374	0.470	0.549	0.603	0.625	0.612	0.563	0.480	0.369	0.246
4.0	0.088	0.175	0.273	0.370	0.456	0.527	0.577	0.601	0.596	0.558	0.486	0.384	0.264
4.5	0.098	0.180	0.268	0.352	0.427	0.488	0.533	0.558	0.561	0.535	0.478	0.390	0.277
5.0	0.102	0.178	0.255	0.326	0.387	0.438	0.478	0.505	0.514	0.500	0.459	0.385	0.283
5.5	0.104	0.172	0.236	0.293	0.342	0.383	0.418	0.444	0.459	0.457	0.430	0.373	0.285
6.0	0.103	0.162	0.214	0.258	0.295	0.328	0.356	0.382	0.401	0.409	0.397	0.356	0.282
6.5	0.101	9.150	0.191	0.224	0.251	0.275	0.299	0.323	0.346	0.361	0.361	0.335	0.276
7.0	0.095	0.135	0.166	0.189	0.207	0.225	0.245	0.267	0.291	0.311	0.321	0.309	0.265
7.5	0.089	0.121	0.143	0.158	0.170	0.182	0.198	0.218	0.242	0.266	0.283	0.283	0.252
8.0	0.083	0.107	0.122	0.130	0.137	0.145	0.157	0.175	0.198	0.224	0.247	0.257	0.238
9.0	0.069	0.081	0.085	0.085	0.086	0.088	0.096	0.109	0.129	0.154	0.183	0.205	0.207
10.0	0.055	0.059	0.059	0.054	0.051	0.052	0.056	0.065	0.080	0.103	0.131	0.159	0.175

TABLE XV.

p	$S(2p\sigma, 2p\sigma)$ $t=0.0$	$t=0.1$	$t=0.2$	$t=0.3$	$t=0.4$	$t=0.5$	$t=0.6$
0.0	−1.000	−0.975	−0.903	−0.790	−0.647	−0.487	−0.328
0.5	−0.927	−0.906	−0.840	−0.738	−0.607	−0.461	−0.312
1.0	−0.736	−0.720	−0.673	−0.599	−0.502	−0.389	−0.271
1.5	−0.483	−0.475	−0.452	−0.416	−0.360	−0.292	−0.214
2.0	−0.226	−0.226	−0.226	−0.222	−0.211	−0.188	−0.150
2.5	0.005	−0.007	−0.029	−0.053	−0.076	−0.092	−0.092
3.0	0.159	0.148	0.121	0.079	0.032	−0.011	−0.040
3.2	0.208	0.196	0.167	0.120	0.067	0.016	−0.022
3.4	0.248	0.237	0.204	0.155	0.097	0.040	−0.005
3.6	0.279	0.268	0.235	0.184	0.123	0.061	0.010
3.8	0.303	0.291	0.258	0.207	0.144	0.079	0.023
4.0	0.319	0.308	0.275	0.224	0.161	0.094	0.035
4.2	0.328	0.318	0.286	0.237	0.175	0.107	0.045
4.4	0.332	0.322	0.293	0.245	0.185	0.117	0.054
4.6	0.332	0.322	0.294	0.250	0.191	0.126	0.062
4.8	0.327	0.318	0.293	0.251	0.196	0.132	0.068
5.0	0.319	0.311	0.288	0.249	0.197	0.136	0.074
5.5	0.289	0.282	0.266	0.236	C.193	0.141	0.083
6.0	0.250	0.247	0.235	0.214	0.181	0.138	0.087
6.5	0.210	0.208	0.201	0.187	0.164	0.131	0.087
7.0	0.171	0.171	0.167	0.160	0.145	0.120	0.085
7.5	0.137	0.137	0.137	0.134	0.126	0.109	0.081
8.0	0.107	0.108	0.109	0.110	0.107	0.096	0.075
8.5	0.083	0.084	0.086	0.089	0.090	0.085	0.069
9.0	0.063	0.064	0.067	0.071	0.075	0.073	0.063
9.5	0.047	0.048	0.052	0.057	0.061	0.063	0.057
10.0	0.035	0.036	0.039	0.045	0.050	0.054	0.051
10.5	0.026	0.027	0.030	0.035	0.041	0.045	0.045
11.0	0.019	0.020	0.021	0.027	0.033	0.038	0.040
11.5	0.013	0.014	0.017	0.021	0.026	0.032	0.035
12.0	0.010	0.010	0.012	0.016	0.021	0.027	0.031
12.5	0.007	0.007	0.009	0.012	0.017	0.022	0.027
13.0	0.005	0.005	0.007	0.009	0.013	0.018	0.023
13.5	0.003	0.004	0.005	0.007	0.010	0.015	0.020
14.0	0.002	0.003	0.003	0.005	0.008	0.012	0.017

TABLE XVI.

p	$S(2p\sigma, 3s)$ $t=-0.6$	$t=-0.5$	$t=-0.4$	$t=-0.3$	$t=-0.2$	$t=-0.1$	$t=0.0$	$t=0.1$	$t=0.2$	$t=0.3$	$t=0.4$	$t=0.5$	$t=0.6$
0.0	0.000	0.000	0.000	0.000	0.000	0.000	0.000	0.000	0.000	0.000	0.000	0.000	0.000
0.5	0.110	0.134	0.142	0.135	0.114	0.085	0.054	0.025	0.002	−0.013	−0.019	−0.018	−0.013
1.0	0.214	0.261	0.278	0.265	0.227	0.173	0.114	0.058	0.013	−0.017	−0.032	−0.032	−0.024
1.5	0.307	0.374	0.399	0.383	0.333	0.261	0.179	0.102	0.034	−0.009	−0.034	−0.039	−0.031
2.0	0.385	0.466	0.497	0.480	0.424	0.340	0.245	0.152	0.071	+0.011	−0.025	−0.038	−0.034
2.5	0.445	0.534	0.569	0.551	0.492	0.405	0.303	0.201	0.111	0.039	−0.008	−0.030	−0.033
3.0	0.486	0.576	0.609	0.591	0.533	0.447	0.346	0.243	0.148	0.069	+0.013	−0.019	−0.028
3.5	0.509	0.594	0.622	0.603	0.547	0.466	0.371	0.272	0.179	0.097	0.035	−0.005	−0.021
4.0	0.515	0.590	0.611	0.589	0.535	0.462	0.376	0.287	0.199	0.120	0.055	+0.010	−0.014
4.5	0.508	0.569	0.580	0.555	0.505	0.439	0.365	0.287	0.208	0.135	0.061	0.023	−0.005
5.0	0.490	0.536	0.537	0.508	0.461	0.404	0.341	0.275	0.208	0.143	0.083	0.034	+0.002
5.5	0.465	0.495	0.485	0.453	0.409	0.360	0.308	0.255	0.199	0.144	0.090	0.043	0.009
6.0	0.434	0.449	0.430	0.395	0.354	0.314	0.271	0.230	0.185	0.139	0.093	0.050	0.015
6.5	0.402	0.403	0.376	0.339	0.301	0.266	0.233	0.201	0.167	0.131	0.093	0.054	0.020
7.0	0.366	0.354	0.322	0.284	0.250	0.220	0.194	0.171	0.147	0.120	0.089	0.055	0.024
7.5	0.331	0.310	0.273	0.236	0.204	0.180	0.160	0.143	0.126	0.107	0.083	0.055	0.027
8.0	0.297	0.268	0.229	0.193	0.165	0.144	0.129	0.118	0.107	0.094	0.076	0.054	0.028
9.0	0.235	0.197	0.157	0.125	0.103	0.089	0.081	0.076	0.073	0.069	0.061	0.048	0.029
10.0	0.182	0.140	0.104	0.078	0.062	0.053	0.048	0.047	0.047	0.048	0.047	0.040	0.028

TABLE XVII.

	$S(2p\sigma, 3p\sigma)$												
p	$t=-0.6$	$t=-0.5$	$t=-0.4$	$t=-0.3$	$t=-0.2$	$t=-0.1$	$t=0.0$	$t=0.1$	$t=0.2$	$t=0.3$	$t=0.4$	$t=0.5$	$t=0.6$
0.0	−0.479	−0.667	−0.826	−0.937	−0.989	−0.979	−0.913	−0.801	−0.659	−0.505	−0.354	−0.222	−0.120
0.5	−0.462	−0.640	−0.790	−0.894	−0.942	−0.934	−0.872	−0.768	−0.636	−0.490	−0.347	−0.220	−0.120
1.0	−0.414	−0.564	−0.687	−0.771	−0.811	−0.805	−0.757	−0.674	−0.567	−0.446	−0.323	−0.211	−0.118
1.5	−0.343	−0.453	−0.537	−0.593	−0.618	−0.615	−0.585	−0.531	−0.460	−0.375	−0.284	−0.194	−0.114
2.0	−0.261	−0.324	−0.365	−0.389	−0.398	−0.396	−0.384	−0.362	−0.330	−0.285	−0.230	−0.168	−0.107
2.5	−0.177	−0.194	−0.195	−0.187	−0.180	−0.178	−0.181	−0.188	−0.192	−0.187	−0.169	−0.136	−0.094
3.0	−0.097	−0.076	−0.042	−0.010	+0.010	+0.013	−0.002	−0.029	−0.062	−0.089	−0.104	−0.101	−0.080
3.5	−0.028	+0.022	+0.081	+0.130	0.159	0.163	+0.142	+0.102	+0.050	−0.002	−0.043	−0.064	−0.063
4.0	+0.027	0.099	0.172	0.230	0.264	0.268	0.245	0.198	0.136	+0.070	+0.012	−0.029	−0.045
4.5	0.070	0.153	0.232	0.292	0.326	0.330	0.307	0.260	0.196	0.124	0.056	+0.003	−0.027
5.0	0.101	0.187	0.264	0.321	0.351	0.355	0.334	0.291	0.231	0.161	0.090	0.029	−0.011
5.5	0.122	0.205	0.275	0.324	0.350	0.353	0.335	0.298	0.246	0.182	0.113	0.050	+0.004
6.0	0.134	0.210	0.270	0.310	0.330	0.331	0.317	0.287	0.244	0.189	0.127	0.066	0.016
6.5	0.139	0.206	0.256	0.286	0.299	0.299	0.288	0.265	0.232	0.187	0.133	0.077	0.026
7.0	0.138	0.194	0.232	0.253	0.261	0.259	0.251	0.235	0.211	0.177	0.133	0.082	0.034
7.5	0.134	0.179	0.207	0.220	0.222	0.220	0.213	0.203	0.186	0.162	0.128	0.085	0.040
8.0	0.127	0.162	0.181	0.186	0.185	0.182	0.177	0.171	0.161	0.145	0.120	0.084	0.044
9.0	0.109	0.127	0.131	0.128	0.122	0.118	0.115	0.115	0.114	0.110	0.099	0.078	0.047
10.0	0.089	0.095	0.091	0.083	0.076	0.072	0.071	0.072	0.075	0.078	0.076	0.067	0.046

TABLE XVIII.

p	$S(3s, 3s)$ $t=0.0$	$S(3p\sigma, 3p\sigma)$ $t=0.0$	$S(3s, 3p\sigma)$ $t=0.0$	p	$S(3s, 3s)$ $t=0.0$	$S(3p\sigma, 3p\sigma)$ $t=0.0$	$S(3s, 3p\sigma)$ $t=0.0$
0.0	1.000	−1.000	0.000	6.4	0.328	0.377	0.371
0.5	0.992	−0.965	0.096	6.6	0.306	0.368	0.352
1.0	0.968	−0.866	0.189	6.8	0.285	0.358	0.333
1.5	0.932	−0.715	0.276	7.0	0.264	0.346	0.314
2.0	0.885	−0.531	0.354	7.2	0.245	0.333	0.295
2.5	0.832	−0.333	0.419	7.4	0.227	0.319	0.277
3.0	0.772	−0.140	0.468	7.6	0.209	0.304	0.259
3.5	0.708	0.031	0.499	7.8	0.193	0.288	0.241
4.0	0.641	0.171	0.511	8.0	0.177	0.273	0.225
4.5	0.572	0.273	0.504	8.5	0.141	0.233	0.185
5.0	0.504	0.342	0.483	9.0	0.114	0.195	0.151
5.2	0.477	0.359	0.470	9.5	0.089	0.161	0.121
5.4	0.451	0.372	0.457	10.0	0.070	0.130	0.096
5.6	0.425	0.380	0.441	10.5	0.054	0.104	0.076
5.8	0.400	0.385	0.425	11.0	0.041	0.082	0.059
6.0	0.375	0.385	0.407	11.5	0.031	0.064	0.045
6.2	0.351	0.382	0.389	12.0	0.024	0.050	0.034

TABLE XIX.

p	$S(5s, 5s)$ $t=0.0$	$S(5p\sigma, 5p\sigma)$ $t=0.0$	$S(5s, 5p\sigma)$ $t=0.0$	p	$S(5s, 5s)$ $t=0.0$	$S(5p\sigma, 5p\sigma)$ $t=0.0$	$S(5s, 5p\sigma)$ $t=0.0$
0.0	1.000	−1.000	0.000	7.0	0.417	0.413	0.445
0.5	0.994	−0.979	0.072	7.5	0.367	0.418	0.411
1.0	0.977	−0.917	0.142	8.0	0.320	0.407	0.374
1.5	0.950	−0.819	0.210	8.5	0.276	0.384	0.334
2.0	0.915	−0.693	0.273	9.0	0.235	0.354	0.294
2.5	0.874	−0.547	0.331	9.5	0.198	0.318	0.255
3.0	0.828	−0.390	0.383	10.0	0.166	0.281	0.218
3.5	0.780	−0.231	0.427	10.5	0.137	0.244	0.185
4.0	0.730	−0.079	0.462	11.0	0.113	0.208	0.154
4.5	0.679	0.059	0.486	11.5	0.091	0.175	0.127
5.0	0.627	0.177	0.499	12.0	0.074	0.145	0.104
5.5	0.574	0.272	0.501	12.5	0.059	0.119	0.084
6.0	0.521	0.343	0.491	13.0	0.047	0.097	0.067
6.5	0.468	0.389	0.472				

TABLE XX.

p	$S(2p\pi, 2p\pi)$ $t=0.0$	$t=0.1$	$t=0.2$	$t=0.3$	$t=0.4$	$t=0.5$	$t=0.6$
0.0	1.000	0.975	0.903	0.790	0.647	0.487	0.328
0.5	0.976	0.951	0.882	0.772	0.633	0.478	0.323
1.0	0.907	0.887	0.823	0.723	0.596	0.453	0.308
1.5	0.809	0.790	0.737	0.652	0.542	0.416	0.287
2.0	0.695	0.680	0.638	0.568	0.477	0.372	0.261
2.5	0.578	0.567	0.535	0.481	0.410	0.325	0.233
3.0	0.468	0.460	0.437	0.398	0.345	0.279	0.205
3.2	0.427	0.420	0.401	0.367	0.320	0.262	0.194
3.4	0:389	0.383	0.366	0.337	0.297	0.245	0.183
3.6	0.352	0.348	0.334	0.309	0.274	0.228	0.173
3.8	0.318	0.315	0.303	0.283	0.253	0.213	0.163
4.0	0.287	0.284	0.275	0.258	0.232	0.198	0.153
4.2	0.258	0.255	0.248	0.234	0.213	0.183	0.144
4.4	0.231	0.229	0.224	0.213	0.195	0.170	0.135
4.6	0.207	0.205	0.201	0.193	0.179	0.157	0.127
4.8	0.184	0.183	0.181	0.174	0.163	0.145	0.119
5.0	0.164	0.163	0.162	0.157	0.149	0.134	0.111
5.2	0.146						
5.4	0.129						
5.5	0.121	0.122	0.122	0.121	0.117	0.109	0.094
5.6	0.114						
5.8	0.101						
6.0	0.089	0.089	0.091	0.092	0.092	0.089	0.079
6.2	0.078						
6.4	0.069						
6.5	0.064	0.065	0.067	0.069	0.072	0.071	0.066
7.0	0.046	0.047	0.049	0.052	0.055	0.057	0.055
7.5	0.033	0.033	0.036	0.039	0.043	0.046	0.046
8.0	0.023	0.024	0.026	0.029	0.033	0.037	0.038
9.0	0.011	0.012	0.013	0.016	0.019	0.023	0.026
10.0	0.005	0.006	0.007	0.008	0.011	0.013	0.018

TABLE XXI.

p	$S(2p\pi, 2p\pi)$ $t=-0.6$	$t=-0.5$	$t=-0.4$	$t=-0.3$	$t=-0.2$	$t=-0.1$	$t=0.0$	$t=0.1$	$t=0.2$	$t=0.3$	$t=0.4$	$t=0.5$	$t=0.6$
0.0	0.479	0.667	0.826	0.937	0.989	0.979	0.913	0.801	0.659	0.505	0.354	0.222	0.120
0.5	0.473	0.658	0.814	0.923	0.973	0.963	0.899	0.786	0.652	0.500	0.352	0.221	0.120
1.0	0.456	0.632	0.779	0.881	0.928	0.920	0.860	0.758	0.628	0.485	0.344	0.219	0.119
1.5	0.431	0.591	0.724	0.816	0.858	0.851	0.798	0.708	0.591	0.461	0.331	0.213	0.118
2.0	0.399	0.541	0.656	0.735	0.771	0.765	0.720	0.643	0.542	0.429	0.313	0.206	0.117
2.5	0.363	0.484	0.581	0.645	0.676	0.672	0.633	0.569	0.486	0.390	0.290	0.195	0.114
3.0	0.325	0.426	0.503	0.554	0.575	0.571	0.542	0.492	0.426	0.348	0.264	0.182	0.109
3.5	0.287	0.369	0.428	0.466	0.480	0.477	0.454	0.416	0.366	0.303	0.237	0.168	0.104
4.0	0.251	0.315	0.359	0.384	0.394	0.389	0.373	0.345	0.308	0.262	0.209	0.153	0.097
4.5	0.218	0.266	0.297	0.313	0.316	0.312	0.301	0.282	0.255	0.222	0.182	0.137	0.090
5.0	0.187	0.222	0.241	0.250	0.251	0.246	0.238	0.225	0.208	0.185	0.157	0.122	0.083
5.2					0.228	0.223	0.216	0.205	0.191	0.172			
5.4					0.206	0.202	0.195	0.187	0.175	0.159			
5.5	0.160	0.184	0.195	0.198	0.196	0.192	0.186	0.178	0.167	0.153	0.133	0.107	0.076
5.6					0.186	0.182	0.176	0.169	0.160	0.147			
5.8					0.168	0.164	0.159	0.153	0.146	0.135			
6.0	0.136	0.151	0.156	0.155	0.151	0.147	0.143	0.139	0.133	0.125	0.112	0.093	0 069
6.5	0.115	0.123	0.124	0.120	0.115	0.112	0.109	0.107	0.104	0.101	0.093	0.081	0.062
7.0	0.097	0.100	0.097	0.092	0.087	0.084	0.082	0.081	0.081	0.080	0.077	0.069	0.055
7.5	0.082	0.081	0.076	0.070	0.065	0.062	0.061	0.061	0.062	0.064	0.063	0.059	0.049
8.0	0.068	0.065	0.059	0.053	0.049	0.046	0.045	0.046	0.048	0.050	0.052	0.050	0.043
9.0	0.048	0.042	0.036	0.030	0.026	0.024	0.024	0.025	0.027	0.030	0.034	0.035	0.033
10.0	0.033	0.027	0.021	0.017	0.014	0.012	0.012	0.013	0.015	0.018	0.022	0.025	0.025

TABLE XXII.

p	$S(2p\pi, 5p\pi)$ $t=0.0$	$t=0.1$
0.5	0.726	0.58
1.0	0.707	0.563
1.5	0.677	0.543
2.0	0.635	0.515
2.5	0.584	0.482
3.0	0.526	0.438
3.5	0.464	0.393
4.0	0.401	0.345
4.5	0.341	0.298
5.0	0.284	0.253
5.5	0.234	0.212
6.0	0.189	0.174
6.5	0.151	0.142
7.0	0.119	0.114
7.5	0.093	0.090
8.0	0.072	0.071
8.5	0.055	0.055
9.0	0.041	0.042
9.5	0.030	0.032
10.0	0.023	0.024
10.5	0.017	0.018
11.0	0.012	0.014
11.5	0.009	0.010
12.0	0.007	0.008
12.5	0.005	0.006
13.0	0.003	0.004
13.5	0.002	0.003
14.0	0.002	0.003

TABLE XXIII.

p	$S(3p\pi, 3p\pi)$ $t=0.0$	$S(5p\pi, 5p\pi)$ $t=0.0$
0.0	1.000	1.000
0.5	0.988	0.993
1.0	0.955	0.972
1.5	0.901	0.938
2.0	0.832	0.893
2.5	0.752	0.839
3.0	0.666	0.778
3.5	0.578	0.712
4.0	0.493	0.644
4.5	0.413	0.575
5.0	0.341	0.508
5.2	0.314	
5.4	0.289	
5.5		0.443
5.6	0.265	
5.8	0.243	
6.0	0.222	0.383
6.2	0.202	0.360
6.4	0.184	0.338
6.6	0.167	0.317
6.8	0.152	0.297
7.0	0.137	0.277
7.2	0.124	0.259
7.4	0.112	0.241
7.6	0.101	0.224
7.8	0.091	0.208
8.0	0.081	0.193
8.5	0.062	0.159
9.0	0.046	0.130
9.5	0.035	0.105
10.0	0.026	0.084
10.5		0.067
11.0		0.053
11.5		0.042
12.0		0.033
12.5		0.025

VII. TABLES OF SELECTED HYBRID SLATER-AO OVERLAP INTEGRALS***

TABLE XXIV.

	$S(1s, 2di\sigma)$							
p	$t=-0.5$	$t=-0.4$	$t=-0.3$	$t=-0.2$	$t=-0.1$	$t=0.0$	$t=0.1$	$t=0.2$
0.0						0.612	0.543	0.461
0.5	0.669	0.757	0.812	0.832	0.818	0.773	0.704	0.611
1.0	0.704	0.807	0.878	0.914	0.913	0.879	0.814	0.723
1.5	0.697	0.804	0.881	0.925	0.936	0.914	0.860	0.778
2.0	0.658	0.758	0.831	0.877	0.895	0.884	0.845	0.778
2.2	0.636	0.730	0.800	0.846	0.865	0.858	0.825	0.765
2.4	0.611	0.699	0.765	0.809	0.829	0.826	0.798	0.746
2.6	0.583	0.665	0.727	0.768	0.789	0.789	0.767	0.722
2.8	0.555	0.630	0.686	0.725	0.746	0.748	0.731	0.694
3.0	0.526	0.593	0.644	0.680	0.700	0.705	0.693	0.662
3.2	0.496	0.556	0.601	0.634	0.654	0.660	0.653	0.629
3.4	0.466	0.519	0.559	0.588	0.607	0.615	0.612	0.594
3.6	0.437	0.483	0.518	0.543	0.561	0.570	0.570	0.558
3.8						0.526	0.529	0.522
4.0	0.381	0.414	0.439	0.457	0.472	0.483	0.489	0.486
4.2						0.442	0.450	0.451
4.4						0.403	0.412	0.417
4.5	0.317	0.336	0.350	0.362	0.372			
4.6						0.366	0.377	0.384
5.0	0.260	0.269	0.274	0.280	0.288	0.299	0.311	0.323
5.5						0.228	0.241	0.256
6.0	0.171	0.166	0.162	0.160	0.163	0.171	0.184	0.200
6.5						0.127	0.138	0.155
7.0	0.110	0.100	0.092	0.088	0.088	0.093	0.103	0.118
7.5						0.067	0.076	0.089
8.0	0.070	0.058	0.050	0.046	0.045	0.048	0.055	0.067
9.0						0.024	0.029	0.037
10.0						0.012	0.014	0.020

TABLE XXV.

	$S(1s, 2tr\sigma)$							
p	$t=-0.5$	$t=-0.4$	$t=-0.3$	$t=-0.2$	$t=-0.1$	$t=0.0$	$t=0.1$	$t=0.2$
0.0						0.500	0.443	0.376
0.5	0.574	0.655	0.708	0.731	0.727	0.689	0.631	0.551
1.0	0.624	0.725	0.797	0.838	0.845	0.821	0.765	0.684
1.5	0.633	0.740	0.821	0.871	0.890	0.876	0.831	0.756
2.0	0.606	0.708	0.788	0.840	0.866	0.862	0.830	0.769
2.2	0.588	0.686	0.762	0.814	0.841	0.841	0.814	0.759
2.4	0.567	0.659	0.732	0.782	0.810	0.813	0.791	0.743
2.6	0.544	0.630	0.697	0.746	0.773	0.779	0.762	0.722
2.8	0.519	0.598	0.661	0.706	0.733	0.741	0.729	0.696
3.0	0.493	0.565	0.622	0.664	0.690	0.701	0.693	0.666
3.2	0.466	0.531	0.583	0.621	0.646	0.658	0.655	0.634
3.4	0.439	0.497	0.543	0.577	0.601	0.614	0.615	0.599
3.6	0.412	0.463	0.504	0.534	0.557	0.571	0.574	0.564
3.8						0.527	0.534	0.529
4.0	0.360	0.399	0.428	0.452	0.471	0.485	0.493	0.493
4.2						0.444	0.455	0.458
4.4						0.405	0.417	0.424
4.5	0.301	0.325	0.343	0.358	0.372			
4.6						0.369	0.381	0.391
5.0	0.247	0.260	0.270	0.278	0.289	0.302	0.316	0.329
5.5						0.231	0.245	0.262
6.0	0.163	0.162	0.160	0.160	0.165	0.174	0.187	0.205
6.5						0.129	0.141	0.158
7.0	0.105	0.097	0.091	0.088	0.089	0.094	0.105	0.121
7.5						0.068	0.077	0.091
8.0	0.067	0.057	0.050	0.046	0.046	0.049	0.056	0.068
9.0						0.025	0.029	0.038
10.0						0.012	0.015	0.020

*** See Section III, especially Eqs. (12), (13), for definitions and conventions, and Eqs. (9), (78), (79) for the hybrid S formulas.

Table XXVI.

p	$S(1s, 2te\sigma)$ $t=-0.5$	$t=-0.4$	$t=-0.3$	$t=-0.2$	$t=-0.1$	$t=0.0$	$t=0.1$	$t=0.2$
0.0						0.433	0.384	0.326
0.5	0.516	0.592	0.643	0.667	0.664	0.635	0.584	0.512
1.0	0.574	0.672	0.744	0.787	0.798	0.779	0.730	0.655
1.5	0.590	0.696	0.778	0.832	0.854	0.845	0.805	0.736
2.0	0.570	0.673	0.754	0.810	0.839	0.840	0.811	0.754
2.2	0.555	0.654	0.732	0.787	0.818	0.822	0.798	0.747
2.4	0.537	0.630	0.705	0.758	0.789	0.796	0.777	0.733
2.6	0.516	0.603	0.673	0.724	0.755	0.765	0.751	0.713
2.8	0.493	0.574	0.639	0.687	0.718	0.729	0.720	0.688
3.0	0.469	0.543	0.603	0.647	0.677	0.690	0.685	0.660
3.2	0.444	0.511	0.565	0.606	0.634	0.649	0.648	0.628
3.4	0.419	0.479	0.527	0.565	0.591	0.606	0.609	0.595
3.6	0.394	0.447	0.490	0.523	0.548	0.564	0.569	0.561
3.8						0.522	0.530	0.526

p	$S(1s, 2te\sigma)$ $t=-0.5$	$t=-0.4$	$t=-0.3$	$t=-0.2$	$t=-0.1$	$t=0.0$	$t=0.1$	$t=0.2$
4.0	0.345	0.385	0.417	0.443	0.464	0.480	0.490	0.491
4.2						0.440	0.452	0.456
4.4						0.402	0.415	0.422
4.5	0.288	0.314	0.335	0.352	0.368			
4.6						0.366	0.380	0.390
5.0	0.238	0.252	0.264	0.274	0.286	0.300	0.314	0.329
5.5						0.229	0.244	0.261
6.0	0.157	0.157	0.157	0.158	0.163	0.173	0.187	0.205
6.5						0.128	0.141	0.158
7.0	0.101	0.094	0.089	0.087	0.088	0.094	0.105	0.121
7.5						0.068	0.077	0.091
8.0	0.064	0.056	0.049	0.046	0.046	0.049	0.056	0.068
9.0						0.025	0.029	0.038
10.0						0.012	0.015	0.020

Table XXVII.

p	$S[1s, 2(\alpha=\frac{1}{3})]$ $t=-0.5$	$t=-0.4$	$t=-0.3$	$t=-0.2$	$t=-0.1$	$t=0.0$	$t=0.1$	$t=0.2$
0.0						0.289	0.256	0.217
0.5	0.386	0.451	0.498	0.524	0.528	0.511	0.475	0.421
1.0	0.459	0.549	0.619	0.665	0.684	0.676	0.641	0.581
1.5	0.490	0.591	0.673	0.731	0.762	0.762	0.733	0.676
2.0	0.485	0.586	0.668	0.730	0.766	0.774	0.755	0.707
2.2	0.476	0.573	0.653	0.714	0.751	0.762	0.747	0.704
2.4	0.462	0.556	0.633	0.691	0.729	0.742	0.731	0.694
2.6	0.447	0.535	0.608	0.664	0.701	0.716	0.709	0.677
2.8	0.429	0.511	0.580	0.633	0.668	0.685	0.682	0.656
3.0	0.410	0.486	0.549	0.598	0.633	0.651	0.651	0.631
3.2	0.389	0.459	0.517	0.562	0.595	0.614	0.617	0.602
3.4	0.369	0.431	0.483	0.525	0.556	0.575	0.581	0.572
3.6	0.347	0.404	0.450	0.488	0.517	0.536	0.545	0.540
3.8						0.497	0.507	0.507

p	$S[1s, 2(\alpha=\frac{1}{3})]$ $t=-0.5$	$t=-0.4$	$t=-0.3$	$t=-0.2$	$t=-0.1$	$t=0.0$	$t=0.1$	$t=0.2$
4.0	0.306	0.349	0.385	0.415	0.440	0.459	0.471	0.471
4.2						0.421	0.435	0.440
4.4						0.385	0.400	0.449
4.5	0.256	0.286	0.311	0.331	0.350			
4.6						0.351	0.366	0.378
5.0	0.212	0.231	0.245	0.259	0.273	0.288	0.304	0.319
5.5						0.221	0.237	0.254
6.0	0.141	0.144	0.146	0.150	0.156	0.167	9.181	0.199
6.5						0.124	0.137	0.154
7.0	0.091	0.087	0.084	0.082	0.085	0.091	0.102	0.118
7.5						0.066	0.075	0.089
8.0	0.058	0.051	0.046	0.044	0.044	0.048	0.055	0.067
9.0						0.024	0.029	0.037
10.0						0.012	0.014	0.020

Table XXVIII.

p	$S(2di\sigma, 2di\sigma)$ $t=0.0$	$S(2tr\sigma, 2tr\sigma)$ $t=0.0$	$S(2te\sigma, 2te\sigma)$ $t=0.0$	$S[2(\alpha=\frac{1}{3}), 2(\alpha=\frac{1}{3})]$ $t=0.0$
0.0	0.000	−0.333	−0.500	−0.778
0.5	0.172	−0.155	−0.325	−0.625
1.0	0.382	+0.086	−0.076	−0.375
1.5	0.590	0.339	+0.195	−0.087
2.0	0.758	0.558	0.436	+0.181
2.5	0.871	0.721	0.622	0.402
3.0	0.907	0.798	0.719	0.532
3.2	0.907	0.813	0.742	0.568
3.4	0.898	0.818	0.753	0.593
3.6	0.882	0.813	0.756	0.608
3.8	0.860	0.802	0.750	0.614
4.0	0.832	0.783	0.738	0.613
4.2	0.801	0.761	0.720	0.606
4.4	0.765	0.733	0.696	0.592

p	$S(2di\sigma, 2di\sigma)$ $t=0.0$	$S(2tr\sigma, 2tr\sigma)$ $t=0.0$	$S(2te\sigma, 2te\sigma)$ $t=0.0$	$S[2(\alpha=\frac{1}{3}), 2(\alpha=\frac{1}{3})]$ $t=0.0$
4.6	0.725	0.699	0.668	0.573
4.8	0.685	0.664	0.637	0.552
5.0	0.645	0.628	0.604	0.527
5.5	0.542	0.534	0.517	0.458
6.0	0.445	0.443	0.430	0.385
6.5	0.358	0.358	0.350	0.316
7.0	0.282	0.284	0.278	0.253
7.5	0.219	0.222	0.218	0.199
8.0	0.168	0.170	0.168	0.154
8.5	0.127	0.129	0.128	0.118
9.0	0.095	0.097	0.096	0.089
9.5	0.070	0.072	0.071	0.066
10.0	0.051	0.053	0.052	0.049

THE MOLECULAR-ORBITAL AND EQUIVALENT-ORBITAL APPROACH TO MOLECULAR STRUCTURE

By J. A. Pople, Ph.D.

(Department of Theoretical Chemistry, University of Cambridge)

Introduction

The electronic theory of chemical valency has to explain a set of facts and empirical rules some of which suggest an interpretation in terms of localised electrons and others require a picture of electrons spread throughout the whole molecule. In the pre-electronic era a chemical bond was regarded as a genuinely local link joining neighbouring atoms in a molecule, and this was associated with a pair of bonding electrons in the early electronic theory developed by Lewis and Langmuir. In accounting for all the electrons some were assigned to atomic inner shells and others were supposed to form inert pairs (or " lone pairs ") on a single atom. The rules of stereochemistry implied certain restrictions about the geometrical arrangement of neighbouring bonds, but, apart from this, there seemed to be considerable evidence that the pairs of electrons in different bonds behaved independently to a large extent. For a great many molecules it was found possible to interpret heats of formation on the assumption that there was a definite energy associated with each type of bond (the bond energy). The refractivity of a large molecule can usually be predicted by assuming that the total is a sum of standard contributions from the various atoms and bonds. Similar additive laws also hold for magnetic susceptibilities. All these facts, which imply the existence of a standard type of bond between two given atoms, are best interpreted in terms of a theory in which a pair of electrons is moving in localised orbits in each bond and is mainly independent of electron pairs in neighbouring bonds.

On the other hand, there are properties of molecules which do not seem to fit this picture. Consider the ionisation of (removal of an electron from) a simple molecule. According to the localised picture, we might expect this process to consist of the removal of an electron from one of the bonds, or possibly from one of the lone pairs. However, in the case of a molecule such as methane, where there are several bonds exactly equivalent to one another, there are various possibilities. There is no *a priori* reason why the electron should be removed from one bond rather than another and, in such circumstances, what actually happens is that the electron is removed partly from them all, or, an equivalent statement, the electron which is removed was moving in an orbit or path extending over the whole molecule. Similar situations arise when we consider the electronic excitation of a molecule. Methane being taken as an example again, instead of exciting the electrons in a single bond, an electron is taken out of one orbit spread over the whole molecule and placed in another excited orbit. It seems, therefore, that in order to interpret spectroscopic properties of molecules such as methane, we ought to treat the electrons as moving in orbits extending over the whole molecule, processes such as ionisation and excitation corresponding to the removal or reallocation of electrons among these paths. Such a procedure is, in fact, a logical extension of the ideas originally used by spectroscopists to interpret atomic spectral lines and it has since proved its value in the theory of the electronic spectra of molecules.

It appears, then, that there are two apparently divergent modes of description of molecular structure, localised electrons in bonds and lone-pair orbits on the one hand and electrons moving in orbits covering the whole molecular framework on the other. But the success of both descriptions in their respective fields of application is so considerable that the two must be more closely related than appears at first sight. When we consider the general quantum-mechanical problem of finding the distribution of electrons in a molecule we find that this is so and that the localised and delocalised pictures are just two different ways of breaking down the same total wave function describing the combined motion of all electrons. The purpose of this Review is to elaborate this transformation and show how it links together alternative descriptions of certain simple molecules.

To do this we begin by considering the general properties that the wave function for the electrons in a molecule must possess. If we consider only one electron moving in the electrostatic field of the nuclei, then it is quite clear that its path or orbital must extend over the whole nuclear framework. Thus the electron in the hydrogen molecule-ion, H_2^+, is equally distributed around both nuclei. When we come to systems of several electrons, however, we also have to take into account the indistinguishability of electrons and, further, the all-important antisymmetry property of the wave-function. The way in which this is incorporated into the molecular-orbital theory is discussed in the next section and its consequences are then illustrated in terms of a simple one-dimensional model. In the remaining sections the transformation between the localised and delocalised descriptions is carried out for certain simple molecules. In this way we can see the relation between the bonding- and antibonding-orbital picture of a diatomic molecule such as F_2 and the alternative description in terms of lone pairs. The relation between the σ–π and the two-bent-bond descriptions of the standard carbon double bond in ethylene also becomes apparent. Similarly a triple bond, as in nitrogen or acetylene, can be regarded as three equivalent bent bonds or as a σ bond and two π bonds.

Quantum-mechanical basis of orbital theories

The basic quantum-mechanical problem is to formulate the wave-like description of an electron moving in the electrostatic field of the nuclei and other electrons. If the potential energy of an electron at a point (x, y, z) is $V(x, y, z)$, this is accomplished by solving the well-known Schrödinger equation for a wave function $\psi(x, y, z)$

$$-\frac{h^2}{8\pi^2 m}\left(\frac{\partial^2\psi}{\partial x^2}+\frac{\partial^2\psi}{\partial y^2}+\frac{\partial^2\psi}{\partial z^2}\right)+V(x, y, z)\psi = E\psi \quad . \quad . \quad (1)$$

where E is the energy of the electron and h and m are Planck's constant and the electronic mass respectively. (For many-electron systems some care has to be taken in obtaining the potential energy V for which a knowledge of other electron distributions is required. The calculations have to be made self-consistent. The details are not relevant to the present topic, however, and we shall not go into them.) The function ψ which depends on the co-ordinates (x, y, z) of the electron in space will be referred to as a *space orbital* or often just as an *orbital*. Its physical interpretation is that $\psi^2\, dx\, dy\, dz$ represents the probability that the electron will be found in a small rectangular element of volume $dx\, dy\, dz$ near the point (x, y, z). Thus ψ^2 is a probability density and the electron is most likely to be found where this density has its largest value.

The other important property of an electron that must be specified besides its spatial distribution is its spin. According to quantum-mechanical arguments, into which we need not go in detail, each electron has a spin which can take one of two values. It is convenient to include this description in the wave function by defining α and β so that $\alpha = 1$ if the spin is in one direction and $\alpha = 0$ if it is in the other. β is defined in a complementary manner. Thus the electron moving in an orbital $\psi(x, y, z)$ may be associated with two functions $\psi(x, y, z)\alpha$ and $\psi(x, y, z)\beta$ according to the direction of its spin. A function such as $\psi(x, y, z)\alpha$ which gives the probability distribution of the spin co-ordinate as well as that of its spatial co-ordinates is sometimes referred to as a *spin orbital*.

All this is very straightforward if we are dealing with a system which contains only one electron such as the hydrogen atom or the hydrogen molecule-ion H_2^+. But when we consider a many-electron molecule we are faced with the problem of combining the orbitals for the individual electrons into a total wave function for the whole system. Suppose we are dealing with two electrons which occupy space orbitals ψ_1 and ψ_2. The simplest compound wave function for both electrons is the product

$$\Psi_{\text{product}} = \psi_1(x_1, y_1, z_1)\psi_2(x_2, y_2, z_2) \quad . \quad . \quad . \quad (2)$$

where (x_1, y_1, z_1) and (x_2, y_2, z_2) are the Cartesian co-ordinates of electrons 1 and 2 respectively. To be complete this should be multiplied by one of the four possible spin functions $\alpha(1)\alpha(2)$, $\alpha(1)\beta(2)$, $\alpha(2)\beta(1)$, or $\beta(1)\beta(2)$. The physical interpretation of this compound wave function is again in terms of probability. Ψ^2 is now proportional to the joint probability of electron 1 being at position (x_1, y_1, z_1) and electron 2 at position (x_2, y_2, z_2) *simultaneously*. If the product form is used this is just the product of the two separate probabilities. Thus the product wave function implies that the two electrons move independently of one another.

Product wave functions can clearly be constructed for any number of electrons. Early wave functions, were constructed on this basis together with the empirical rule that not more than two electrons could be assigned to a single orbital, one of each spin. Further, electrons tend to occupy the orbitals with lowest possible energy in the absence of other factors.

The major disadvantage of the product function is that it fails to satisfy another important quantum-mechanical principle, namely that of antisymmetry. This is really a consequence of the indistinguishability of electrons. If we consider the operation of interchanging the positions of two electrons the probability of the new configuration must be just the same as previously. Thus the square of the total wave function must be unaltered, and consequently the wave function itself can only be multiplied by $+1$ or by -1. It is found that the second choice is demanded for electrons so that we formulate the antisymmetry principle by requiring that the wave function changes sign if we interchange the co-ordinates of any two electrons. Clearly the product function (2) does not satisfy this condition, for if we interchange the co-ordinates of electrons 1 and 2 we obtain $\psi_1(x_2, y_2, z_2)\psi_2(x_1, y_1, z_1)$ which is not a direct multiple of its previous form.

The next step is to construct a wave function from products of the type (2) which satisfies this further condition. This can be done in terms of what is called an *antisymmetrised product.* Let us consider, first of all, the case of two electrons in the same space orbital ψ_1 with two different spins. The simple product function is

$$\psi_1(x_1, y_1, z_1)\psi_1(x_2, y_2, z_2)\alpha(1)\beta(2) \quad . \quad . \quad . \quad (3)$$

The antisymmetrised product is obtained by subtracting from this the corresponding product with the suffixes 1 and 2 interchanged. This gives

$$(1/\sqrt{2})\psi_1(x_1, y_1, z_1)\psi_1(x_2, y_2, z_2)\{\alpha(1)\beta(2) - \alpha(2)\beta(1)\} \quad . \quad . \quad (4)$$

The factor $(1/\sqrt{2})$ is inserted so that the total probability added over all configurations is unity. This function may be said to be symmetric in the space co-ordinates but antisymmetric in the spins. For an overall interchange it is antisymmetric.

Next, suppose we have two electrons in different space orbitals ψ_1 and ψ_2 but with the same spin α. Then the simple product function is

$$\psi_1(x_1, y_1, z_1)\psi_2(x_2, y_2, z_2)\alpha(1)\alpha(2) \quad . \quad . \quad . \quad (5)$$

and the antisymmetrised product constructed in the same way is

$$(1/\sqrt{2})\{\psi_1(x_1, y_1, z_1)\psi_2(x_2, y_2, z_2) - \psi_1(x_2, y_2, z_2)\psi_2(x_1, y_1, z_1)\}\alpha(1)\alpha(2) \quad (6)$$

Both the antisymmetric functions (4) and (6) can be written as 2×2 determinants. Thus (4) is

$$\frac{1}{\sqrt{2}} \begin{vmatrix} \psi_1(1)\alpha(1) & \psi_1(2)\alpha(2) \\ \psi_1(1)\beta(1) & \psi_1(2)\beta(2) \end{vmatrix} \quad . \quad . \quad . \quad . \quad (7)$$

and (6) is

$$\frac{1}{\sqrt{2}} \begin{vmatrix} \psi_1(1)\alpha(1) & \psi_1(2)\alpha(2) \\ \psi_2(1)\alpha(1) & \psi_2(2)\alpha(2) \end{vmatrix} \quad . \quad . \quad . \quad . \quad (8)$$

Here we have written $\psi_1(1)$ as a short form of $\psi_1(x_1, y_1, z_1)$.

These simple determinantal functions for two electrons suggest that we can construct antisymmetric wave functions for any number of electrons in a similar manner. Thus if we have a set of orbitals $\psi_1, \psi_2 \ldots \psi_n$ each containing two electrons, one of each spin (this applies to most molecules), an antisymmetric wave function can be constructed as a determinant with a different spin orbital in each row.

$$\Psi = \sqrt{\frac{1}{(2n)!}} \begin{vmatrix} \psi_1(1)\alpha(1) & \psi_1(2)\alpha(2) & \dots & \psi_1(2n)\alpha(2n) \\ \psi_1(1)\beta(1) & \psi_1(2)\beta(2) & \dots & \\ \psi_2(1)\alpha(1) & \psi_2(2)\alpha(2) & \dots & \\ \cdot & \cdot & \cdot & \cdot \\ \psi_n(1)\beta(1) & \psi_n(2)\beta(2) & \dots & \psi_n(2n)\beta(2n) \end{vmatrix} \quad . \quad . \quad (9)$$

The interchange of the co-ordinates and spins of two electrons corresponds to interchanging two columns of this determinant. This leads to a change of sign, so that the antisymmetry property is satisfied. This type of total wave function is that used in molecular-orbital theory.

Another well-known property of determinants is that they vanish if they have two identical rows. This means that it is not possible to construct a non-vanishing antisymmetrised product in which two electrons in the same orbital have the same spin. Thus the rule that not more than two electrons must be assigned to any one space orbital follows as a direct consequence of the antisymmetry principle ; for product wave functions it had to be introduced as an extra postulate.

Another important physical interpretation of the molecular-orbital determinant follows from an application of a similar argument to the columns. The elements of two columns become identical if two electrons have the same spin (α or β) and are at the same point (x, y, z). The determinant then vanishes and consequently the probability of such a configuration is zero. Such an argument does not apply to electrons of different spin, however. The antisymmetry principle operates, therefore, in such a way that electrons of the same spin are kept apart. We shall see in later sections that this is an important factor in determining stereochemical valence properties.

The antisymmetry principle is also of great importance in understanding the dualism between localised and delocalised descriptions of electronic structure. We shall see that these are just different ways of building up the same total determinantal wave functions.[1] This can be developed mathematically from general properties of determinants, but a clearer picture can be formed if we make a detailed study of the antisymmetric wave function for some highly simplified model systems.

Simple models illustrating the effects of antisymmetry

The simplest system that can be used for illustrative purposes is one in which electrons are free to move in one dimension along a wire of length l. The potential energy will be constant and can be taken as zero. If the position of a point on the wire is measured by the distance x from one end, the Schrödinger equation is

$$-\frac{h^2}{8\pi^2 m}\frac{d^2\psi}{dx^2} = E\psi \quad . \quad . \quad . \quad . \quad (10)$$

[1] Lennard-Jones, *Proc. Roy. Soc.*, 1949, *A*, **198**, 1, 14.

where E is the energy. This is just the simple harmonic equation which has the general solution

$$\psi = A \sin\left\{\sqrt{\left(\frac{8\pi^2 mE}{h^2}\right)}x\right\} + B \cos\left\{\sqrt{\left(\frac{8\pi^2 mE}{h^2}\right)}x\right\} \quad . \quad . \quad (11)$$

where A and B are constants of integration. The wave function must also satisfy the boundary conditions of being zero at both ends (that is at $x = 0$ and $x = l$). The first condition requires that $B = 0$ and the second that $\sqrt{(8\pi^2 mE/h^2)}l$ is an integral multiple of π. The lowest orbitals that electrons can occupy (those with least nodes) are therefore

$$\psi_1 = \sqrt{(2/l)} \sin (\pi x/l)$$
$$\psi_2 = \sqrt{(2/l)} \sin (2\pi x/l) \quad . \quad . \quad . \quad . \quad (12)$$

ψ_1 is positive over the whole length of the segment while ψ_2 is zero at the centre of the wire (Fig. 1).

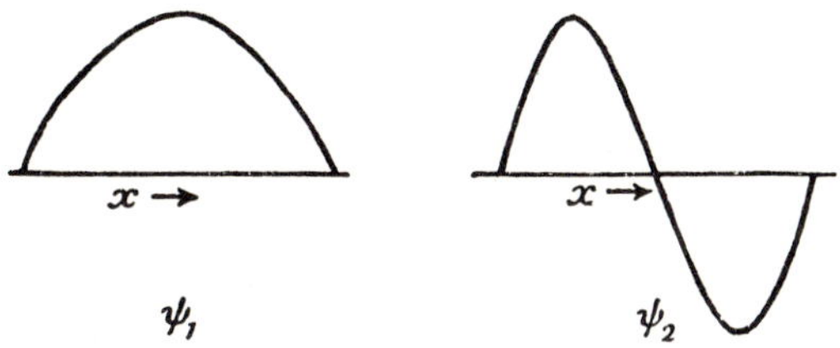

FIG. 1

Lowest occupied molecular orbitals in model system.

Now suppose two electrons are placed one in each of these orbitals. The distribution of these electrons in their individual orbitals will simply be given by ψ_1^2 and ψ_2^2. If we wish to examine the probability of various simultaneous positions of the two electrons, we have to consider the total wave function Ψ, which will be an antisymmetric product with a form depending on the spins of the electrons. If we wish to investigate the effect of the antisymmetry principle on the spatial arrangement of the electrons, it is convenient to examine the case in which they both have the same spin α. Then the wave function will be of the form given in equations (6) and (8). If the factor $\alpha(1)\alpha(2)$ is omitted,

$$\Psi = \begin{vmatrix} \psi_1(1) & \psi_1(2) \\ \psi_2(1) & \psi_2(2) \end{vmatrix} = (2/l)\{\sin (\pi x_1/l) \sin (2\pi x_2/l) - \sin (\pi x_2/l) \sin (2\pi x_1/l)\}$$

$$= (4/l) \sin (\pi x_1/l) \sin (\pi x_2/l)\{\cos (\pi x_1/l) - \cos (\pi x_2/l)\} \quad . \quad . \quad (13)$$

Ψ^2 is then proportional to the probability of electron 1 being found at position x_1 and electron 2 at x_2. The significant points to be noted are (1) that if $x_1 = x_2$ the wave function vanishes so that the configuration has zero probability and (2) that there are two equivalent most probable configurations in which the electrons are in different halves of the wire. These two configurations differ only in the numbering of the electrons and are otherwise indistinguishable.

The effect of using an antisymmetrised wave function instead of an ordinary product, therefore, is to cause the electrons to move in two different regions in the two halves of the segment, the probability of configurations in which both are in the same segment at the same instant being relatively small. This suggests that the system could be alternatively described in terms of two localised orbitals, one in either segment with one electron in each.

This alternative description in terms of localised orbitals can indeed be set up by taking linear combinations of the orbitals ψ_1 and ψ_2 and using these in the determinant instead. If the linear combinations are suitably chosen, the value of the determinant is unaltered, although the individual rows change. Let us consider, therefore, how we can construct localised orbitals from our two starting orbitals ψ_1 and ψ_2. As we have already noted, ψ_1 is positive everywhere while ψ_2 is positive in the left-hand part of the segment and negative in the right. If we consider $\psi_1 + \psi_2$, the two components will add on the left, but partly cancel on the right. This, therefore, can be used as a localised orbital mainly concentrated in the left-hand part. Similarly $\psi_1 - \psi_2$ is mainly concentrated on the right. We therefore define two new localised orbitals χ_a and χ_b by

$$\begin{aligned}\chi_a &= (\psi_1 + \psi_2)/\sqrt{2} = \sqrt{(4/l)} \cos(\pi x/2l) \sin(3\pi x/2l) \\ \chi_b &= (\psi_1 - \psi_2)/\sqrt{2} = \sqrt{(4/l)} \sin(\pi x/2l) \cos(3\pi x/2l)\end{aligned} \qquad (14)$$

The factor $(1/\sqrt{2})$ is included to keep the total probability equal to unity. These functions are illustrated in Fig. 2. They are mirror images in the

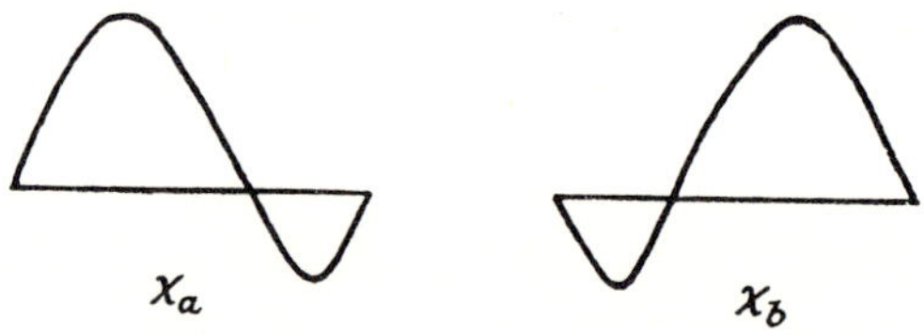

FIG. 2
Equivalent orbitals in model system.

mid-point of the line-segment. They are sometimes called *equivalent orbitals*.[1, 2]

The total wave function can now be written in terms of the equivalent orbitals

$$\Psi = -\begin{vmatrix} \chi_a(1) & \chi_a(2) \\ \chi_b(1) & \chi_b(2) \end{vmatrix} \qquad (15)$$

If we substitute for χ_a and χ_b and expand the expression, it is easily confirmed that the value of this determinant is identical with the original total wave function (13). This is a particular example of what is known as an orthogonal transformation of the rows of the determinant.

It appears, therefore, that we have two possible descriptions of this system. We can describe it as two electrons, each of which occupies one of the delocalised (or molecular) orbitals which are solutions of the

[2] Lennard-Jones and Pople, *Proc. Roy. Soc.*, 1950, *A*, **202**, 166.

Schrödinger equation. Or alternatively, we may say, equally accurately, that the two electrons occupy two localised orbitals χ_a and χ_b, one at each end of the segment. These are just two different ways of interpreting the same total wave function.

The two descriptions are useful in rather different contexts. If we are interested in the relative positions of the two electrons, then the interpretation in terms of localised orbitals gives a clearer description of the qualitative features of the overall probability distribution. On the other hand, if we are interested in the removal of an electron, the first description is more appropriate, for the remaining electron must occupy an orbital which is a solution of the original Schrödinger equation. Thus the electron must be removed from ψ_1 or ψ_2.

This sort of model can easily be generalised to deal with more than two electrons and other assignments of the spins. The case of most interest in molecular studies is that in which a set of molecular orbitals are all occupied by two electrons. Thus if there were two electrons, one of either spin, in both orbitals ψ_1 and ψ_2, the total wave function would be a 4×4 determinant. But most of the features of the two-electron model are retained. The system could be alternatively described as consisting of two electrons in each of the equivalent orbitals. The effect of the antisymmetry principle is then to keep electrons of the same spin apart, the motion of the two opposite spin-types being uncorrelated.

Although the one-dimensional model bears little resemblance to any real molecular system, many of its features carry over to cases of practical interest. Suppose we consider three-dimensional motion in a central field as in atoms. The orbitals or single-electron functions now become atomic orbitals and can be classified in the usual manner as $1s$, $2s$, . . . $2p$, $3p$, . . . $3d$, ... Suppose we are dealing with an atom in which there are two electrons of the same spin (α, say) occupying the $2s$ and $2p$ orbitals (inner shells being ignored for the present). Then the antisymmetric product function is

$$\Psi = \begin{vmatrix} \psi_{2s}(1) & \psi_{2s}(2) \\ \psi_{2p}(1) & \psi_{2p}(2) \end{vmatrix} \alpha(1)\alpha(2) \qquad . \quad . \quad . \quad (16)$$

This wave function has many features in common with that of the previous model. While ψ_{2s} is spherically symmetric, ψ_{2p} has a nodal plane through the centre of symmetry. A similar transformation can be applied and we can use two equivalent orbitals

$$\begin{aligned} \chi_1 &= \sqrt{(1/2)}(\psi_{2s} + \psi_{2p}) \\ \chi_2 &= \sqrt{(1/2)}(\psi_{2s} - \psi_{2p}) \end{aligned} \qquad . \quad . \quad . \quad . \quad (17)$$

Atomic orbitals of this mixed type are usually referred to as hybrids (or more specifically digonal s–p hydrids). As with the one-dimensional model, they reinforce on one side of the nucleus and partly cancel on the other. Hence the transformation is from the delocalised s and p description to a description in terms of two equivalent orbitals, localised on opposite sides of the nucleus. Again a similar transformation may be applied to a 4×4 determinant describing a system with two electrons in each of these orbitals.

For the next example, consider a system of three electrons of the same spin occupying atomic orbitals $2s$, $2px$, and $2py$. Here the antisymmetric wave function is

$$\Psi = \begin{vmatrix} \psi_{2s}(1) & \psi_{2s}(2) & \psi_{2s}(3) \\ \psi_{2px}(1) & \psi_{2px}(2) & \psi_{2px}(3) \\ \psi_{2py}(1) & \psi_{2py}(2) & \psi_{2py}(3) \end{vmatrix} \alpha(1)\alpha(2)\alpha(3) \quad . \qquad (18)$$

In this case we can transform these into three equivalent orbitals which are s–p hybrids (called trigonal hybrids) pointing towards the vertices of an equilateral triangle, so that the angle between neighbouring directions is 120°. The actual transformation is

$$\begin{aligned} \chi_1 &= \sqrt{(1/3)}\psi_{2s} + \sqrt{(2/3)}\psi_{2px} \\ \chi_2 &= \sqrt{(1/3)}\psi_{2s} - \sqrt{(1/6)}\psi_{2px} + \sqrt{(1/2)}\psi_{2py} \\ \chi_3 &= \sqrt{(1/3)}\psi_{2s} - \sqrt{(1/6)}\psi_{2px} - \sqrt{(1/2)}\psi_{2py} \end{aligned} \quad . \quad . \qquad (19)$$

It is not immediately clear from the form in which these are written that they are equivalent functions, that is, differ only in their orientation, but it is easily confirmed that they do transform into each other if the axes are rotated through 120°. Again it can be shown that the determinant of χ-functions has the same value as (18). One other point about this set of equivalent orbitals is that there appears to be no preferential direction in which any one of the vertices of this triangle may be chosen. The choice is, in fact, arbitrary and any set of three equivalent directions perpendicular to the z direction would suffice. This only applies for an atomic wave function, of course. In molecules (such as planar XY_3) there may be a preferred choice of axes on account of symmetry. This will be clear from some examples considered in the next section.

The case of four electrons in the atomic orbitals $2s$, $2px$, $2py$, and $2pz$ can be handled in a similar manner. Here we can transform the expression into four equivalent orbitals given by

$$\begin{aligned} \chi_1 &= \tfrac{1}{2}(\psi_{2s} + \psi_{2px} + \psi_{2py} + \psi_{2pz}) \\ \chi_2 &= \tfrac{1}{2}(\psi_{2s} + \psi_{2px} - \psi_{2py} - \psi_{2pz}) \\ \chi_3 &= \tfrac{1}{2}(\psi_{2s} - \psi_{2px} + \psi_{2py} - \psi_{2pz}) \\ \chi_4 &= \tfrac{1}{2}(\psi_{2s} - \psi_{2px} - \psi_{2py} + \psi_{2pz}) \end{aligned} \quad . \quad . \quad . \qquad (20)$$

which are directed towards the vertices of a regular tetrahedron. These equivalent orbitals are usually called tetrahedral s–p hybrids. If we have eight electrons (two of each spin in each orbital) this description can be applied directly to the outermost shell of electrons in the neon atom. The neon atom is not usually described in terms of localised tetrahedral orbitals, but such a description is just as valid as the more conventional s^2p^6. We shall see in the next section that the localised picture is useful in discussing the structure of molecules isoelectronic with neon.

The orbital description of molecules

We now turn to the description of actual molecules in terms of molecular orbitals. The usual procedure is to find orbital functions $\psi_1, \psi_2, \ldots$ which are solutions of a suitable Schrödinger equation, assign the electrons in pairs to those orbitals of lowest energy, and then construct an antisymmetric determinantal wave function [as in eqn. (9)]. We can then consider possible alternative descriptions obtained by transformations of the rows of the determinant as with the models of the previous section.

In the complex electrostatic field of a molecule, it is usually impracticable to obtain accurate molecular orbitals, so it is customary to express them approximately as linear combinations of atomic orbitals belonging to the constituent atoms. This is called the "linear combination of atomic orbital" or LCAO form. Although they are only approximate, the LCAO functions do show most of the properties of the precise orbitals. Both molecular and localised equivalent orbitals can be expressed in this manner.

Diatomic Molecules.—We shall begin by discussing diatomic molecules, which bear some relation to the models discussed in the previous section. To begin with, the hydrogen molecule has two electrons which both occupy the lowest molecular orbital whose LCAO form is

$$\psi_1 = \lambda(1s_A + 1s_B) \qquad . \quad . \quad . \quad . \quad (21)$$

$1s_A$ and $1s_B$ are the two hydrogen $1s$ atomic orbitals. The factor λ is introduced so that the total probability adds up to unity. If the overlap between the atomic orbitals is small, λ is approximately $1/\sqrt{2}$. Since there is only one space orbital in the determinantal wave function [eqn. (7)] no transformation of the orbitals is possible.

If we now go to a pair of interacting helium atoms, there will be four electrons of which the first pair will go into the corresponding orbital ψ_1 and the second pair into the next lowest orbital for the system whose LCAO form will be

$$\psi_2 = \mu(1s_A - 1s_B) \qquad . \quad . \quad . \quad . \quad (22)$$

This function is zero for all points equidistant from the two nuclei (that is, it has a nodal plane). The orbital ψ_1 is large in the region between the nuclei (where $1s_A$ and $1s_B$ overlap and the electrostatic potential is low) and is generally referred to as a *bonding* orbital. Similarly, ψ_2, which keeps its electron away from the internuclear region, is *antibonding*. The two functions ψ_1 and ψ_2 are analogous to the symmetric and antisymmetric orbitals for the one-dimensional model. A similar transformation can be applied and two equivalent orbitals constructed. These are

$$\left.\begin{aligned} \chi_A &= (1/\sqrt{2})(\psi_1 + \psi_2) = (1/\sqrt{2})(\lambda + \mu)1s_A + (1/\sqrt{2})(\lambda - \mu)1s_B \\ \chi_B &= (1/\sqrt{2})(\psi_1 - \psi_2) = -(1/\sqrt{2})(\lambda - \mu)1s_A + (1/\sqrt{2})(\lambda + \mu)1s_B \end{aligned}\right\} \quad . \quad (23)$$

If the overlap of the functions is not large, λ and μ are both nearly $1/\sqrt{2}$ and so the equivalent orbitals approximate to the atomic orbitals for the isolated atoms. The complete equivalence of the two configurations $\psi^2_{\text{bonding}}\psi^2_{\text{antibonding}}$ and $\chi_A{}^2\chi_B{}^2$ is the simplest example of the dual description of a molecular system.

Proceeding further along the series of homonuclear diatomic molecules, the $1s$ inner shells can still be described in either manner. Since the $1s$

electrons do not play any appreciable part in bonding, it is usually most convenient to treat them as localised. The lithium molecule Li_2 can be described in terms of a pair of inner shells and a bonding orbital which is similar to that in H_2. There is a difference, however, in that there is now a possibility of appreciable hybridisation between the $2s$ and $2p$ atomic orbitals which have comparable energies. The best LCAO representation of the bonding orbital will be a sum of two hybrid orbitals of the form

$$\alpha(2s) + \beta(2p\sigma)$$

where $2p\sigma$ represents an atomic $2p$ orbital with its axis along the internuclear line. Again, since this is the only occupied orbital formed from valence shell atomic orbitals, no transformation to localised orbitals is possible.

Proceeding further along the Periodic Table, let us next consider the nitrogen molecule N_2. Here we have to consider molecular orbitals constructed from all the $2p$ functions for each atom. (The $2p$ orbitals with axes perpendicular to the molecular axis are usually called $2p\pi$ functions.) To begin with, four electrons are assigned to the inner shells, represented by equivalent orbitals $1s_A$ and $1s_B$. Secondly, there will be two molecular orbitals, bonding and antibonding, formed from the next s orbitals, $2s_A$ and $2s_B$. These can be transformed into two equivalent orbitals in a similar manner and correspond to lone-pair or inert electrons. Then there will be a bonding orbital formed from $2p\sigma$ functions

$$\psi_{\sigma\text{-bonding}} = (1/\sqrt{2})(2p\sigma_A + 2p\sigma'_B) \quad . \quad . \quad . \quad (24)$$

and two bonding orbitals whose LCAO forms are sums of the $2p\pi$ atomic orbitals

$$\left.\begin{aligned} \psi_{\pi x\text{-bonding}} &= (1\sqrt{2})(2p\pi x_A + 2p\pi x_B) \\ \psi_{\pi y\text{-bonding}} &= (1/\sqrt{2})(2p\pi y_A + 2p\pi y_B) \end{aligned}\right\} \quad . \quad . \quad . \quad (25)$$

If two electrons are assigned to each of these orbitals, all fourteen in the molecule are accounted for. This set of orbitals would be slightly modified if hybridisation between the $2s$ and $2p\sigma$ electrons is allowed.

This description of the triple bond represents it as an axially symmetric σ bond together with two perpendicular π bonds. This is appropriate for spectroscopy and must be used if we are discussing excited N_2 or N_2^+. But for N_2 in its ground state, another description in terms of three equivalent bonding orbitals can be obtained by applying the trigonal transformation to (24) and (25). Thus if we write

$$\left.\begin{aligned} \chi_{1\text{-bonding}} &= \frac{1}{\sqrt{3}}\psi_{\sigma\text{-bonding}} + \frac{\sqrt{2}}{\sqrt{3}}\psi_{\pi x\text{-bonding}} \\ \chi_{2\text{-bonding}} &= \frac{1}{\sqrt{3}}\psi_{\sigma\text{-bonding}} + \frac{1}{\sqrt{6}}\psi_{\pi x\text{-bonding}} + \frac{1}{\sqrt{2}}\psi_{\pi y\text{-bonding}} \\ \chi_{3\text{-bonding}} &= \frac{1}{\sqrt{3}}\psi_{\sigma\text{-bonding}} + \frac{1}{\sqrt{6}}\psi_{\pi x\text{-bonding}} - \frac{1}{\sqrt{2}}\psi_{\pi y\text{-bonding}} \end{aligned}\right\} \quad . \quad (26)$$

the three new orbitals will be turned into one another by a rotation through 120° about the axis of the molecule. They represent three bent bonds concentrated in three different azimuthal planes. If we square the orbitals and add to get the total electron density

$$\rho = (\chi_{1\text{-bonding}})^2 + (\chi_{2\text{-bonding}})^2 + (\chi_{3\text{-bonding}})^2$$

this is found to be axially symmetric. Nevertheless, the existence of the three equivalent orbitals implies that the pairs of electrons dispose themselves relative to each other in such a way that their distributions are similar and interrelated by a 120° rotation.

Although the nitrogen molecule represents the standard type of triple bond, the bond in O_2 is in no way typical of a double bond. There are two extra electrons and the next orbitals to be filled are the antibonding π orbitals

$$\begin{aligned} \psi_{\pi x\text{-antibonding}} &= (1/\sqrt{2})(2p\pi x_{\mathrm{A}} - 2p\pi x_{\mathrm{B}}) \\ \psi_{\pi y\text{-antibonding}} &= (1/\sqrt{2})(2p\pi y_{\mathrm{A}} - 2p\pi y_{\mathrm{B}}) \end{aligned} \quad . \quad . \quad . \quad (27)$$

These both have the same energy so that, in the absence of other determining factors, the electrons go one into each with the same spin (or an equivalent state). This means that they are kept apart by the antisymmetry principle and so the energy is lowered by the reduction of Coulomb repulsion. In this rather exceptional case, therefore, the orbitals are not all doubly occupied and we cannot carry out any simple transformation into localised orbitals.

If we now proceed further to the fluorine molecule, both the π-antibonding orbitals will be doubly occupied. As with the s functions, the configuration $(\psi_{\pi x\text{-bonding}})^2(\psi_{\pi x\text{-antibonding}})^2$ can be transformed into two π-lone-pair orbitals, one on each atom. Similarly with the πy orbitals. The localised description of F_2, therefore, has four localised π-lone-pairs, there being only one single bonding orbital.

Molecules Isoelectronic with Neon.—Another set of molecules whose structure is typical of many standard chemical environments is the set of ten-electron first row hydrides Ne, HF, H_2O, NH_3, and CH_4. On p. 281 we saw how the outer electrons of the neon atom could be described either as being in the configuration $(2s)^2(2p)^6$ or, alternatively, as occupying four tetrahedral orbitals χ_1, χ_2, χ_3, and χ_4 orientated relative to one another in a tetrahedral manner, the orientation of the tetrahedron in space being arbitrary. The electronic structures of the other molecules of the series can now be discussed in terms of this basic system if we imagine unit positive charges to be removed successively from the nucleus.

If a single positive charge is removed to give HF, a preferred direction is established and the orbitals have to be referred to the internuclear line. Suppose we take this as the z axis. The orbitals will be somewhat distorted but their general arrangement will not be radically altered. One of the four localised neon orbitals will be pulled out into a localised bonding orbital ; it can probably be expressed fairly accurately in the LCAO form as

$$\chi_{\text{bonding}} = \lambda(2s)_{\mathrm{F}} + \mu(2pz)_{\mathrm{F}} + \nu(1s)_{\mathrm{H}} \quad . \quad . \quad . \quad (28)$$

where λ, μ, and ν are numerical coefficients. This is a linear combination of an s–p hybrid on the fluorine atom directed along the z axis and the

1s hydrogen orbital (Fig. 3). The other three neon-like localised orbitals will not be distorted as much, so they will remain as three equivalent tetrahedral hybrids pointing in directions making an approximately tetrahedral angle with the bond. They are three equivalent lone pairs. It is interesting that the most important lone-pair direction (where the negative charge is most likely to be found) may not be directly at the back of the fluorine atom. As the lone-pair electrons play an important rôle as the negative end of hydrogen bonds, this is probably closely connected with the non-linear structure of hydrogen fluoride polymers.

The molecular-orbital description of HF can be obtained if we note that the three equivalent lone pairs can be obtained from a σ orbital and two π orbitals by a transformation similar to that used for obtaining the bent-bond description of N_2. In the LCAO form the σ lone pair will be another s–p hybrid and the two lone pairs will be $(2px)^2$ and $(2py)^2$. It is generally found that π lone pairs are less firmly bound than σ lone pairs so that the lowest ionisation potential would correspond to the removal of an electron from one of the last two orbitals.

We can now consider the structure of the water molecule by supposing

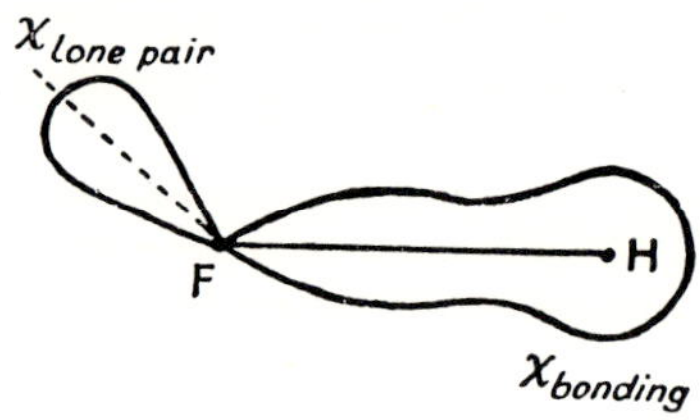

FIG. 3
Localised orbitals for hydrogen fluoride.

a further positive charge removed from the nucleus. The localised description gives some insight into the reason for the non-linear structure. Given that one positive charge has been removed, as in HF, the second charge will prefer to be pulled out in the directions where the remaining electrons are most likely to be found. As we have seen above this is in a direction at an approximately tetrahedral angle to the first bond. In the localised-orbital picture, therefore, the outer electrons of the water molecule occupy two sets of two equivalent orbitals. The first two are bonding orbitals concentrated mainly along the O–H bonds, and the other two are localised lone pairs which point in two equivalent directions towards the back of the molecule, above and below the plane of the nuclei.[3]

Once again this is a very useful description for understanding molecular interaction. The normal form of the ice crystal, for example, is held together by hydrogen bonds in such a way that each molecule is surrounded tetrahedrally by four others.[4] This is completely consistent with the electrostatic theory of the hydrogen bond according to which a proton is

[3] Pople, *Proc. Roy. Soc.*, 1950, *A*, **202**, 323.
[4] Barnes, *ibid.*, 1929, *A*, **125**, 670.

attracted by a localised lone pair of electrons on another molecule.[5] There is also considerable evidence that this structure persists to a large extent in the liquid.[6, 7]

To deal with the molecular orbitals for water it is useful to examine the effect of certain symmetry operations on the molecule. We choose a set of rectangular Cartesian axes (Fig. 4) so that the x axis bisects the angle between the bonds and the z axis is perpendicular to the nuclear plane.

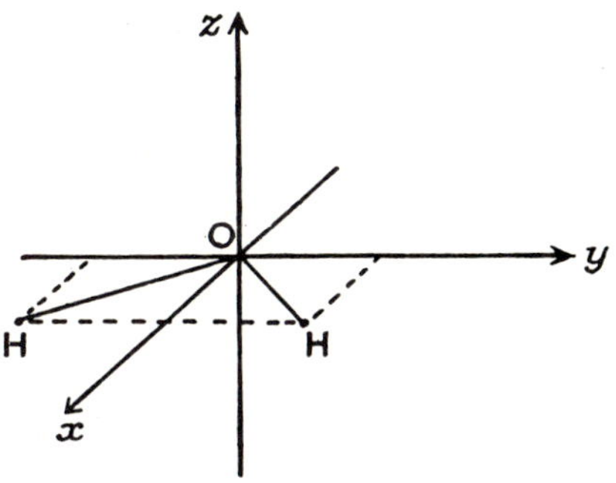

FIG. 4
Cartesian axes for the water molecule.

Then we can classify the molecular orbitals according to whether they are antisymmetric or not in the planes of symmetry. These are Oxy and Oxz. The molecular orbitals are summarised in the Table, together with LCAO forms.

TABLE. *Molecular orbitals for the water molecule.*

	Symmetry	Description	LCAO form
ψ_1	Totally symmetric .	Oxygen inner shell	$(1s)_O$
ψ_2	Totally symmetric .	Symmetric bonding orbital	Mixture of oxygen hybrid of $(2s)_O$ and $(2px)_O$ with $(1s)_H + (1s)_{H'}$
ψ_3	Antisymmetric in plane Oxz	Antisymmetric bonding orbital	Mixture of oxygen $(2py)_O$ with $(1s)_H - (1s)_{H'}$
ψ_4	Totally symmetric .	Symmetric lone pair	Hybrid of $(2s)_O$ and $(2px)_O$
ψ_5	Antisymmetric in plane Oxy	Antisymmetric lone pair	$(2pz)_O$

The localised equivalent orbitals are connected with these by the transformations

$$\begin{aligned} \chi_{\text{bonding}} &= (1/\sqrt{2})(\psi_2 \pm \psi_3) \\ \chi_{\text{lone pair}} &= (1/\sqrt{2})(\psi_4 \pm \psi_5) \end{aligned} \quad . \quad . \quad . \quad . \quad (29)$$

It is interesting that the molecular-orbital functions give an alternative

[5] Lennard-Jones and Pople, *Proc. Roy. Soc.*, 1951, *A*, **205**, 155.
[6] Bernal and Fowler, *J. Chem. Phys.*, 1933, **1**, 515.
[7] Pople, *Proc. Roy. Soc.*, 1951, *A*, **205**, 163.

description of the lone-pair electrons which still distinguishes them from bonding electrons. There are two distinct lone-pair molecular orbitals, one of which, ψ_4, is an s–p hybrid on the oxygen atom directed along the negative x-axis, that is, backwards along the line bisecting the two O–H bonds. The other (ψ_5) is antisymmetric in the HOH plane and approximates to an atomic $2p$-function. This is the orbital of lowest energy for the water molecule and corresponds to the lone-pair electron removed in the first ionisation process.

The structure of ammonia, the next molecule in the series, can be considered in a similar manner. If a further unit positive charge is removed from the nucleus in H_2O, the most favourable direction energetically will be towards one of the localised lone pairs. The ammonia molecule, therefore, will have a tetrahedral-like structure with three equivalent localised bonding orbitals and a hybrid lone-pair orbital in the fourth direction. The three bonding orbitals can be transformed into three delocalised orbitals, but here the lone pair is already symmetrical and approximates to a molecular orbital. It is interesting to consider the behaviour of the lone pair during the inversion of the molecule (this is known to occur with relatively high frequency). In the equilibrium configuration, the orbital is close to a tetrahedral s–p hybrid. As the molecule flattens, the amount of s character decreases until, in the intermediate planar configuration, the lone-pair orbital is a pure p function. After passing through this position, s character reappears, causing the lone pair to project in the opposite direction.

The final molecule of this series is methane, the tetrahedral structure of which follows if a fourth unit positive charge is removed from the nucleus in the ammonia lone-pair direction. There are now four equivalent bonding orbitals, which may be represented approximately as linear combinations of carbon s–p hybrid and hydrogen $1s$ functions. The transformation from molecular orbitals into equivalent orbitals or *vice versa* is exactly the same as for the neon atom.

Molecules with Multiple Bonds.—The double bond in a molecule such as ethylene provides a striking example of the transformation between equivalent and molecular orbitals.[8] The nuclear configuration of ethylene is known to be planar, so the molecular or symmetry orbitals can be divided into two classes according to whether they are symmetrical or antisymmetrical in this plane. By analogy with the classification for diatomic molecules, these are referred to as σ and π orbitals respectively. If we take the z direction to be normal to the plane, the LCAO forms of the σ molecular orbitals (apart from the carbon inner shells) will be constructed from the hydrogen $1s$ and carbon $2s$, $2px$, and $2py$ atomic orbitals. The only low-lying π atomic orbitals are $2pz$. Two types of transformation are possible. In the first place, the σ orbitals may be transformed among themselves, so that all orbitals will retain the property of symmetry or antisymmetry in the nuclear plane. The occupied σ molecular orbitals could be transformed in this way into a set of localised σ orbitals which correspond to bonds of single axially-symmetric type. There will be five

[8] Lennard-Jones and Hall, *Proc. Roy. Soc.*, 1951, *A*, **205,** 357.

in all, four local C–H bonds and one C–C bond whose LCAO form will be approximately

$$\psi_{\text{C-C }\sigma\text{-bond}} = (1/\sqrt{2})(\text{tr}_A + \text{tr}_B) \quad . \quad . \quad . \quad (30)$$

where tr_A and tr_B are trigonal s–p hybrids. The remaining two electrons will occupy the π-bonding orbital

$$\psi_{\text{C-C }\pi\text{-bond}} = (1/\sqrt{2})(2pz_A + 2pz_B) \quad . \quad . \quad . \quad (31)$$

which will be antisymmetric in the nuclear plane. These two orbitals

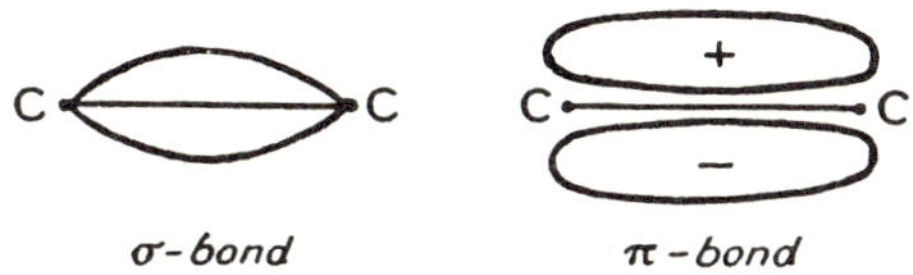

FIG. 5

σ and π Bonding orbitals in ethylene.

constitute the σ–π representation of the double bond (Fig. 5). If we now carry out a further transformation by writing

$$\begin{aligned} \chi_1 &= (1/\sqrt{2})(\psi_{\text{C-C }\sigma\text{-bond}} + \psi_{\text{C-C }\pi\text{-bond}}) \\ \chi_2 &= (1/\sqrt{2})(\psi_{\text{C-C }\sigma\text{-bond}} - \psi_{\text{C-C }\pi\text{-bond}}) \end{aligned} \quad . \quad . \quad . \quad (32)$$

we get two equivalent orbitals, concentrated one above and one below the plane. This description corresponds to two bent bonds (Fig. 6). Each carbon atom takes part in four bonds in directions which are approximately tetrahedral, two being bent round towards the other carbon atom.

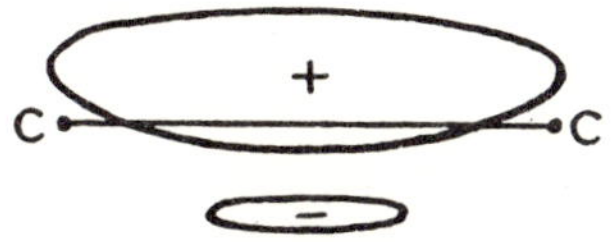

FIG. 6

Equivalent or bent bonding orbitals in ethylene.

Actually the HCH bond angle in ethylene is rather larger than the tetrahedral value. According to the equivalent-orbital picture, this can be attributed to the closing up of one pair of bonds leading to the opening of the other pair.

The carbon–oxygen double bond in aldehydes and ketones is similar and can be described in either of these two ways. If we adopt the localised-orbital description, formaldehyde will have two directed lone pairs in place of two of the C–H bonds in ethylene. In this case the axes of these hybrid orbitals will be in the molecular plane (unlike the oxygen lone pairs in water). Either the components of the double bond or the lone pairs can be transformed back into symmetry forms. The alternative description of the lone pairs would be one σ-type along the C–O direction and one π-type with axis perpendicular to the C–O bond but in the molecular plane. It is the latter orbital which has the highest energy, so that an electron is removed from it in ionisation or excitation to the lowest excited state.

The carbon–carbon triple bond in acetylene can be treated in a similar way to that in the nitrogen molecule.[8] The details of hybridisation may

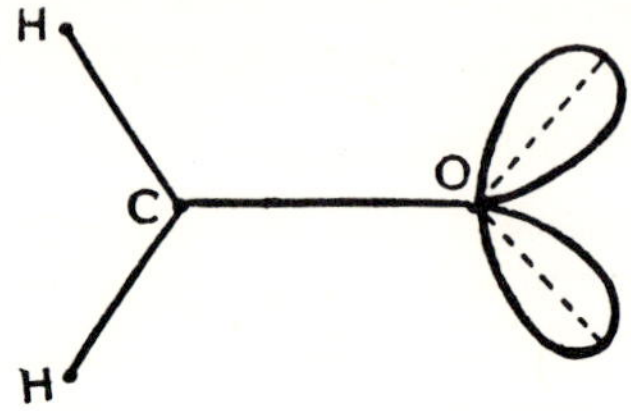

FIG. 7
Equivalent lone pairs in formaldehyde.

differ somewhat, but there will be a C–C σ bond and two perpendicular C–C π bonds. The alternative description is in terms of three equivalent bent bonds. The triple bond in hydrogen cyanide HC≡N is similar.

Resonance and Conjugation.—All the molecules described so far have been simple ones which can be described in terms of a single classical valence structure. For such systems we have seen how the molecular-orbital wave function can be expressed in terms of a set of localised equivalent bonding orbitals, each such orbital corresponding to a chemical bond or to a lone pair of electrons. In many more complex molecules, it is generally recognised that a single valence structure is insufficient and that the ground state should be represented as a mixture of several structures. This raises the question of what happens to the localised bonding orbitals when such mixing occurs.

This can be illustrated by the π electrons of buta-1 : 3-diene as an example (inset). This molecule is planar and its principal structure has two ethylenic-type double bonds. The correct equivalent-orbital description of this structure would be in terms of two localised bonding orbitals

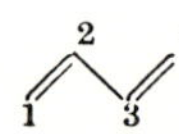

$$\begin{aligned}\chi_A &= (1/\sqrt{2})(\phi_1 + \phi_2)\\ \chi_B &= (1/\sqrt{2})(\phi_3 + \phi_4)\end{aligned} \quad . \quad . \quad . \quad . \quad (33)$$

where ϕ_1, ϕ_2, ϕ_3, and ϕ_4 are the $2p\pi$ atomic orbitals. The corresponding symmetrical orbitals are obtained by taking the sum and difference of these two :

$$\begin{aligned}\psi_{sym} &= \tfrac{1}{2}(\phi_1 + \phi_2 + \phi_3 + \phi_4)\\ \psi_{antisym} &= \tfrac{1}{2}(\phi_1 + \phi_2 - \phi_3 - \phi_4)\end{aligned} \quad . \quad . \quad . \quad (34)$$

Now actual calculations based on a simple model of a hydrocarbon such as this suggest that these molecular orbitals are better approximated by

$$\begin{aligned}\psi_{sym} &= 0{\cdot}3717\phi_1 + 0{\cdot}6015\phi_2 + 0{\cdot}6015\phi_3 + 0{\cdot}3717\phi_4\\ \psi_{antisym} &= 0{\cdot}6015\phi_1 + 0{\cdot}3717\phi_2 - 0{\cdot}3717\phi_3 - 0{\cdot}6015\phi_4\end{aligned} \quad . \quad . \quad (35)$$

The equivalent orbitals corresponding to these are obtained by applying the reverse transformation and are

$$\begin{aligned}\chi_A &= (1/\sqrt{2})(\psi_{sym} + \psi_{antisym}) = 0{\cdot}6882(\phi_1 + \phi_2) + 0{\cdot}1625(\phi_3 - \phi_4)\\ \chi_B &= (1/\sqrt{2})(\psi_{sym} - \psi_{antisym}) = -0{\cdot}1625(\phi_1 - \phi_2) + 0{\cdot}6882(\phi_3 + \phi_4)\end{aligned} \quad . \quad (36)$$

Thus it appears that the best equivalent orbitals in this molecule are not completely localised in the two double bonds but are to some extent distributed over the whole system. This failure to obtain localisation is the molecular-orbital analogue of resonance between valence structures.

Discussion

The main result that emerges from the discussions of particular cases is that it has proved possible to give a description of a molecule in terms of equivalent orbitals which are approximately localised, but which can be transformed into delocalised molecular orbitals without any change in the value of the total wave function. The equivalent orbitals are closely associated with the interpretation of a chemical bond in the theory, for, in a saturated molecule, the equivalent orbitals are mainly localised about two atoms, or correspond to lone-pair electrons. Double and triple bonds in molecules such as ethylene and acetylene are represented as bent single bonds, although the rather less localised σ–π description is equally valid.

Another property of these equivalent orbitals is that they include in themselves effects of delocalisation. Such effects are most important in conjugated molecules, although they are present in all molecules to a greater or lesser extent. In a highly conjugated system such as benzene only a limited amount of localisation can be achieved by transforming the orbitals.

For large molecules, the equivalent-orbital analysis is the most convenient starting point for a molecular-orbital treatment. In a molecule such as a long-chain paraffin it is possible to write approximate equivalent orbitals corresponding to each bond and then to apply a transformation to obtain the delocalised molecular orbitals. Simple assumption about the interaction of neighbouring bonds will then lead to estimates of the relative stability of the various energy levels.[9]

[9] Hall, *Proc. Roy. Soc.*, 1951, *A*, **205**, 541.

466. *The Electronic Orbitals, Shapes, and Spectra of Polyatomic Molecules. Part I.* AH_2 *Molecules.*

By A. D. Walsh.

The electronic orbitals possible for an AH_2 molecule when linear are correlated with those possible for the molecule when non-linear. Qualitative curves of binding energy *versus* $\angle$HAH are given for the various orbitals. These curves are then used to explain and predict (i) the shapes and (ii) the electronic spectra and associated characteristics of AH_2 molecules. AH_2 molecules containing 4 valency electrons should be linear in their ground states. AH_2 molecules containing 5—8 valency electrons should be bent in their ground states.

The purpose of this paper is first to correlate the electronic orbitals of a bent with a linear AH_2 molecule, next to decide whether a given orbital becomes more or less tightly bound with increase of the angle HAH, and then to use the resulting graphs of binding energy *versus* apex angle to explain and predict the shapes and spectra of AH_2 molecules.

Linear AH_2 *Molecules*.—The lowest-energy intra-valency-shell orbitals of a linear AH_2 molecule, may, on the assumption that these are built solely from s and p atomic orbitals, be described as follows: (i) Two orbitals binding the hydrogen atoms to the central atom. These may be thought of as each formed by the overlap of an sp hybrid atomic orbital on A with the 1s orbital of hydrogen. If so regarded they will be orbitals of σ-type predominantly localized one in each A–H distance. If, however, one were discussing a transition involving excitation of an electron from one of the bonding orbitals, one could not regard the electron as coming from one A–H link or the other—the two are indistinguishable. If one first conceives of the orbitals as completely localized one has to take combinations of them in order to express this indistinguishability. These combinations are either in-phase or out-of-phase and may be labelled σ_g and σ_u respectively. (ii) A π_u orbital. This is simply a p orbital localized on the central atom and pointing in a direction at 90° to the HAH line. It is non-bonding. Since there are two such directions that are independent but equivalent (except for a rotation by 90°), the orbital is two-fold degenerate. If the apex angle were changed from 180°, the degeneracy must become split. The order of decreasing binding energy of these orbitals is σ_g, σ_u, π_u.

Bent AH_2 *Molecules.*—A bent AH_2 molecule belongs to the symmetry class C_{2v}. The definitions of the symbols appropriate to the non-localized orbitals of such a molecule are given below. The z axis bisects the HAH angle and lies in the molecular plane. The y axis also lies in the molecular plane and is parallel with the H · · · H line. $C_2(z)$ means a rotation by 180° about the z axis. $\sigma_v(y)$ means a reflection in the xz plane. + and − mean respectively that the wave function does not or does change sign when one of the symmetry operations $C_2(z)$ or $\sigma_v(y)$ is carried out.

	$C_2(z)$	$\sigma_v(y)$
a_1	+	+
a_2	+	−
b_1 *	−	+
b_2 *	−	−

* Certain authors reverse the definitions of b_1 and b_2.

The lowest-energy intra-valency-shell orbitals of an AH_2 molecule whose apex angle is 90° are then as follows: (i) Two orbitals binding the hydrogen atoms to the central atom. These may be thought of, in the first place, as each formed by the overlap of a pure p atomic orbital on A with the 1s orbital of H. This would be to think of the orbitals as predominantly localized, one in each A–H distance. As with the bonding orbitals of the linear molecule, however, for discussions of spectroscopic transitions involving these orbitals they must be regarded as non-localized. The equivalent non-localized orbitals are simply the in-phase or out-of-phase combinations of the two localized orbitals, and, being no longer localized, may, by the above table of definitions, be labelled a_1 and b_2 respectively. (ii) A p orbital on atom A pointing in the x direction. The foregoing Table shows that the molecular symbol to be applied to this orbital is b_1. (iii) An s orbital on atom A. It is non-bonding. The Table shows that the molecular symbol to be applied to this orbital is a_1.

Correlation of the Orbitals for Bent and Linear Molecules.—Clearly, if the HAH angle is gradually increased, the a_1 and the b_2 bonding orbitals must eventually become the σ_g and the σ_u orbitals respectively of the linear molecule. At the same time the binding energy of the orbitals must increase. The reasons for this may be seen by thinking of the orbitals in either their localized or their non-localized forms. For the localized forms, the reasons are: (1) the orbitals are built solely from p atomic orbitals of A in the 90° molecule but partly from an s atomic orbital of A in the linear molecule; (2) an sp hybrid valency gives rise to a stronger bond than does a pure p valency (see Walsh, *Discuss. Faraday Soc.*, 1947, **2**, 18).

When considering the non-localized forms, it is convenient to form first the two possible group orbitals of the H_2 group, *viz.*, 1s + 1s and 1s − 1s. To do this automatically takes account of the symmetry of the molecule. The components from which the non-localized molecular orbitals are built may then be written as follows:

Bent molecule		Linear molecule	
1s + 1s, a_1	p_z, a_1	1s + 1s, σ_g	$p_z p_x$, π_u
1s − 1s, b_2	p_y, b_2	1s − 1s, σ_u	p_y, σ_u
s_A, a_1	p_x, b_1	s_A, σ_g	

Components of the same species may then be " mixed " to form the actual molecular orbitals. This procedure leads to the same end as taking combinations of the localized orbitals. We now introduce three principles: (i) in the 90° molecule the s_A orbital does not mix (" hybridize ") with the other orbitals *; (ii) whether or not an orbital becomes more tightly bound with change of angle is determined primarily by whether or not it changes from being built from a p orbital of A to being built from an s orbital of A; and (iii) if no change of A valencies from which the orbital is built occurs when the angle is changed, the following subsidiary effect determines whether the orbital becomes more

* Introduced to make the description of the non-localized bond orbitals consistent with that of the localized ones.

or less tightly bound: if the orbital is anti-bonding between the end atoms it is most tightly bound when the latter are as far apart as possible (*i.e.*, in the linear molecule); if it is bonding between the end atoms it is most tightly bound when the latter are as near together as possible (*i.e.*, in the 90° molecule). The a_1–σ_g bond orbital is thus built from a combination of $1s + 1s$ and p_z in the 90° molecule, from a combination of $1s + 1s$ and s_A in the linear molecule, and from a combination of $1s + 1s$, s_A, and p_z in the intermediate molecule. It follows that the orbital is more tightly bound in the linear than in the bent molecule. The b_2–σ_u orbital is built from a combination of the $1s - 1s$ and p_y components in both the bent and the linear molecule; but from principle (iii), it is more tightly bound in the linear molecule since it is anti-bonding between the end atoms.

Clearly, as the HAH angle is increased, the b_1 orbital of the bent molecule must become one of the two π_u orbitals of the linear molecule. To a first approximation, the orbital is the same when the apex angle is 90° as when it is 180°. We may therefore suppose its binding energy to remain approximately constant as the angle changes.

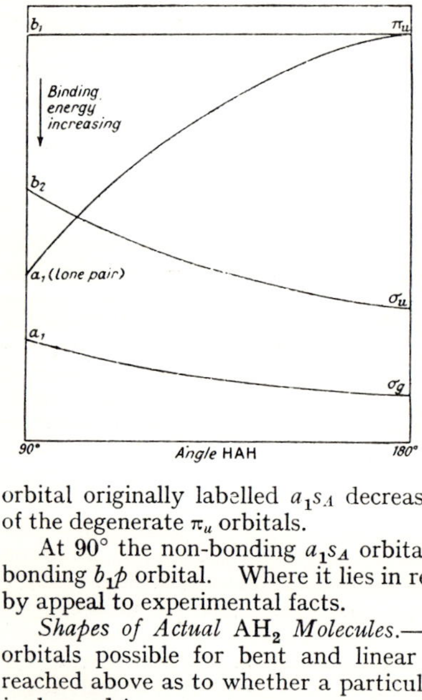

Increase of the apex angle from 90° implies mixing of the a_1s_A orbital with the a_1p_z orbital on A. The a_1 bond orbital, instead of being built from a pure p valency of A, becomes more and more built also from the s orbital of A; while at the same time the non-bonding orbital, instead of being built solely from the pure s orbital of A, becomes more and more built from a p_z orbital on A. In the linear molecule, the bonding orbital has become built from an orbital of A that is solely s, while the non-bonding orbital has become built solely from a p_z orbital of A. In other words, as the HAH angle increases from 90° to 180°, the bond orbital increases in binding energy, while the non-bonding orbital originally labelled a_1s_A decreases in binding energy until at 180° it becomes one of the degenerate π_u orbitals.

At 90° the non-bonding a_1s_A orbital must be much more tightly bound than the non-bonding b_1p orbital. Where it lies in relation to the b_2 and a_1 bond orbitals is best decided by appeal to experimental facts.

Shapes of Actual AH_2 *Molecules.*—The Figure is a correlation diagram between the orbitals possible for bent and linear AH_2 molecules. It incorporates the conclusions reached above as to whether a particular orbital rises or falls in energy as the HAH angle is changed.*

The Figure does not include *all* the possible intra-valency-shell orbitals, but only the lowest-lying ones. Two others are possible and are referred to below. There is no particular reason to make the Figure include all and only all the intra-valency-shell orbitals; for an infinite number of extra-valency-shell orbitals is also possible, the lowest of which (see the discussion below of the spectrum of water vapour) will be comparable in energy with the highest intra-valency-shell orbitals. It is just as arbitrary to make the Figure show six orbitals as to make it show four; and, indeed, for the purpose of interpreting observed spectra it is less desirable since it is the highest intra-valency-shell orbitals that are most likely to lie above or mix with the lowest extra-valency-shell orbitals. There is no doubt about the four lowest-lying orbitals—they are intra-valency-shell in type; but the fifth and sixth orbitals in order of energy are probably not both simple intra-valency-shell orbitals. A similar point should be borne in mind in reading all the

* *Added in Proof.*—The drop from left to right of the a_1–σ_g orbital curve, due to principle (ii), is offset by a smaller rise due to principle (iii). The rise from left to right of the a_1–π_u orbital curve is not offset. The Figure has therefore been drawn with the a_1–σ_g and b_2–σ_u curves falling from left to right by similar amounts; and both falling considerably less than the a_1–π_u curve rises.

papers of this series, where the correlation diagrams show what are thought to be the lowest energy orbitals and do not necessarily include *all* the intra-valency shell orbitals.

Each of the curves must be a maximum or a minimum on the 180° line, since from 180° to 270° the curves must repeat their behaviour from 180° to 90°. The two most tightly bound orbitals have a minimum on the 180° line. One therefore expects that all AH_2 molecules containing only four valency electrons will be linear in their ground states. As far as is known, this is true. As examples, the BeH_2 and HgH_2 molecules are expected to be linear in their ground states. On the other hand, the ground states of AH_2 molecules containing 5, 6, 7, or 8 valency electrons are expected to be bent because at least one electron has to be placed in the a_1 non-bonding orbital. Similarly, the first excited state of HgH_2 or BeH_2 should be bent. This a_1 orbital rises steeply from left to right in the Figure because, as already explained, it changes from a pure *s* orbital to a pure *p* orbital. The ground state of the H_2O molecule, with eight valency electrons, has therefore a bent nuclear configuration. The CH_2 molecule, with six valency electrons, is similarly expected to be bent. The actual value of the apex angle cannot be predicted, while the Figure remains merely qualitative. If the angle is close to 180°, the ground state of the molecule may be a triplet, the configuration

$$(a_1)^2(b_2)^2(a_1)(b_1),{}^3B_1 \quad . \; . \; . \; . \; . \; . \; . \; . \; . \quad (1)$$

lying lower in energy than

$$(a_1)^2(b_2)^2(a_1)^2,{}^1A_1 \quad . \; . \; . \; . \; . \; . \; . \; . \; . \quad (2)$$

because the existence of electron repulsion offsets the small energy difference of the (a_1) and (b_1) orbitals. If, on the other hand, the apex angle is considerably less than 180°, the ground state will be a singlet, having the configuration (2). By analogy with H_2O and in view of the fact that the two most weakly bound electrons of the H_2O molecule lie in the b_1 orbital and should therefore have comparatively little effect on the apex angle, the second possibility seems the more likely.

The small decrease that has been observed in the apex angle in the series H_2O, H_2S, H_2Se, H_2Te may be due either to the a_1–π_u curve rising even more steeply or to the bonding orbital curves falling less steeply as one passes from H_2O to H_2Te. However, the total range of the decrease is only 14°.

Spectra of AH_2 *Molecules*.—(i) *Spectrum of* H_2O. In the ground state of the H_2O molecule all the orbitals represented in the Figure are fully occupied. These are the only low-lying orbitals of the molecule. No electronic spectrum of H_2O is therefore expected until comparatively short wave-lengths are reached. In agreement, the first absorption of the molecule occurs as a continuum between 1830 and 1500 Å ($\lambda_{max.}$ *ca.* 1675 Å; Hopfield, *Phys. Review*, 1950, **77**, 560).

In the Figure the b_2 and the upper a_1 curve have been drawn to cross at an angle of *ca.* 110°. The reason for this is as follows. Three ionization potentials of H_2O are known (Price and Sugden, *Trans. Faraday Soc.*, 1948, **44**, 108, 116), *viz.*, 12·61 v, 14·5 ± 0·3 v, 16·2 ± 0·3 v. A good Rydberg series is known leading to the first limit (Price, *J. Chem. Phys.*, 1936, **4**, 147). A cruder Rydberg series is also known leading to the third limit (Henning, *Ann. Physik*, 1924, **13**, 599). Rydberg series are only likely for the excitation of lone-pair electrons (or, sometimes, of weakly bonding or anti-bonding electrons); excitation of an electron from a strongly bonding orbital is hardly likely to give a series of discrete transitions. For this reason it seems best to interpret the first and third limits as due respectively to ionization from the (b_1) and (a_1) lone-pair orbitals; leaving the second limit to represent ionization of the (b_2) bonding electrons. The ground state configuration of H_2O is therefore written as $(a_1)^2(a_1)^2(b_2)^2(b_1)^2$ instead of $(a_1)^2(b_2)^2(a_1)^2(b_1)^2$. The main conclusions of this paper would not, however, be altered if the order of the (a_1) and (b_2) orbitals were reversed.*

Two intra-valency-shell orbitals other than those shown in the Figure are possible. These are highly anti-bonding in nature, being obtained by considering out-of-phase overlap of an oxygen valency with the hydrogen 1*s* valency. The lower of the two is of

* Price and Sugden (*loc. cit.*) assume that (b_2) is more tightly bound than (a_1).

species $\bar{a}_1$–σ_g, where the bar indicates the anti-bonding nature. For spectroscopic purposes it must be considered as non-localized and therefore built by the in-phase overlap of the two hydrogen 1*s* atomic orbitals, these interacting out-of-phase with an oxygen valency that is 2*s* in the linear molecule and $2p_z$ in the 90° molecule. Because of its anti-bonding nature, it may well lie so high that transition to it requires as much energy as transition to the lowest extra-valency-shell (Rydberg) orbital. In other words, the longest wave-length allowed transition of H_2O may be formulated

$$(a_1)^2(a_1)^2(b_2)^2(b_1)(a_1),{}^1B_1 \longleftarrow (a_1)^2(a_1)^2(b_2)^2(b_1)^2,{}^1A_1$$

where the uppermost (a_1) orbital refers either to the anti-bonding intra-valency-shell orbital or to the 3*s* orbital of the oxygen atom. The 1830—1500-Å absorption has already been interpreted as due to transition of an electron from (b_1) to $(3sa_1)$ (Mulliken, *J. Chem. Phys.*, 1935, **3**, 506; Price, Teegan, and Walsh, *Proc. Roy. Soc.*, 1950, *A*, **201**, 600). Its continuous nature, however, makes it best considered as involving also transition to a repulsive upper state containing an electron in the anti-bonding $(\bar{a}_1)$ orbital (cf. Walsh, unpublished work).

According to Wilkinson and Johnston (*J. Chem. Phys.*, 1950, **18**, 190) the absorption actually consists of three diffuse bands superimposed on a continuum. Rathenau (*Z. Physik*, 1934, **87**, 32) had earlier found structure in the continuum, stating that a frequency difference ~1300 cm.$^{-1}$ was present. The diffuse bands found by Wilkinson and Johnston are at 1608, 1648, and 1718 Å. It is possible that they represent the Rydberg transition while the continuum represents the intra-valency-shell transition. The maximum molecular extinction coefficient is about 1000.* It seems therefore that the transition(s) must be regarded as allowed, in agreement with the present assignments. Because of this, one expects either or both of the two totally symmetrical vibrational frequencies of the upper state to appear in the spectrum. It is possible that the separation of the 1608 and the 1648 Å band (1520 cm.$^{-1}$) represents the bending vibration ν_2 (1595 cm.$^{-1}$ in the ground state), and that the separation of the 1608 and the 1718 Å band (3388 cm.$^{-1}$) represents the stretching vibration ν_1 (3655 cm.$^{-1}$ in the ground state). The smallness of the reductions in these frequencies from the ground-state values would accord with only one quantum of each vibration appearing. It seems that comparatively little change of the molecular dimensions occurs on excitation, in agreement with the transition being of an electron from one non-bonding orbital (whose binding energy changes little with change of HOH angle) to another.

(ii) *Spectrum of* NH_2. In the ground state of the NH_2 radical only one electron lies in the (b_1) orbital. A long-wave-length transition

$$\cdots(a_1)(b_1)^2,{}^2A_1 \longleftrightarrow \cdots(a_1)^2(b_1),{}^2B_1 \quad . \quad . \quad . \quad . \quad . \quad . \quad (3)$$

should therefore be possible. It represents an allowed transition. The further, long-wave-length, transition

$$\cdots(b_2)(b_1)^2,{}^2B_2 \longleftrightarrow \cdots(b_2)^2(b_1),{}^2B_1$$

is forbidden. The allowed transition should be polarized in the *x* direction, *i.e.*, perpendicularly to the molecular plane. Since the transition involves transfer of an electron from the orbital represented in the Figure by the steep a_1–π_u curve to the orbital represented by the approximately horizontal b_1–π_u line, it should result in an increase of the apex angle in the equilibrium form of the upper state. On the other hand, since the (a_1) and the (b_1) orbital are non-bonding, there should be little change of N–H length.

The so-called " ammonia α " band, lying in the visible region, has long been attributed (though without proof) to the NH_2 radical. It was first observed by Eder (*Denkschr.*

* Wilkinson and Johnston record intensities in terms of the atmospheric absorption coefficient at 30° c. Their values have been multiplied by (22·4 × 303)/(2·30 × 273) in order to obtain molecular extinction coefficients. The present discussion assumes that the bands observed by them in the continuum are not due to impurities. There is some doubt of this, however, since Rathenau's findings are not entirely in agreement with those of Wilkinson and Johnston.

Wien. Akad., 1893, **60**, 1) in the ammonia–oxygen flame. It was later found to occur (more weakly) in the spectrum of an uncondensed discharge through ammonia (Rimmer, *Proc. Roy. Soc.*, 1923, *A*, **103**, 696). The latter method of obtaining it supports the identification of the emitter as a decomposition product of NH_3. The band also occurs in the spectra of the hydrogen–nitrous oxide (Fowler and Badami, *ibid.*, 1931, *A*, **133**, 325; Gaydon, *ibid.*, 1942, *A*, **181**, 197) and the methane–nitrous oxide flames (Gaydon, *loc. cit.*). Infra-red bands observed in the spectrum of the former flame have been provisionally assigned to an extension of the band (Gaydon, *loc. cit.*). Rimmer's high-resolution photographs showed the great complexity of the band and the many lines (some 3000) present. The fine structure appears much too complex for the emitter to be the diatomic radical NH. Its attribution to NH_2 therefore seems very plausible. If the attribution is correct, then the interpretation of the band can hardly be other than as the emission transition (3). Analysis of the observed rotational structure should proceed in the light of the expectations formulated above.* That the structure should be so complex is not surprising (i) if the HNH angle is such that the molecule is not even an approximate symmetric top † and (ii) if, as expected here, the apex angle changes markedly in the transition.

(iii) *Spectrum of* CH_2. If the ground state of the molecule is a singlet, then the following is the longest-wave-length absorption transition :

$$K^2(a_1)^2(b_2)^2(a_1)(b_1),{}^1B_1 \longleftarrow K^2(a_1)^2(b_2)^2(a_1)^2,{}^1A_1$$

Mulliken (quoted by Venkateswarlu, *Phys. Review*, 1950, **77**, 676) has earlier made the same assignment. The transition is allowed, with polarization perpendicular to the main symmetry axis. It should probably occur at quite long wave-lengths (cf. the supposed NH_2 transition above). It would result in an increase of apex angle in the equilibrium form of the upper state. On the other hand, there should be little change of C–H length. The Figure thus enables very specific statements to be made about the expected spectrum of CH_2, which should be helpful in the search currently being undertaken in various laboratories for that spectrum.

It is worth noting what the expectations would be if the ground state of the CH_2 molecule should turn out to be a triplet. This implies that the molecule is nearly linear in the ground state and we shall therefore use the σ, π, $\cdots$ nomenclature for its orbitals. Consider only transitions with no change of multiplicity : the lowest energy transition, *viz.*,

$$(\sigma_g)^2(\sigma_u)(\pi_u)^2(\pi_u),{}^3\Pi_g \longleftarrow (\sigma_g)^2(\sigma_u)^2(\pi_u)(\pi_u),{}^3\Sigma_g^-$$

is forbidden. There will be no allowed transitions until comparatively short wave-lengths (probably, by analogy with H_2O, not until wave-lengths <2500 Å). The longest wave-length allowed transition will be ${}^3\Pi_u \longleftarrow {}^3\Sigma_g^-$ whose interpretation may be

$$(\sigma_g)(\sigma_u)^2(\pi_u)^2(\pi_u) \longleftarrow (\sigma_g)^2(\sigma_u)^2(\pi_u)(\pi_u) \quad . \; . \; . \; . \; . \; (4)$$

or

$$(\sigma_g)^2(\sigma_u)^2(\pi_u)(\sigma_g') \longleftarrow (\sigma_g)^2(\sigma_u')^2(\pi_u)(\pi_u)$$

where (σ_g') stands for either the 3*s* carbon atom orbital or the anti-bonding orbital referred to above in the discussion of the H_2O spectrum.

If the CH_2 molecule is found to have a strong transition at wave-lengths considerably longer than 2500 Å, it should be safe to conclude that (*a*) the apex angle in the ground state is considerably less than 180° and (*b*) the ground state is a singlet.

* Since this paper was written, Dr. D. A. Ramsay has informed me that Professor Herzberg and he have now (*a*) proved that the ammonia α bands are due to NH_2, (*b*) obtained the system in absorption between 4500 and 7400 Å, and (*c*) shown, in agreement with the present expectations, that the upper and the lower state differ considerably in the geometrical arrangement of the nuclei (see Herzberg and Ramsay, *J. Chem. Phys.*, 1952, **20**, 347; *Discuss. Faraday Soc.*, 1953, **14**, 11). Dr. Ramsay has also pointed out to me that the same assignment of the bands as made here has been given earlier by Mulliken and quoted as a personal communication in a paper by Swings, McKellar, and Minkowski (*Astrophys. J.*, 1943, **98**, 142).

† According to the Figure the apex angle in the ground state should be close to that of the H_2O molecule.

Relation to Other Work.—In the following papers we shall show that, as with AH_2 molecules, the shape of a molecule in its ground state depends primarily on the number of valency electrons. In general terms, the recognition of this is of course not new (see, *e.g.*, Cassie, *Nature*, 1933, **131**, 438; Penney and Sutherland, *Proc. Roy. Soc.*, 1936, *A*, **156**, 654), though the newer data now available frequently enable the older generalizations to be extended and made more precise. There seems, however, to have been only one previous attempt to plot a correlation diagram between the orbitals possible for a polyatomic molecule in each of two nuclear configurations, and to use such a diagram to discuss both the shapes (in excited as well as ground states) and electronic spectra of molecules. This was by Mulliken (*Rev. Mod. Phys.*, 1942, **14**, 204) who plotted a correlation diagram for the limited case of AB_2 molecules. He did not specifically apply his diagram to AH_2 molecules. He also said that he was unable to give any simple explanation of why particular curves should rise or fall with increase of angle. His diagram was either empirical or based upon unpublished computations.

UNIVERSITY OF LEEDS. [*Received, May 15th*, 1952.]

---

467. *The Electronic Orbitals, Shapes, and Spectra of Polyatomic Molecules. Part II.* Non-hydride* AB_2 *and* BAC *Molecules.*

By A. D. WALSH.

A correlation diagram is plotted for the orbitals of linear and non-linear triatomic molecules. Simple reasons are given why particular orbitals become more or less tightly bound as the apex angle changes. The diagram is used to interpret and predict the shapes, reactivities, and spectra of non-hydride triatomic molecules. As regards the shapes, the following facts (most of which have been recognised for some time) become understandable: that molecules with not more than 16 valency electrons are linear in their ground states; that molecules with 17, 18, 19, or 20 valency electrons are bent in their ground states, the apex angle decreasing markedly from 16- to 17- and from 17- to 18-electron molecules and less markedly from 18- to 19- and from 19- to 20-electron molecules; and that 22-electron molecules are linear or nearly linear in their ground states. The paper has much in common with an earlier paper by Mulliken.

THE purpose of this paper is to apply to AB_2 and BAC molecules the procedure which in Part I * was applied to AH_2 molecules. We consider first symmetrical, non-hydride, triatomic molecules AB_2.

AB_2 *Molecules.*—The lowest-energy orbitals of a linear AB_2 molecule may, on the assumption that they are built solely from *s* and *p* atomic orbitals, be described as follows: (i) Two lone-pair *s* orbitals, one on each B atom. (ii) Two bond orbitals analogous to the bond orbitals discussed in Part I. For discussion of molecular shape, they may be thought of as built from *sp* hybrid valencies of A plus $p\sigma$ valencies of the two B atoms and largely localized one to each A–B distance. In discussion of spectroscopic transitions involving them, they would have to be regarded as non-localized combinations of the localized orbitals and be labelled σ_g and σ_u. (For simplicity, hybridization of the *s* and the *p* valencies of the B atoms is here neglected. It is unlikely that its inclusion would alter the general form of the correlation diagram plotted in the Figure. A similar comment applies to later papers of this series.) (iii) A π_u orbital built by the in-phase overlap of a $p\pi$ atomic orbital on each of the three atoms. It will be bonding and two-fold degenerate. (iv) A π_g orbital built by the out-of-phase overlap of a $p\pi$ atomic orbital on each of the B atoms. It has zero amplitude at the central atom, being localized entirely on the B atoms. It is two-fold degenerate, weakly B$\leftarrow\rightarrow$B anti-bonding and A$\leftarrow\rightarrow$B non-bonding. (v) A $\bar{\pi}_u$ orbital built by the in-phase overlap of a $p\pi$ atomic orbital on each of the B atoms,

* Part I, preceding paper.

these overlapping out-of-phase with a $p\pi$ orbital on the central atom. It is two-fold degenerate, B<-->B bonding and A<-->B anti-bonding. A bar is placed over the symbol to indicate the A<-->B anti-bonding nature.

The π_u orbitals are both built from atomic orbitals on each of the three atoms. Nevertheless they will be partly localized in a way that may be deduced as follows. The sharing of an electron between two atomic orbitals is only likely when those two atomic orbitals have equal, or almost equal, binding energy. If the two atomic orbitals differ widely in binding energy no sharing will take place. It follows that if the $p\pi$ atomic orbitals of A are *much* less tightly bound than those of B the possible π orbitals in the molecule AB_2 are as follows: two orbitals (each two-fold degenerate) practically entirely localized on the B atoms and one higher-energy orbital (also two-fold degenerate) practically entirely localized on the A atom. In most actual AB_2 molecules (*e.g.*, CO_2, SO_2, OF_2, NO_2), B has considerably greater electronegativity (*i.e.*, has considerably more tightly bound valency orbitals) than A. While therefore it would be too extreme to speak of complete localization of any of the π orbitals, it follows that the three two-fold degenerate π-orbitals fall into two groups: (*a*) the two lower-energy ones (π_u and π_g) which should be more localized on the B than on A atoms, and (*b*) the higher-energy $\bar{\pi}_u$ orbital which should be more localized on A than on the B atoms. In the case of the π_g orbital, this argument merely reinforces what was already evident from the nature of the orbital [see (iv) above]. In the non-linear molecule the degeneracy of each of the orbitals is split. The π_u orbitals each yield an a_1' and a b_1'' orbital while the π_g orbital yields an a_2'' and a b_2' orbital. The primes are added for clarity to the usual group-theory symbols. A single prime means that the orbital is symmetric with respect to reflection in the plane of the molecule, a double prime that it is anti-symmetric. (vi) A $\bar{\sigma}_g$ orbital which is similar to the $\bar{\sigma}_g$ orbital described in Part I in the discussion of the spectrum of H_2O. It is built by the in-phase overlap of a $p\sigma$ valency on each of the B atoms, these overlapping out-of-phase with an *s* atomic orbital on the central atom. It is A<-->B anti-bonding, though B<-->B bonding.

In AH_2 molecules, there is general agreement that the π_u lies below the $\bar{\sigma}_g$ orbital. This is likely because the former is non-bonding whereas the latter is A<-->H anti-bonding. In non-hydride, AB_2 molecules, however, the orbital corresponding to π_u of AH_2 is $\bar{\pi}_u$, *i.e.*, an orbital that is A<-->B anti-bonding. A major reason for the orbital's lying lower than $\bar{\sigma}_g$ is therefore removed. We shall see in discussing the spectra of AB_2 molecules, however, that there is no compelling evidence for $\bar{\sigma}_g$ not still lying above $\bar{\pi}_u$.

By arguments similar to those given above, the $\bar{\sigma}_g$ orbital is more localized on the A than on the B atom.

The lowest-energy orbitals of a 90° AB_2 molecule may then be described as follows.

(i) Two lone-pair *s* orbitals, one on each B atom. It is assumed that these play little part in the binding of the B to the A atoms and that, being rather more tightly bound than any of the other valency orbitals of B or than any of the valency orbitals of A, they vary little in energy as the apex angle changes from 90° to 180°. They are therefore represented by horizontal straight lines on the correlation diagram (see Figure).

(ii) Two bond orbitals. Considered as localized orbitals, they are built from pure *p* valencies of the A atoms overlapping with *p* valencies of the B atoms. Considered as non-localized orbitals, their symbols are a_1' and b_2'. As described in Part I they both become more tightly bound as the apex angle changes from 90° to 180°.

(iii) An $a_1 s_A$ orbital. This, in first approximation, is a pure *s* orbital localized on atom A. As the apex angle changes from 90° to 180° this must, as described in Part I, tend to go over into a $p\pi$ orbital localized on atom A. It must therefore tend to correlate with the upper rather than with the lower π_u orbital. The localization on atom A was complete for AH_2 molecules. For AB_2 molecules the orbital will be largely, but not completely, localized on A. The curve representing the orbital rises steeply from left to right on the correlation diagram (Figure). The steep rise may be accentuated because the orbital becomes markedly A<-->B anti-bonding in the linear molecule. The Figure thus incorporates largely unchanged three of the curves of the Figure in Part I.

(iv) Three orbitals anti-symmetrical with respect to reflection in the plane of the molecule. The first of these (b_1'') is built from a p atomic orbital on each of the three atoms overlapping in-phase. It is A←→B and (weakly) B←→B bonding, and becomes one of the lower π_u orbitals in the linear molecule. The A atomic orbital concerned in it is pure p in both bent and linear molecules. According to the principles enunciated in Part I any change of binding energy as the apex angle is changed is therefore governed by the fact that it is weakly B←→B bonding. This means it will be rather more stable (lead to more evolution of energy) when the two B atoms are as close together as possible (*i.e.*, in the 90° molecule). The curve representing the orbital in the Figure has therefore been drawn to rise slightly from left to right. The second of these orbitals (a_2'') is built from a p atomic orbital on each of the B atoms overlapping out-of-phase. It is A←→B non-bonding and B←→B anti-bonding, and becomes one of the π_g orbitals in the linear molecule. Because it is B←→B anti-bonding it will be more stable when the two B atoms are as far apart as possible (*i.e.*, in the 180° molecule). The curve representing the orbital in the Figure has therefore been drawn to decrease from left to right. The third orbital (b_1'') is built from a p atomic orbital on A overlapping out-of-phase with each of two p orbitals, one on each of the B atoms. In a 60° B_3 molecule this b_1'' orbital would be degenerate with the a_2'' orbital. It is the analogue of the b_1 orbital shown in the Figure of Part I. Because of the presence of low-lying p atomic orbitals on the B atoms it is now, however, A←→B anti-bonding and B←→B bonding, instead of non-bonding and localized solely on A. Because it is B←→B bonding it will be most stable when the two B atoms are as close together as possible. It is represented in the Figure therefore, not by an horizontal line (as in Part I), but by a line that rises from left to right.

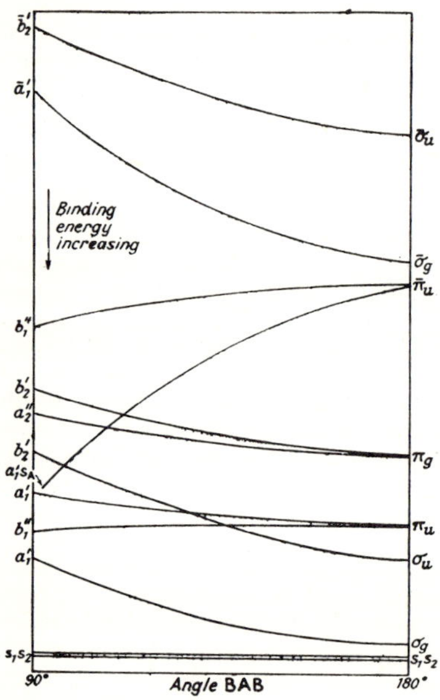

(v) Two orbitals symmetrical with respect to reflection in the plane of the molecule and primarily built from two p atomic orbitals lying one on each B atom with axes in the molecular plane and at right angles to the direction of the adjacent B–A line. That which involves in-phase overlap of these two orbitals is of species a_1'. That which involves out-of-phase overlap is of species b_2'. The a_1' orbital becomes one of the lower π_u orbitals in the linear molecule. It is B←→B bonding, which supplies a reason why it should decrease in binding energy with increase of apex angle. On the other hand, because the s orbital on A is expected to interact little with the p orbitals on the B atoms from which a_1' is built, in the 90° molecule the a_1' orbital largely loses the A←→B bonding character which it has in the linear molecule. This supplies a reason why the a_1' orbital should become more tightly bound as the apex angle increases. The a_1' orbital has therefore been inserted in the Figure as showing comparatively little dependence on angle.* The b_2'–π_g orbital is B←→B anti-bonding and is therefore represented in the Figure by a line that falls from left to right.

(vi) An $\bar{a}_1'$ orbital that correlates with the $\bar{\sigma}_g$ orbital of the linear molecule. This

* As a net effect the a_1' orbital curve probably falls from left to right in the Figure for the following reason. In a 60° B_3 molecule the a_1' orbital leading to the upper π_u should become degenerate with the b_2' orbital leading to π_g. This might happen if the a_1' orbital leading to the lower π_u curls upwards from right to left. For if orbital curves of the same species cannot cross, the a_1' orbital curve leading to the upper π_u may contain a minimum to the left of which it curves steeply upward to meet the b_2' orbital. It should be emphasized, however, that it is the part of the diagram from 90° to 180° that concerns real molecules.

orbital consists largely of an A atomic orbital. According to the principles given in Part I, it is built from a p_z orbital of A in the 90° molecule but from an *s* orbital of A in the 180° molecule; and therefore is represented in the Figure by a line that falls steeply from left to right.

If it should happen that the $\bar{\sigma}_g$ orbital lies *below* $\bar{\pi}_u$, then the curve from the $a_1's_A$ orbital will lead, not to the $\bar{\pi}_u$, but to the $\bar{\sigma}_g$ orbital. Orbital curves of the same species cannot cross.* In that case, the steep rise over most of its course of the curve from the $a_1's_A$ orbital could still be regarded as due to the reasons already described, *i.e.*, to its "trying" to reach $\bar{\pi}_u$; but an "avoided crossing" would have to be drawn, with the result that the curve from the $\bar{a}_1'$ orbital which "tended" to proceed to $\bar{\sigma}_g$ would be drawn in fact to proceed to the $\bar{\pi}_u$ orbital. However, we shall see below that the balance of evidence is not unfavourable to $\bar{\sigma}_g$ lying above $\bar{\pi}_u$, *i.e*, no such "avoided crossing" is necessary.

(vii) A total of twelve intra-valency-shell orbitals can be built from one *s* and three *p* atomic orbitals on each of the three atoms of AB_2. For completeness therefore a twelfth orbital curve has been added to the Figure. It represents a $\bar{b}_2'$–$\bar{\sigma}_u$ orbital which is built by the out-of-phase overlap of valencies on the B atoms, these themselves overlapping out-of-phase with a *p* valency of A. It is A$\leftarrow\!\!\rightarrow$B and B$\leftarrow\!\!\rightarrow$B anti-bonding. Because it involves no change of A valency as the apex angle is changed and is B$\leftarrow\!\!\rightarrow$B anti-bonding, it is represented in the Figure by a curve that falls from left to right. It may well lie so high in energy, however, that it is above the lowest extra-valency-shell orbital.

It should be emphasized (1) that the actual forms of the curves in the Figure are uncertain, except that each must be a maximum or minimum on the 180° line, (2) that though the energy order of the orbitals on the 180° ordinate is probably definite (except perhaps for $\bar{\sigma}_g$ and $\bar{\pi}_u$) † that of the orbitals on the 90° ordinate is uncertain, (3) that no stress should be put upon the quantitative energy differences shown even on the 180° ordinate,‡ and (4) that though the above arguments provide plausible support for the rise or fall from left to right of a given orbital they are not always conclusive. In other words, some of the support for much in a correlation diagram such as that illustrated must be empirical. We therefore proceed below to examine how far the Figure can explain known facts.

BAC *Molecules.*—A diagram very similar to the Figure for AB_2 molecules should also hold for BAC molecules. The absence of a centre of symmetry now means that the *g* and *u* suffixes for the orbitals in the linear molecule must be abandoned. With the exception of the primes, the symbols for the bent molecule (C_s symmetry) also become inappropriate. The only appropriate symbols are now *a'* and *a''*.

Shapes of Non-hydride, Triatomic Molecules.—Of the lowest eight orbital curves on the right-hand side of the Figure, all but one either decrease from left to right or are horizontal. The one exception rises only slightly from left to right. One therefore expects that AB_2 or BAC molecules containing 16 or less valency electrons will be linear in their ground states. So far as is known there are no exceptions to this rule. CO_2, COS, CS_2, N_2O, ClCN, $HgCl_2$, NCO^-, and N_3^- supply well-known examples. Of less well-known molecules, the ions CO_2^+ (see below) and NO_2^+ (see Gillespie and Millen, *Quart. Reviews*, 1948, **2**, 277) are linear, while the radicals C_3 and N_3 would be expected to be linear. The uranium atom has six valency electrons, the configuration in the ground state being $\cdots(5f)^4(7s)^2$, where the $(5f)$- and the $(7s)$-orbitals have closely similar binding energies. One would expect, therefore, that the ion UO_2^{++} would be linear in the ground state. This is true (Zachariasen, *Acta Cryst.*, 1948, **1**, 277, 281; Sutton, *Nature*, 1952, **169**, 235). It seems

* The point is the same as that involved in the familiar diagrams correlating the orbitals of two separated atoms with those of the united atom. "The lowest σ orbital to the right goes into the lowest σ orbital to the left, the second lowest σ orbital to the right goes into the second lowest to the left and so on" (Herzberg, "Molecular Spectra and Molecular Structure," Van Nostrand, New York, 1950, p. 327).

† For CO_2 the ionization energies for electrons in the σ_u, π_u, and π_g orbitals are known to be 18·00, 17·32, and 13·79 ev.; *i.e.*, the σ_u lies only just below the π_u orbital, the π_g orbital lying considerably higher, in accord with the Figure.

‡ In any case these energy differences must vary from molecule to molecule.

that the implications of the Figure are not limited to light atoms but apply throughout the Periodic Table. The ions $AgCl_2^-$ and $AuCl_2^-$ are linear.* One may deduce, either directly by arguments similar to those of this paper or by analogy with the isoelectronic CO_2 that the ground states of keten and allene molecules will have linear CCO or CCC chains respectively.

The ground state of a molecule such as NO_2, however, containing 17 valency electrons, has to have its outermost electron in the $a_1's_A-\bar{\pi}_u$ orbital which rises steeply from left to right in the Figure. The ground state of the NO_2 molecule is therefore bent,† its apex angle showing a considerable drop from 180°. When NO_2^+ reacts, a donation of electrons to it may be considered to occur. These electrons occupy the $a_1's_A-\bar{\pi}_u$ orbital, resulting in the NO_2 group's becoming triangular. NO_2 is the only 17-electron molecule for which an apex angle has been reported. Two rather widely different values for this angle have been given (Table 1). The mean drop from 180° is 37° ± 11°. One would expect the apex

TABLE 1. *Apex angles of ground states of triatomic molecules.*

No. of valency electrons	Molecule	Apex angle	Ref.	Mean value	Method of determn.
16	Numerous	180°		180°	—
17	NO_2	154 ± 4	*a*	143 ± 11	Electronic spectrum
		132 ± 3 *	*b*		Electron diffraction
18	NOCl	116	*c*	117	" "
	NOBr	117	*c*		" "
	NO_2^- (in crystal)	114 ± 4	*d*		*X*-Ray diffraction
		115	*e*		" "
	O_3	127 ± 3	*f*		Electron diffraction
		116·5—117	*g*		Micro-waves
	SO_2	119° 2′	*h*		"
19	ClO_2	116·5°	*i*	116·5	Electron diffraction
20	F_2O	101	*j*	103	" "
	Cl_2O	110·8	*i*		" "
	Cl_2S	101 ± 4	*k*		" "
		103 ± 3	*l*		" "
	Br_2Te	98 ± 3	*m*		" "
22	$BrIBr^-$	180° or *ca.* 180°	*n*	180	*X*-Ray diffraction
	$ClICl^-$		*o*		" "
	I_3^-		*p*		" "
	$ClIBr^-$		*q*		" "

* Spurr, using the electron diffraction data of Maxwell and Mosley (*J. Chem. Phys.*, 1940, **8**, 738), found 141° (quoted by Jost and Russell, "Systematic Inorganic Chemistry," Prentice Hall, New York, 1944).

[a] Harris and King, *J. Chem. Phys.*, 1940, **8**, 775. [b] Claesson, Donohue, and Schomaker, *ibid.*, 1948, **14**, 207. [c] Ketelaar and Palmer, *J. Amer. Chem. Soc.*, 1937, **59**, 2629. [d] Truter, personal communication; Mrs. Truter (*Nature*, 1951, **168**, 344) originally reported the angle as 132° 48′, but has since revised her estimate. [e] Carpenter, *Acta Cryst.*, 1952, **5**, 132. [f] Shand and Spurr, *J. Amer. Chem. Soc.*, 1943, **65**, 179. [g] Hughes, *Bull. Amer. Phys. Soc.*, 1951, **26**, 21, No. 6. [h] Crable and Smith, *J. Chem. Phys.*, 1951, **19**, 502. [i] Dunitz and Hedberg, *J. Amer. Chem. Soc.*, 1950, **72**, 3108. [j] Brockway, *Rev. Mod. Phys.*, 1936, **8**, 231. [k] Palmer, *J. Amer. Chem. Soc.*, 1938, **60**, 2360. [l] Stevenson and Beach, *ibid.*, p. 2872. [m] Rogers and Spurr, *ibid.*, 1947, **69**, 2102. [n] Bozorth and Pauling, *ibid.*, 1925, **47**, 1561. [o] Wyckoff, *ibid.*, 1920, **42**, 1100. [p] Mooney, *Z. Krist.*, 1935, **90**, 143. [q] Mooney, *Phys. Review*, 1935, **47**, 807; *Z. Krist.*, 1938, **98**, 324; 1939, **100**, 519.

angle to show a further marked drop if an eighteenth electron were added. The exact value of the apex angle in 18-electron molecules must vary from molecule to molecule because the heights of the various orbitals and the steepness of their change with angle must vary from molecule to molecule. In other words, if it ever becomes possible to plot the Figure in a completely quantitative way, a separate figure really needs to be plotted for each molecule. It is remarkable, however, that the apex angles reported for various

* The hybrid valencies giving linearity may not be *sp* when heavy elements are considered. The general ideas of the present argument should, however, still be valid.

† *Added in Proof*.—It is not obvious that *one* electron in the $a_1's_A-\bar{\pi}_u$ orbital should outweigh the effect of, *e.g.*, the filled $a_1-\sigma_g$ and $b_2-\sigma_u$ orbitals. However, the fall of the $a_1-\sigma_g$ curve from left to right in the Figure is offset by the B ←→ B bonding nature of the orbital; whereas the rise of the $a_1's_A-\bar{\pi}_u$ curve is accentuated by the B ←→ B bonding nature of the orbital and by its becoming A ←→ B anti-bonding in the linear molecule. It is plausible therefore that the latter curve should rise much more steeply than the $a_1-\sigma_g$ curve falls. See also the footnote concerning this point on p. 2263.

18-electron molecules only vary over a range (*ca.* 12° at most and probably only 5°) that is small compared with the total range of the abscissa in the Figure.* These angles are given in Table 1. The mean value implies a drop of *ca.* 26° from the mean value for NO_2. A 19-electron molecule has to place its outermost electron in the b_1'–$\bar{\pi}_u$ orbital. Since the curve of this orbital rises from left to right in the Figure, but not as steeply as that of the $a_1's_A$–$\bar{\pi}_u$ orbital, one would expect the apex angle of such a molecule to show a further, but smaller, drop from the value characteristic of 18-electron molecules. The only 19-electron molecule for which an apex angle has been reported is ClO_2. The value of the angle reported for this molecule probably has a considerable uncertainty, but is not incompatible with the expected small drop from the mean value for the 18-electron molecules. One would expect a further small drop in the apex angle of a 20-electron molecule. One expects and finds the actual angle reported for such molecules to cover a range of values, but again the range is small (*ca.* 13° : see Table 1). The mean value represents a drop of about 13° from the value for ClO_2.

To sum up, all 17-, 18-, 19-, and 20-electron molecules have ground states that are bent; the apex angle drops considerably from 16- to 17-electron molecules; and also from 17- to 18-electron molecules; the apex angle also drops, but less markedly, from 18- to 19-electron molecules and from 19- to 20-electron molecules; and these facts fit very well with expectations from the Figure. It is remarkable that mere number should determine, for AB_2 and BAC as for AH_2 molecules, the crude magnitude of the apex angle (cf. Cassie, *Nature,* 1933, **131**, 438; Penney and Sutherland, *Proc. Roy. Soc.*, 1936, *A*, **156,** 654; Sidgwick and Powell, *ibid.*, 1940, *A*, **176**, 153).

The first excited state of the CO_2 molecule has its most weakly bound electron in the $a_1's_A$–$\bar{\pi}_u$ orbital. It has the same reason as the ground state of NO_2 for being bent, and at the same time has one less electron in the orbitals stabilizing the linear form. One strongly expects therefore the first excited state of CO_2 to be bent. We discuss evidence in favour of this below. The keten and the allene molecule are isoelectronic with CO_2. It follows either directly by arguments similar to those of this paper, or by analogy with CO_2, that the CCO or CCC chains in the equilibrium forms of the lowest-lying excited states of these molecules should be non-linear (cf. Sutcliffe and Walsh, *J.*, 1952, 899). Further examples of molecules which have linear ground states but probably bent upper states are given in a discussion below of the spectra of triatomic molecules. It is probably a very common phenomenon for polyatomic molecules to have different shapes in their excited and in their ground states (cf. the following papers).

The ground state of a 22-electron molecule must have two electrons in the $\bar{a}_1' - \bar{\pi}_u$ orbital. The trend of the apex angle in 16—20-electron molecules should therefore be sharply reversed on proceeding further to 22-electron molecules. Indeed, if the $\bar{a}_1'$–$\bar{\pi}_u$ curve falls sufficiently steeply, 22-electron molecules should be linear in their ground states. As Table 1 records, the trihalide ions are reported, experimentally, to be linear or nearly so. Pimentel (*J. Chem. Phys.*, 1951, **19**, **446**) has earlier discussed the molecular orbitals involved in the ground states of trihalide ions, accepting these states as linear, but he made no attempt to discuss the more difficult problem of why this was so. In essentials Pimentel's conclusions agree with the orbitals that would be predicted from the Figure for the ground state of a 22-electron molecule. In particular agreement with Pimentel, the Figure shows that there is no need to invoke the use of *d* orbitals to explain (qualitatively) the bonding (and the linearity) of the trihalide ions.

Alternative Statement of the Factors determining the Apex Angle.—It is striking that the apex angles of the ground states of H_2O and F_2O are nearly the same (104° and 101° respectively).† It presumably means that H–H and F–F repulsions are of minor importance in determining the angle. This is readily interpreted in terms of the Figure. The major factors determining the apex angle lie in the a_1'–σ_g, b_2'–σ_u, and a_1s_A–$\bar{\pi}_u$, curves, *i.e.*, in the curves largely common to the Figures of the preceding and of the present paper. It can be seen from the Figure that the sum of the b_1''–$\bar{\pi}_u$, b_2'–π_g, a_2''–π_g, a_1'–π_u, and

* Similarly, it was pointed out in Part I that the known angles for 8-electron AH_2 molecules vary only over a range of 14°.

† I am indebted to Dr. P. Torkington for stressing this to me

b_1''–π_u curves changes comparatively little with the angle. The major changes in binding energy with angle may therefore be regarded as due to the three remaining curves which, for F_2O, must be substantially the same as for H_2O. In other words, the major factor determining the apex angle is a property of the central atom. It is the interaction between the valencies of this atom used for the bonding orbitals and the lone pair a_1' orbital.

So far we have expressed this interaction in terms of hybridization of the valencies of the central atom. One could express the interaction in alternative terms of electrostatic repulsions between the electrons in the bond orbitals and those in the " lone-pair " orbital (cf. Pople, *Proc. Roy. Soc.*, 1950, *A*, **202**, 323). Thus the fundamental reason why 17- and 18-electron molecules have ground states that are bent is because they have 1 and 2 electrons respectively which are in orbitals more localized on the central than on the end atoms and repel the bond electrons. As a consequence of the molecule's bending, the extra electrons come to occupy an orbital whose contribution from the central atom is no longer pure *p*, but an *sp* hybrid pointing in the $+z$ direction (axes as in Part I). At the same time the valencies of the central atom used for the bond orbitals become more nearly pure *p*, *i.e.*, become less electronegative (Walsh, *Discuss. Faraday Soc.*, 1947, No. 2, p. 18). The hybridization changes thus imply that the electrons in the lone-pair orbital move away from the bond orbitals in the *z* direction (the orbital no longer having its centre of gravity at the nucleus of A as it had in the linear molecule) while the bond electrons move away from atom A towards atoms B. These are just the changes expected to follow from repulsion between lone-pair electrons on A and the bond electrons.

In the ground states of 22-electron molecules, two electrons are placed in a further orbital which is more localized on atom A than on atoms B. This orbital is built from a p_z orbital of A in the 90° molecule and from a pure *s* orbital of A in the linear molecule. At intervening angles the orbital represents the second *sp* hybrid that can be formed from the *s* and p_z orbitals of A. This second *sp* hybrid points in the $-z$ direction and so restores symmetry to the electron cloud around A. The bond electrons are now subject to repulsions from the electrons in the *sp* hybrids which tend to cause bending in opposite directions; thus the molecule resumes a linear or nearly linear form.

The key curves causing the major changes of shape are: (a) the bond orbital curves; (b) the $a_1's_A$–$\bar{\pi}_u(p_z)$ curve; and (*c*) the $\bar{a}_1'(p_z)$–$\bar{\sigma}_g(s_A)$ curve.

Reactivity of Triatomic Molecules.—The outermost, unpaired electron of the ground state of the NO_2 molecule lies in the $a_1's_A$–$\bar{\pi}_u$ orbital. From the arguments above, this electron is more localized on the N atom than on the O atoms. Reaction of NO_2 with a free radical is therefore likely to form a nitro-compound rather than a nitrite. Inhibition of certain free-radical reactions by NO_2 may plausibly be attributed to such reaction: the inhibition occurs under conditions where addition of a nitro-compound has no effect, but addition of a nitrite catalyses the reaction. Similarly, initiation of chains by H-abstraction by NO_2 (see, *e.g.*, McDowell and Thomas, *Trans. Faraday Soc.*, 1950, **46**, 1030) is likely to form $H{\cdot}N{<}^{O}_{O}$ rather than HO·NO. One might expect that, since NO_2 molecules readily associate to form N_2O_4, the latter molecule has the structure ${}^{O}_{O}{>}N{\cdot}N{<}^{O}_{O}$ (cf. Ingold and Ingold, *Nature*, 1947, **159**, 743; Walsh, *J. Chem. Phys.*, 1947, **15**, 688; Claesson, Donohue, and Schomaker, *ibid.*, 1948, **16**, 207; Broadley and Robertson, *Nature*, 1949, **164**, 915). On arguments similar to those given here, the outermost, unpaired electron of the ground state of the NO molecule is more localized on the N than on the O atom. Since NO and NO_2 associate to form N_2O_3, a certain presumption is raised in favour of the structure $O{-}N{\cdot}N{<}^{O}_{O}$ for N_2O_3 (see also Part V of this series).

The nitrite ion has its two most weakly bound electrons in the $a_1's_A$–$\bar{\pi}_u$ orbital, *i.e.*, more localized on the N atom than on the O atoms. One might therefore expect reaction with an acceptor entity to take place at the N atom rather than at the O atoms. In agreement, the nitrite ion reacts with a carbonium ion to form a nitro-compound (Austin, Thesis, London, 1950).

The ClCO radical has its outermost, unpaired electron more localized on the C than on the Cl or O atom. One therefore expects it to react most readily at the C atom. This is in agreement with the reaction that has been postulated for the radical (Rollefson and Montgomery, *J. Amer. Chem. Soc.*, 1933, **55**, 142, 4025), namely, formation of carbonyl chloride : $Cl_2 + ClCO = Cl_2CO + Cl$.

The first excited state of the SO_2 molecule contains one electron in the $a_1's_A$–$\bar{\pi}_u$ orbital and one in the b_1''–$\bar{\pi}_u$ orbital. Both these orbitals are more localized on the S atom than on the O atoms. It is interesting that photochemical reactions of SO_2, presumably involving this excited state, have been shown to form two bonds to the *S* atom (Dainton and Ivin, *Trans. Faraday Soc.*, 1950, **46**, 382). Moffitt (*Proc. Roy. Soc.*, 1950, *A*, **200**, 414) has previously noted that the most weakly bound electrons of the ground state of the SO_2 molecule lie in an (a_1) orbital (here called $a_1's_A$) which is predominantly localized on the S atom; and that the oxidation (" donor ") reactions of SO_2 to form SO_3 or SO_2Cl_2 can therefore be readily understood.

The discussion in this section is somewhat naive in that reactivity is a complicated matter; but it is sufficiently successful to suggest that the ideas involved are useful in considering not merely the ground state, but also the photochemical, reactions of molecules. Combined with the conclusions of the preceding section concerning the shapes of excited states, they should be of particular use in considering possible photochemical syntheses.

Relation to Other Work.—As mentioned in Part I, Mulliken (*Rev. Mod. Phys.*, 1942, **14**, 204) has earlier plotted a correlation diagram for the orbitals of a linear and a bent triatomic molecule. Mulliken, however, was unable to give any simple reason why particular curves rose or fell with change of apex angle. In many respects, our Figure agrees with that of Mulliken. There are two major differences. The first is that Mulliken supposed $\bar{\sigma}_g$ to lie below $\bar{\pi}_u$ (for his reasons, see *J. Chem. Phys.*, 1935, **3**, 739). As already explained, this results in the curve from $a_1's_A$ leading to $\bar{\sigma}_g$ instead of to $\bar{\pi}_u$ and in the curve from $\bar{a}_1'$ leading to $\bar{\pi}_u$ instead of to $\bar{\sigma}_g$. The second major difference is that Mulliken drew the curve from $\bar{a}_1'$ as rising from left to right, whereas in the present Figure it has the opposite curvature. Mulliken's diagram made the reported linearity of the 22-electron molecules incomprehensible

Mulliken particularly used his diagram to interpret observed spectra of triatomic molecules. In the next section we give a fuller discussion of these observed spectra, indicating our agreement with, or changes from, Mulliken's assignments.

Spectra of Non-hydride, Triatomic Molecules.—(i) CO_2 *and* N_2O. It is striking that, whereas the N_2O and CO_2 molecules are isoelectronic, ultra-violet absorption by the former begins at *ca.* 3000 Å whereas that by CO_2 does not occur until *ca.* 1700 Å.* A simple explanation is as follows. The lowest-energy transition of these molecules, according to the Figure, is of an electron from the π_g to the $\bar{\pi}_u$ orbital. In other words, it is from an orbital largely localized on the end atoms to one largely localized on the central atom. The orbital localized on the end atoms will be less tightly bound in NNO than in OCO since the electronegativity of N is less than that of O.† On the other hand, the orbital largely localized on the central atom will be more tightly bound in NNO than in OCO since the electronegativity of N is greater than that of C. We have therefore the situation represented diagrammatically below and the transition in N_2O should require considerably less energy than that in CO_2.

The first electronic transitions of N_2O *and* CO_2.

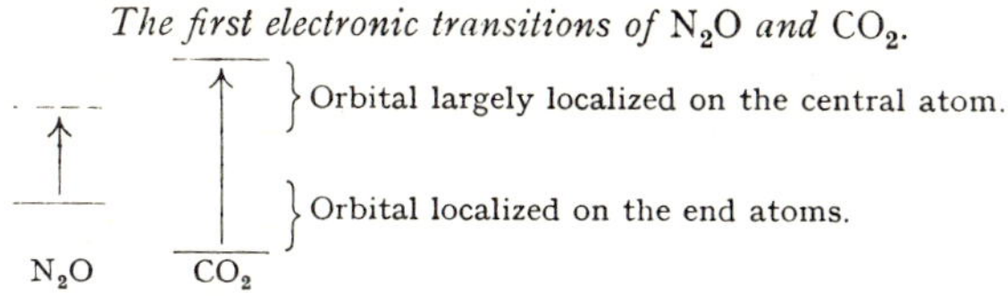

According to our Figure (p. 2268) the upper state of the longest-wave-length-absorption

* Even in liquid CO_2 no absorption occurs until at least 2150 Å.

† It is easily seen that in N_2O the analogue of the lower π_u orbital is predominantly localized on the O atom, while the analogue of the π_g orbital is predominantly localized on the end N atom.

system of CO_2 should be related to the configuration $(\pi_g)^3(\bar{\pi}_u)$. This gives rise to $^1\Delta_u$ (two-fold degenerate), $^1\Sigma_u^-$, and $^1\Sigma_u^+$ states. By analogy with atoms and diatomic molecules one would expect $^1\Delta_u$ to be the lowest of these states. The transitions

$$\cdots(\pi_g)^3(\bar{\pi}_u),\ ^1\Delta_u \longleftarrow \cdots(\pi_g)^4,\ ^1\Sigma_g^+ \qquad (1)$$

and

$$\cdots(\pi_g)^3(\bar{\pi}_u),\ ^1\Sigma_u^- \longleftarrow \cdots(\pi_g)^4,\ ^1\Sigma_g^+ \qquad (2)$$

are forbidden; while

$$\cdots(\pi_g)^3(\bar{\pi}_u),\ ^1\Sigma_u^+ \longleftarrow \cdots(\pi_g)^4,\ ^1\Sigma_g^+ \qquad (3)$$

is allowed. Now, (3) should be a particularly intense transition since it belongs to the class of so-called " $V \longleftarrow N$ " transitions which, compared with other intra-valency-shell transitions, are expected to have particularly high intensities (Mulliken, *J. Chem. Phys.*, 1939, **7**, 20). The different possible states correspond, in C_{2v} symbols, to the four configurations (see Figure)

$$(a_2'')^2(b_2')(a_1'),\ ^1B_2$$
$$(a_2'')(b_2')^2(a_1'),\ ^1A_2$$
$$(a_2'')^2(b_2')(b_1''),\ ^1A_2$$
$$(a_2'')(b_2')^2(b_1''),\ ^1B_2$$

The first two of these configurations represent states that are expected to be strongly bent, and the second two states that are expected to be slightly bent or linear. $^1\Delta_u$ correlates with $^1A_2$ and $^1B_2$; $^1\Sigma_u^+$ and $^1\Sigma_u^-$ correlate with $^1B_2$ and $^1A_2$ respectively. For bent upper states, the degeneracy of $^1\Delta_u$ is split, so that (1) has to be replaced by the two transitions

$$^1A_2 \longleftarrow \cdots(\pi_g)^4,\ ^1\Sigma_g^+ \qquad (1A)$$

and

$$^1B_2 \longleftarrow \cdots(\pi_g)^4,\ ^1\Sigma_g^+ \qquad (1B)$$

Both transitions would be weak, partly because forbidden by the Franck–Condon principle and partly because both are related to the forbidden transition (1). In addition, (1A) is forbidden by the symmetry-selection rules and has therefore an extra reason to be weak compared with (1B). Similarly, for bent upper states, (2) and (3) have to be replaced respectively by

$$^1A_2 \longleftarrow \cdots(\pi_g)^4,\ ^1\Sigma_g^+ \qquad (2A)$$

and

$$^1B_2 \longleftarrow \cdots(\pi_g)^4,\ ^1\Sigma_g^+ \qquad (3A)$$

(2A) should be very weak because it is forbidden by the symmetry-selection rules and is related to the forbidden transition (2). Provided, however, that the upper state is not *strongly* bent, (3A) should be strong since it is allowed by the selection rules and is related to the intense transition (3).

The lowest-energy absorption of CO_2 falls into two regions, 1700—1400 and 1390—1240 Å. The latter region has $\lambda_{max.}$ at *ca.* 1335 Å (Price and Simpson, *Proc. Roy. Soc.*, 1939, *A*, **169**, 501) and is much stronger than the 1700—1400-Å region, though not as strong as the Rydberg bands at shorter wave-lengths. There is no doubt that it represents an allowed intra-valency-shell transition. Indeed its intensity makes it only plausibly identified as transition (3) or (3A). In the latter case the upper state cannot be *strongly* bent. The appearance pressure of the 1390—1240-Å absorption, relative to the Rydberg bands, appears to be of the order found in other molecules for $V \longleftarrow N$, relative to Rydberg, transitions (*e.g.*, acetylene; see Part III). The 1700—1400-Å absorption contains a peak of $\varepsilon_{max.}$ ~110 * (*ca.* 1495 Å; Wilkinson and Johnston, *J. Chem. Phys.*, 1950, **18**, 190). This is too strong for the transition to be plausibly interpreted as a triplet $\longleftarrow$ singlet transition (which would be expected to have an extinction coefficient < 1; Kasha, *Discuss. Faraday Soc.*, 1950, **9**, 72). It is also too strong to be plausibly interpreted as (2) or (2A). Further, it is not plausibly interpreted simply as (1), partly because of its intensity and partly because of the strong expectations that the first absorption of CO_2 should lead to a bent upper state. The only plausible interpretation is as transition (1B). In view of the expectation (see Figure) that the lowest-energy

* See footnote, p. 2264

absorption should lead to a *strongly* bent upper state, the 1495-Å peak may be interpreted, in more detail, as

$$\cdots (a_2'')^2(b_2')(a_1'),\ {}^1B_2 \text{ (correlating with } {}^1\Delta_u) \longleftarrow \cdots (\pi_g)^4,\ {}^1\Sigma_g^+ \quad . \quad \text{(1B')}$$

This leaves the interpretation of the 1390—1240-Å region to be given in more detail as

$$\cdots (a_2'')(b_2')^2(b_1''),\ {}^1B_2 \text{ (correlating with } {}^1\Sigma_u^+) \longleftarrow \cdots (\pi_g)^4,\ {}^1\Sigma_u^+ \quad . \quad \text{(3A')}$$

raising the expectation that the upper state may be *slightly* bent. That the expectation is correct is strongly supported by the fact that in the analogous region of CS_2 (see below) there is evidence that the upper state is bent through a few degrees. In view of the remarkable dependence of apex angle upon number of occupied orbitals (irrespective of the particular nuclei concerned) we would expect the apex angle in the upper state of the ~1495-Å absorption to be less than the ground state angle of NO_2 (143° ± 11°) by a small amount which represents the effect of losing an electron from the linear-stabilizing π_g orbital. A reasonable expectation for the apex angle would be 135° ± 10°.

It is interesting that Wilkinson and Johnston find, besides the peak at *ca.* 1495 Å, three still weaker bands at 1690, 1673, and 1662 Å, which appear to represent a separate system. These bands, which are very diffuse, are also mentioned by Price and Simpson (*loc. cit.*). It is possible that they represent transition (1A).

The 3000—1760-Å absorption of N_2O is also known to subdivide into at least two regions ($\lambda_{max.}$ *ca.* 2900 and *ca.* 1900 Å), requiring different appearance pressures (see summary given by Sponer and Teller, *Rev. Mod. Phys.*, 1941, **13**, 75).

As stated above, Mulliken (*loc. cit.*; *J. Chem. Phys.*, 1935, **3**, 720) supposes the $\bar{\sigma}_g$ to lie below the $\bar{\pi}_u$ orbital. Since if that were true we should need to draw our Figure with an "avoided crossing" and the $a_1's_A$ curve leading to $\bar{\sigma}_g$ instead of to $\bar{\pi}_u$, Mulliken's interpretation of the longest-wave-length absorption is identical with the present interpretation, provided that the upper state is formulated in the C_{2v} symbols given above. The difference is that Mulliken's interpretation implies that the ${}^1B_2$ upper state correlates with the $(\pi_g)^3(\bar{\sigma}_g)$, ${}^1\Pi_g$ linear state instead of with the $(\pi_g)^3(\bar{\pi}_u)$, ${}^1\Delta_u$ linear state. [In a bent molecule, the degeneracy of the ${}^1\Pi_g$ state would be split, leading to transitions (1A) and (1B) above.] Now, if the upper state of the first absorption correlates with ${}^1\Pi_g$, one might expect the analogous absorption in N_2O to be much stronger, since it should be related to the *allowed* transition

$$\cdots (\pi)^3(\bar{\sigma}),\ {}^1\Pi \longleftarrow \cdots (\pi)^4,\ {}^1\Sigma^+ \quad . \quad . \quad . \quad . \quad . \quad . \quad . \quad (4)$$

The first absorption of N_2O, however, remains weak. It seems better, therefore, to regard the first absorption as related to the transition

$$\cdots (\pi)^3(\bar{\pi}),\ {}^1\Delta \longleftarrow \cdots (\pi)^4,\ {}^1\Sigma^+ \quad . \quad . \quad . \quad . \quad . \quad . \quad . \quad (5)$$

which is forbidden in N_2O, as is (1) in CO_2, *i.e.*, at least with N_2O, to regard $\bar{\sigma}$ as lying above $\bar{\pi}$. [Doubtless altogether the 3000—1760-Å absorption of N_2O involves absorption to the three bent upper states which correspond to ${}^1B_2$, ${}^1A_2$ (correlating with ${}^1\Delta_u$) and ${}^1A_2$ (correlating with ${}^1\Sigma_u^-$) of CO_2.] It is true that with COS the first absorption becomes stronger than in CO_2, but (as we explain below) this is not compelling evidence in favour of $\bar{\sigma}_g$ lying below $\bar{\pi}_u$. Further, in other molecules [*e.g.*, diatomic molecules, C_2H_2 (see Part III) and HCHO] the details of the spectra are only plausibly interpreted by supposing that the orbital analogous to $\bar{\sigma}_g$ lies above that analogous to $\bar{\pi}_u$. With no molecule does there appear to be any compelling evidence that $\bar{\sigma}_g$ (or its analogue) lies below $\bar{\pi}_u$. At least pending further evidence, therefore, we shall proceed on the assumption that $\bar{\sigma}_g$ lies above $\bar{\pi}_u$, since this implies a simpler form for our Figure. If later work should show that after all $\bar{\sigma}_g$ lies below $\bar{\pi}_u$, it would not be difficult to make the necessary changes in the present assignments.

The continuous nature of the absorption in CO_2 and N_2O does not seem surprising in view of the fact that the energy absorbed is sufficient to cause dissociation into the ground states of CO and O or N_2 and O respectively. It is clear, however, that the equilibrium form of the upper state may possess much less energy than the form reached in absorption at $\lambda_{max.}$ from the ground state. The upper state may therefore be stable in the equilibrium form and a discrete emission spectrum may be possible. If the absorption spectrum of CO_2 at high temperatures were studied, so that transitions from high vibrational levels of the ground state were possible at sufficiently long wave-lengths, the absorption spectrum would presumably become banded.

Returning now to the 1390—1240-Å region of CO_2, we should expect from the interpretation given above that both the ν_2 bending frequency and (more strongly) the ν_1 stretching frequency of the upper state would appear in the absorption. The latter expectation follows from the fact that the π_u orbital is C<-->O anti-bonding. For an allowed transition, only frequencies that represent vibrations which are totally symmetrical should appear strongly; if the upper state remained linear, ν_2, being then non-totally symmetrical, would not appear strongly. The bands in the region are diffuse and shaded to the red. Their separation is $\sim$500—600 cm.$^{-1}$, but is too irregular for the structure to be interpreted in terms of a single frequency of this magnitude. Price and Simpson suggest that two progressions are present, one starting at 1380 Å and the other at 1368 Å, each with frequency intervals of *ca.* 1225 cm.$^{-1}$ and each rising to a maximum intensity at the same wave-length. However, the measured intervals of these supposed progressions vary from 1080 to 1350 cm.$^{-1}$. No doubt the difficulty of measuring the diffuse bands can partly explain this irregularity, but the variation is so great that it seems more probable that two different frequencies are involved. In this connection it should be noted that, for a reason that will be made clear in Part III, it would be possible for the bending frequency to be slightly *increased* in the upper state relative to the value of 667 cm.$^{-1}$ in the ground state (cf. N_2O below).* The stretching frequency, on the other hand, is expected to be decreased in the upper state from its value of *ca.* 1340 cm.$^{-1}$ in the ground state.

The analogue of the CO_2 1390—1240-Å system in N_2O appears to be the absorption occurring between 1520 and 1425 Å (Duncan, *J. Chem. Phys.*, 1936, **4**, 638), although the appearance pressure of this absorption is only about a tenth of that for the CO_2 system. The absorption consists of nine diffuse bands. Duncan represents their spacing by a frequency of 621 cm.$^{-1}$ in the upper state. This might represent the totally symmetrical ν_1 vibration of 1285 cm.$^{-1}$ in the ground state, though if so the drop is very great. Alternatively, 621 cm.$^{-1}$ might represent the bending vibration which has a frequency of 589 cm.$^{-1}$ in the ground state; the small increase is not implausible (see Part III). If the latter interpretation is correct, then it implies that the upper state is (slightly) bent. Sponer and Teller (*loc. cit.*, p. 109) suggest the further interpretation that two separate progressions, each with a spacing of *ca.* 1240 cm.$^{-1}$ (representing the 1285-cm.$^{-1}$ ground-state frequency), are present; but if so the drop in frequency seems surprisingly small. However, the spacing between the bands is irregular and, though this is doubtless partly due to the difficulty of measurement, it is not improbable that *two* frequencies are involved, one representing ν_2' and the other ν_1'. It is unfortunate that the diffuseness of the bands precludes resolution of the rotational fine structure to decide whether the bands are parallel or perpendicular. Whatever the correct interpretation, if ν_2' is involved the progression in it is fairly short (maximum intensity occurs at the fifth band), so that the upper state is slightly, rather than strongly, bent.

An extensive emission band system believed to be due to the CO_2 molecule is known. It appears in the carbon monoxide flame spectrum (see Gaydon and Wolfhard, *Nature*, 1949, **164**, 22). The bands lie between 5500 and 3000 Å, with maximum intensity around 4200 Å. Smyth (*Phys. Review*, 1931, **38**, 2000) was unable to find emission bands that could be ascribed to CO_2 over the range 2700—1400 Å. In other words, there appeared to be no bands corresponding to those found in absorption around 1500 Å. However, as Gaydon (*Proc. Roy. Soc.*, 1940, *A*, **176**, 505; " Spectroscopy and Combustion Theory," Chapman and Hall, London, 2nd Edn., 1948) pointed out, if the upper state is considerably bent one would expect the emission system to lie at wave-lengths much longer than those of the corresponding absorption bands. This follows from the Franck–Condon principle. The extensive range covered by the emission spectrum confirms that the molecular

* In the first absorption of CS_2, however, $\nu_2' < \nu_2''$.

dimensions change considerably in the transition. Excited CO_2 molecules in the carbon monoxide flame almost certainly arise by the reaction

$$O(^3P) + CO(^1\Sigma^+) \longrightarrow CO_2^* \quad . \quad . \quad . \quad . \quad . \quad . \quad . \quad . \qquad (6)$$

It follows that the excited molecules are in a triplet state (Laidler, *J. Chem. Phys.*, 1949, **17**, 221) and that this triplet state is not more than *ca.* 130 kcal./mole above the ground state. If the excited state arises from the first excited configuration, then the system is due to the transition

$$\cdots (b_2')(a_1'),\ ^3B_2 \longrightarrow \cdots (\pi_g)^4,\ ^1\Sigma_g^+ \quad . \quad . \quad . \quad . \quad . \quad . \quad . \qquad (7)$$

i.e., involves a change of multiplicity. Gaydon assumes that the transition is to the ground state but does not point out that this must mean the transition is forbidden because of the change of multiplicity. If the ground state is not involved, the lower state must be strongly bent and (probably) triplet. Only so will it be possible for the height of the lower state above the ground state to be sufficiently low for an emission stretching to *ca.* 3000 Å to arise from an upper state produced by (6). The lower state should not be more than 30—40 kcal. above the ground state. If the ground state is not involved, then, for obvious reasons apart from one of the states' being strongly bent, the emission system does not correspond to any known absorption system; and the vibrational frequencies present are not, as Gaydon supposes, those of the ground state.

In fact, the evidence that the lower state is the ground state is unconvincing. Thus Gaydon interprets frequency differences of 2065 and 565 cm.$^{-1}$ as the 2349 cm.$^{-1}$ (ν_3) and 667 cm.$^{-1}$ (ν_2) frequencies of the ground state. The discrepancies are very large if they are due to anharmonic factors concerned in transitions to high vibrational levels of the ground state. The anharmonic factors for the lower ground-state levels are known to be very small [*e.g.*, $(x_e\omega_e)_2 = -1{\cdot}3$ cm.$^{-1}$]. The observed progressions are so long that the bands can hardly ***all*** be to very high levels of the ground state. In the ground state Fermi resonance occurs between the ν_2 and ν_1 levels, the resonance splitting being $\sim$102 to 144 cm.$^{-1}$. This separation does not appear in the emission spectrum. Further, the upper state is expected to involve an excited orbital that, in comparison with the ground state, is strongly C$\longleftrightarrow$O anti-bonding. In other words, if the ground state is concerned, the transition should show progressions in ν_1 of the ground state. These progressions should be represented by pairs of bands separated by the Fermi resonance splitting and with $\sim\nu_1$ of the ground state between the pairs. The ν_1 frequency in the ground state is known to have the approximate value of 1340 cm.$^{-1}$. The anharmonicity $(x_e\omega_e)$ is known to be very small ($-0{\cdot}3$ cm.$^{-1}$), but, if the other two frequencies are markedly reduced by anharmonic factors, the ν_1'' progressions should be represented by pairs of bands separated by markedly less than 1340 cm.$^{-1}$. Now the most commonly occurring intervals in the flame spectrum (Gaydon, *loc. cit.*) are 565, 1130, and 1500 cm.$^{-1}$. Intervals that occur less frequently are 345, 925, 1355, 1700, 2065, and 2260 cm.$^{-1}$; intervals that occur still less frequently are 160, 230, 515, 780, 1865, and 1915 cm.$^{-1}$. In addition Gaydon refers to the bands' tending to occur in pairs with a separation of *ca.* 60 cm.$^{-1}$ between the members of a pair. None of these frequency differences is plausibly identified as ν_1 of the ground state. The only two that are close to the expected magnitude are 1130 and 1355 cm.$^{-1}$; but the first of these is rather low and is much more plausibly interpreted as 2×565 cm.$^{-1}$, while the second is too high. Gaydon makes no attempt to identify any of the observed frequency differences as ν_1. If the lower state is the ground state, non-appearance of progressions in ν_1 of the ground state is difficult to understand.

Let us see what interpretation of the observed frequency differences can be given if the ground state is not involved. Gaydon assumes long progressions in ν_3. However, there is no precedent in the known spectra of triatomic molecules for long progressions in the unsymmetrical stretching frequency ν_3. Further, such progressions would imply that one of the low-lying states of CO_2 has the C atom nearer to one O atom than to the other; and there is no theoretical expectation of such a lack of symmetry. In fact, if the emission involves a single upper state and a single lower state, we may confidently expect the system to be capable of interpretation in terms of four fundamental frequencies alone, *viz.*, ν_1', ν_2', ν_1'', and ν_2''. ν_1' and ν_1'' should both be less than 1340 cm.$^{-1}$. ν_2' and ν_2'' are likely to be less than or close to 667 cm.$^{-1}$ (the value of ν_2 in the ground state). *All* the frequency differences greater than 1340 cm.$^{-1}$ must be interpreted as overtones or combination bands of the fundamentals. 60, 160, 230, and probably 345 cm.$^{-1}$ are too small to be plausibly interpreted as fundamentals. The very commonly occurring 565-cm.$^{-1}$ frequency is an obvious choice for one of the fundamentals. It immediately enables 1130, 1700, and 2260 cm.$^{-1}$ to be interpreted as 2×565, 3×565 and 4×565 cm.$^{-1}$ respectively. Moreover, the magnitude 565 cm.$^{-1}$ makes it practically certain that this frequency is a bending one; it is too low to be plausibly chosen as a ν_1 frequency. $\nu_2'' =$ 565 cm.$^{-1}$ is the obvious choice, since in an emission transition ν_2'' is likely to be more prominent than ν_2'. The existence of long progressions in ν_2'' then implies that the two states concerned differ markedly in apex angle. Since we have argued that the lower state must be strongly bent, the upper must therefore be linear or only slightly bent. This is in contrast to Gaydon's interpretation that the transition is from a strongly bent upper to a linear lower (ground) state. Since the emission takes place from a hot source ($\sim$600° c in the " glow " of oxidizing carbon monoxide and a much higher temperature in the ordinary flame of carbon monoxide), we expect ν_2' also to appear in the system, but probably less prominently than ν_2''. $\nu_2' = 515$ cm.$^{-1}$ is the obvious choice to try. Such a choice may at once explain the tendency of bands to occur in pairs with *ca.* 60 cm.$^{-1}$ between the members of a pair since 565—515 is close to 60. On account of Boltzman factors, we should not expect the frequency 2×515 cm.$^{-1}$ to be prominent; and, in agreement, no frequency of *ca.* 1030 cm.$^{-1}$ appears. The only likely choices for the ν_1 frequencies (since they must be less than *ca.* 1340 and since 1130 cm.$^{-1}$ has already been identified) are 925 and 780 cm.$^{-1}$. Of these, it is 925 cm.$^{-1}$ which, being the commoner of the two, is likely to represent ν_1''. Such a choice probably accounts also for (*a*) *ca.* 1865 cm.$^{-1}$, since $2 \times 925 = 1850$, (*b*) *ca.* 345 cm.$^{-1}$, since $925 - 565 = 360$, (*c*) *ca.* 1500 cm.$^{-1}$, since $925 + 565 = 1490$, and (*d*) *ca.* 2065 cm.$^{-1}$, since $925 + (2 \times 565) = 2055$. The commonness of occurence of, *e.g.*, 1500 cm.$^{-1}$ is understandable, since it is composed from a very commonly and a fairly commonly occurring interval. This leaves $\nu_1' = 780$ cm.$^{-1}$. Just as 2×515 cm.$^{-1}$ did not occur, so we should not expect 2×780 cm.$^{-1}$ to appear; nor is it found. The choice, however, plausibly identifies (*a*) *ca.* 1355 cm.$^{-1}$, since $780 + 565 =$ 1345, (*b*) *ca.* 1915 cm.$^{-1}$, since $780 + (2 \times 565) = 1910$, (*c*) *ca.* 160 cm.$^{-1}$, since $925 - 780 =$ 145, and (*d*) *ca.* 230 cm.$^{-1}$, since $780 - 565 = 215$. This accounts for all the observed frequency intervals and moreover the whole forms an entirely reasonable, consistent, interpretation. The magnitudes of $2\nu_2''$ and $2\nu_2'$ are such that no Fermi resonance is expected with the ν_1 frequencies; and in agreement no further spearation remains to be accounted for by such resonance. Moreover, it is very reasonable that ν_2'' should be less than ν_2', and ν_1'' less than ν_1'; for the lower, strongly bent state (*a*) is likely, for a reason that will be referred to again in Part III, to have more resistance to bending than the upper linear or nearly linear state and (*b*) almost certainly involves the $a_1's_A\text{–}\bar{\pi}_u$ orbital which tends to lose its C$\leftrightarrow$O anti-bonding character as the molecule bends.

Summarizing, $\nu_2'' = 565$, $\nu_1'' = 925$, $\nu_2' = 515$, $\nu_1' = 780$ cm.$^{-1}$ and it is clear that the ground state is not involved and that the system consists of a single electronic transition. All the vibrations concerned are totally symmetrical with respect to the bent molecule and therefore there is no symmetry restriction on the number of quanta of each that may occur. The O(^{3}P) atom in (6) is expected to approach the C atom of the CO molecule along the C–O line, in such a way that an unpaired $p\sigma$ electron of the O atom interacts with a σ lone pair on the C atom. It appears probable therefore that the resulting state of CO_2 is $^3\Pi$. Having regard to what are the low-lying states (see the Figure), this must be $^3\Pi_g$. Our analysis of the flame bands requires this state to be linear or nearly linear; it is satisfactory that our Figure (though not Mulliken's correlation diagram) predicts linearity for the $^3\Pi_g$ state. The lowest excited state of CO_2, according to the Figure, should be the strongly bent $^3B_2$ (or less probably $^3A_2$) state. The flame bands are therefore most probably to be identified as due to the allowed transition

$$^3\Pi_g \longrightarrow {}^3B_2 \qquad \ldots\ldots\ldots\ldots \qquad (8)$$

Further, there is a very low probability that a triplet, strongly bent (apex angle 135° ± 10°), excited state of CO_2 would emit radiation and pass to the ground state. It would behave almost like a separate chemical species, a fact that may have something to do with the " latent energy " that has been reported for carbon monoxide flames (David, Leah, and Pugh, *Phil. Mag.*, 1941, **31**, 156).

(ii) CO_2^+. Discharges through CO_2 give rise to a pair of strong bands at 2882 and 2896 Å. It is known (Bueso-Sanllehí, *Phys. Review*, 1941, **60**, 556) that these are due to the CO_2^+ molecule; that they represent the transition $\cdots(\sigma_u)(\pi_u)^4(\pi_g)^4,\ {}^2\Sigma_u^+ \longrightarrow \cdots(\sigma_u)^2(\pi_u)^4(\pi_g)^3,\ {}^2\Pi_g$; that (as would be expected from the present considerations) the molecule is linear in both upper and lower states; and that the transition causes little change of the molecular dimensions.

The negative glow of discharges through streaming CO_2 or excitation of CO_2 by electron impact gives rise to an extensive system of weaker bands lying between 5000 and 2900 Å. The bands were first observed by Fox, Duffendack, and Barker and have been studied by Smyth, by Schmid, and especially by Mrozowski (*Phys. Review*, 1941, **60**, 730; 1942, **62**, 270; *Rev. Mod. Phys.*, 1942, **14**, 216). According to Pearse and Gaydon (" The Identification of Molecular Spectra," Chapman and Hall, London, 2nd Edn., 1950) these bands are believed to be due to the neutral CO_2 molecule, but this neglects the work of Mrozowski which appears to show conclusively that the bands are due to CO_2^+ and that they represent the transition $\cdots(\pi_u)^3(\pi_g)^4,\ {}^2\Pi_u \longrightarrow \cdots(\pi_u)^4(\pi_g)^3,\ {}^2\Pi_g$. As would be expected from the present arguments, rotational analysis shows that the molecule is linear in both upper and lower states. In accord with this, only ν_1 vibrations appear strongly. The degradation of the bands to the red and the extensive range covered by the spectrum are understandable in view of the properties of the π_u and π_g orbitals.

(iii) CS_2. From our Figure we should expect the first absorption of the CS_2 molecule to be due to transition to a configuration correlated to $(\pi_g)^3(\bar{\pi}_u)$. The first absorption consists of a large number of bands between 3980 and 2767 Å ($\lambda_{max.}$ ~3200 Å) (Wilson, *Astrophys. J.*, 1929, **69**, 34). It is weak, requiring several cm. pressure with a 1-m. absorption path, and is, like the first absorption peak of CO_2, plausibly interpreted as (1B′). That the CS_2 spectrum, unlike that of CO_2, consists of discrete bands is understandable since the excitation energy for CS_2 lies below that required for dissociation. One would expect the upper state to be considerably bent and to possess a considerably increased C–S distance over the ground state. Mulliken (*J. Chem. Phys.*, 1935, **3**, 720) early concluded that the upper state of the system was bent in its equilibrium form. Liebermann (*Phys. Review*, 1941, **60**, 496), from a partial vibrational analysis, showed that long progressions in ν_2' occur and also that the intensity of bands originating from $v_2'' = 2$, relative to bands originating from $v_2'' = 0$ but proceeding to the same upper level, is considerably greater than would be expected from the Boltzmann factor alone. Both these facts strongly suggest a considerably bent upper state. In addition, one-quantum jumps of ν_2' occur, which would be expected for a bent upper state but not for a linear one. Rotational analysis of six of the bands (at 3673, 3637, 3601, 3535, 3501, and 3468 Å) by Liebermann (*ibid.*, 1940, **58**, 183; 1941, **59**, 106A; 1941, **60**, 496) has shown them to be of the parallel type with simple P and R branches, closely resembling ${}^1\Sigma_u^+ \longleftarrow {}^1\Sigma_g^+$ bands of a molecule in which both states were linear. If, however, the upper state *were* linear, alternate bands of the ν_2' progression should be of a $\Pi \longleftarrow \Sigma$ type, which they are not. In fact, Liebermann has shown that the observed simple rotational structure is equally compatible with any apex angle between 125° and 180°. (As estimated above for the first excited state of CO_2, 135° ± 10° would be a reasonable value for the apex angle in the excited state.) If the upper state is bent, the parallel nature of the bands identifies them as ${}^1B_2 \longleftarrow {}^1\Sigma_g^+$, in agreement with the expected interpretation (1B′). Mulliken (*ibid.*, 1941, **60**, 506) has earlier shown how the observed characteristics are compatible with interpretation (1B′). The only difference between his and the present interpretation is that he supposed the upper state correlated with ${}^1\Pi_g$ rather than with ${}^1\Delta_u$. The rotational structure, on the assumption of a markedly bent molecule, indicates a considerable increase of C–S length on excitation. This is as expected for transition (1B′) and implies that progressions in ν_1' should be present. These have not yet been identified. It may be noted that ν_1', when found, is expected to be somewhat greater than its value (410 cm.$^{-1}$) in the $^1\Sigma_u^+$ state (see below) since the $a_1's_A-\pi_u$ orbital is expected to lose some of its anti-bonding character as the molecule bends; on the other hand, ν_1' should be less than ν_1'' which is 657 cm.$^{-1}$. The magnitude of the angular restoring forces is considerably reduced by the excitation, the frequency ν_2 being 275 cm.$^{-1}$ in the upper state as against 401 cm.$^{-1}$ in the ground state. Transition (1A) has not yet been identified, but may be present in the 3980—2767-Å system.

The second absorption system of CS_2 consists of a much more intense set of bands with $\lambda_{max.}$ *ca.* 1970 Å. Almost certainly this system corresponds to the CO_2 system of $\lambda_{max.}$ *ca.* 1335 Å (Price and Simpson, 1939, *loc. cit.*) and may be interpreted as (3A'). Mulliken has previously briefly suggested that the absorption represents (3). According to the present considerations, the upper state would be expected to be slightly bent, while the C–S length would be expected to be considerably increased relatively to the ground state. That the transition covers an extensive range of the spectrum (2300—1800 Å) is not therefore surprising. Price and Simpson have already concluded that the upper state is bent through a few degrees. As a result the molecule acquires one low moment of inertia and this explains the many heads, separated by about 40 cm.$^{-1}$, observed in each band under high dispersion. The bands as a whole are shaded to the red, as expected. Only one upper-state vibration appears as a long progression. This has a frequency of *ca.* 410 cm.$^{-1}$ (390—430 cm.$^{-1}$), and undoubtedly represents the symmetrical valency-stretching frequency ν_1 which is 657 cm.$^{-1}$ in the ground state.

(iv) COS. The first absorption of COS (2550—1600 Å) is much stronger than that of CO_2 or CS_2. This might be taken as evidence that the first absorption of CO_2 or CS_2 is to an upper state that correlates with $^1\Pi_g$ rather than with $^1\Delta_u$, *i.e.*, that $\bar{\sigma}_g$ lies below $\bar{\pi}_u$. The corresponding absorption of COS might then be expected to be much stronger because of the absence of a centre of symmetry in linear COS; the transition $^1\pi \longleftarrow {}^1\Sigma^+$ is allowed, whereas $^1\Pi_g \longleftarrow {}^1\Sigma_g^+$ is forbidden. An alternative explanation is equally valid, however, *viz.*, that, though the transitions $^1A_2 \longleftarrow {}^1\Sigma_g^+$ are symmetry-forbidden for CO_2 and CS_2, the analogous transitions $^1A'' \longleftarrow {}^1\Sigma^+$ are allowed for COS (apart from the Franck–Condon restriction on intensity). There should be two of these $^1A'' \longleftarrow {}^1\Sigma^+$ transitions, corresponding to the $^1A_2$ upper states of CO_2 or CS_2 correlating with $^1\Delta$ and $^1\Sigma_u^-$. In addition there should be a third transition, to a bent upper state, corresponding to the $^1B_2$ state of CO_2 or CS_2 correlating with $^1\Delta_u$. It is significant that at very low pressures (~0·01 mm. in a 1-m. path length) the absorption sub-divides into several wide diffuse bands (2380—2150, 2120—2080, and 2050—1860 Å) apparently involving three separate electronic transitions (Price and Simpson, 1939, *loc. cit.*). [That the first absorption of N_2O remains weak, while that of COS becomes comparatively strong, may be because the electronegativities of N and O are closer than those of S and O (Walsh, *Proc. Roy. Soc.*, 1951, *A*, **207**, 13), so that a bent state of NNO is not so far removed from C_{2v} symmetry as a bent state of COS.]

The second absorption region consists of some seven diffuse bands extending from *ca.* 1550 to *ca.* 1410 Å with $\lambda_{max.}$ at 1510 Å (Price and Simpson, 1939, *loc. cit.*). It undoubtedly corresponds to the CO_2 absorption of $\lambda_{max.}$ 1335 Å and the CS_2 absorption of $\lambda_{max.}$ 1970 Å. A progression probably involving ν_1' is present with a difference of *ca.* 760 cm.$^{-1}$. The corresponding frequency in the ground state is 859 cm.$^{-1}$. Price and Simpson refer to the 1510-Å system as a possible first member of a Rydberg series, but this would not accord with the present interpretation as an intra-valency-shell transition and as corresponding to the CO_2 and the CS_2 system discussed above.

(v) $HgCl_2$, $HgBr_2$, HgI_2. According to analyses by Wehrli (*Helv. Phys. Acta*, 1938, 339; 1940, **13**, 153) and Sponer and Teller (*J. Chem. Phys.*, 1939, **7**, 382; *Rev. Mod. Phys.*, 1941, **13**, p. 106), the upper states of the 1731—1670, 1862—1813, and 2108—2066 Å absorption regions of $HgCl_2$, $HgBr_2$, and HgI_2 respectively are probably $^1\Sigma_u^+$. The absorption regions are therefore to be interpreted as (3). As expected, the bands are degraded to the red; and the stretching frequencies in the upper states are considerably reduced relative to the ground state. The excited states do not depart appreciably from linearity. This of course is quite possible according to the present arguments. Although the $b_1''-\bar{\pi}_u$ curve falls from right to left in the Figure its curvature is not necessarily sufficient to cause the ${}^1\Sigma_u^+$ states to be bent. Indeed, since the b_1''–$\bar{\pi}_u$ curve falls from right to left because the orbital is bonding between the end atoms, it is not unlikely that the curvature would be less when the central atom is very large (as Hg) and the end atoms are therefore far apart, whatever the molecular shape. It is significant that ν_2' does occur in the spectra though, in accord with the selection rules for a strictly linear upper state, always as two quanta; its occurrence indicates that the forces controlling bending have changed as a result of the transition. ν_3' does not occur in the spectra. ν_2' is less than ν_2'' by a few cm.$^{-1}$ for each molecule (65, 36, and 30 cm.$^{-1}$, compared with 70, 41, and 33 cm.$^{-1}$ for $HgCl_2$, $HgBr_2$, HgI_2 respectively). In accord with the above absorption representing transitions to ${}^1\Sigma_u^+$ upper states a longer-wave-length absorption continuum has been observed for each molecule (Wieland, *Z. Physik*, 1932, **76**, 801; **77**, 157). The maxima of these continua are at *ca.* 1810 Å ($HgCl_2$), *ca.* 1950 Å ($HgBr_2$), and *ca.* 2660 Å (HgI_2). The continua should represent transitions to strongly bent ${}^1B_2$ upper states, correlating with ${}^1\Delta_u$.

(vi) NO_2. That a Rydberg series exists in the spectrum of NO_2 (Price and Simpson, *Trans. Faraday Soc.*, 1941, **37**, 106), the members of which are accompanied by very little vibrational structure, implies that a state of NO_2^+ exists which is non-linear and has an angle not very different from that in the ground state of NO_2. As Price and Simpson have already remarked, the Rydberg series can hardly proceed to the ground state of NO_2^+ since there is strong evidence that the latter is linear and the Franck–Condon principle would forbid photo-ionization to such a state without a large amount of accompanying vibration. The magnitude of the ionization limit obtained from the Rydberg series (12·3 v) makes it very probable that the series leads to the *first* excited state of NO_2^+ (certainly to a very low-lying excited state). We have, therefore, evidence that the first (or at least a very low-lying) excited state of NO_2^+ is considerably bent, as the present arguments would lead one to expect.

(vii) NO_2. From our Figure, for a ground-state angle of 143° ± 11°, the allowed transitions of longest wave-length for NO_2 should be

$$\cdots(a_2'')^2(b_2')^2(b_1''),\ {}^2B_1 \longleftarrow \cdots(a_2'')^2(b_2')^2(a_1'),\ {}^2A_1 \quad . \quad . \quad . \quad (10)$$

$$\cdots(a_2'')^2(b_2')^2(\bar{a}_1'),\ {}^2A_1 \longleftarrow \cdots(a_2'')^2(b_2')^2(a_1'),\ {}^2A_1 \quad . \quad . \quad . \quad (11)$$

$$\cdots(a_2'')^2(b_2')(a_1')^2,\ {}^2B_2 \longleftarrow \cdots(a_2'')^2(b_2')^2(a_1'),\ {}^2A_1 \quad . \quad . \quad . \quad (12)$$

Transitions of the types

$$\cdots(a_2'')(b_2')^2(a_1')^2,\ {}^2A_2 \longleftarrow \cdots(a_2'')^2(b_2')^2(a_1'),\ {}^2A_1 \quad . \quad . \quad . \quad (13)$$

$$\cdots(a_2'')^2(b_2')(a_1')(b_1''),\ {}^2A_2 \longleftarrow \cdots(a_2'')^2(b_2')^2(a_1'),\ {}^2A_1 \quad . \quad . \quad . \quad (14)$$

would be forbidden. By Mulliken's reasoning (*Rev. Mod. Phys.*, 1942, **14**, 204), the observed absorption bands of NO_2 between 9000 and 3200 Å (strongest between 5750 and 3520 Å) may be attributed to one or more of the transitions (10), (11), and (12). Pearse and Gaydon (*loc. cit.*) comment that more than one electronic system may be involved in the bands. Expressions (10), (11), and (12) supply examples of transitions that might be polarized in any of the three possible axes, might lead to an increase or a decrease of apex angle, and might markedly increase or hardly change the N–O distance. Transitions that cause a marked change of angle or N–O distance would be expected to occupy an extensive region of the spectrum.

Towards shorter wave-lengths should come the transitions

$$\cdots(a_2'')^2(b_2')(a_1')(\bar{a}_1'),\ {}^2B_2 \longleftarrow \cdots(a_2'')^2(b_2')^2(a_1'),\ {}^2A_1 \quad . \quad . \quad (15)$$

and

$$\cdots(a_2'')(b_2')^2(a_1')(b_1''),\ {}^2B_2 \longleftarrow \cdots(a_2'')^2(b_2')^2(a_1'),\ {}^2A_1 \quad . \quad . \quad (16)$$

followed at still shorter wave-lengths by

$$\cdots(b_2')(a_2'')^2(b_2')^2(a_1')^2,\ {}^2B_2 \longleftarrow \cdots(b_2')^2(a_2'')^2(b_2')^2(a_1'),\ {}^2A_1 \quad . \quad (17)$$

Both these are allowed and they are of the same symmetry type.

NO_2 has a second absorption system between 2600 and 2270 Å. Certain of the observed bands of this system are known to be of the parallel type with red-degraded

K-structure (Mulliken, 1942, *loc. cit.*), *i.e.*, polarized parallel to the axis of least moment of inertia and so to the O···O line, NO_2 being an approximately symmetric top molecule. This implies that the bands are part of a $B_2 \leftarrow A_1$ system, the transition being assumed to be electronically allowed. Mulliken chooses either (15) or (16) to represent the system. (17) might be considered as another possibility, but, whereas (15) should lead to a small *increase* of apex angle and (16) to only a small decrease, (17) should cause a very marked decrease. Rotational analysis of the 2491-Å band has yielded angles in the ground and the excited state (154° ± 4° and 154° ± 6° respectively) which differ little (Harris and King, *J. Chem. Phys.*, 1940, **8**, 775). This favours assignment to (15) or (16) rather than to (17). However, some doubt must attach to the conclusions from the rotational analysis since (*a*) the ground state angle obtained is very different from the angle reported from electron-diffraction studies (132°), and (*b*) it is not based on a complete resolution of the *J*-structure. That the angle has in fact changed from the ground state to the upper state is indicated by the appearance in the spectrum of vibrational bands probably representing the upper-state bending frequency ν_2 (Harris, King, Benedict, and Pearse, *ibid.*, p. 765). The frequency is 523 cm.$^{-1}$, compared with 648 cm.$^{-1}$ in the ground state. Expressions (15) and (16) should lead to a considerable increase of N–O distance. In agreement, from the rotational analysis of the band at 2491 Å, it has been concluded that this distance is 1·41 ± 0·06 Å in the upper state, compared with 1·28 ± 0·03 Å in the ground state. One would expect the totally symmetrical stretching vibration (probably 1320 or 1373 cm.$^{-1}$ in the ground state) to appear in the upper state. A frequency of 714 cm.$^{-1}$ possibly represents this. However, the vibrational analysis of the system is as yet far from satisfactory, one difficulty being that the separations to long wave-lengths of the apparent origin of the system do not obviously correspond to the frequencies found in the infra-red spectrum. (In order to avoid absorption by N_2O_4, the spectrum has been studied at temperatures above 100° C, so that a number of bands due to transitions from vibrational levels of the ground state appear.) The difficulty is enhanced by the fact that the Raman spectrum of NO_2 has not yet been observed (owing to the molecules absorbing throughout the visible and the near ultra-violet regions) with the result that the ν_1'' frequency is not definitely known. It should also be noted that predissociation of the bands of shorter wave-length shows that a second electronic upper state is concerned in the absorption.

A further absorption system of NO_2 occurs between 1600 and 1350 Å with $\lambda_{max.}$ at 1467 Å (Price and Simpson, *Trans. Faraday Soc.*, 1941, **37**, 106). This contains a very long vibrational progression, covering at least some fifty bands. The frequency difference is 200 cm.$^{-1}$ and almost certainly represents the $\nu_2 a_1$ deformation vibration. It follows that the equilibrium form of the upper state has an apex angle very different from that of the ground state, and also has considerably reduced angular restoring forces. This suggests that the upper orbital concerned is a_1'–π_u, whose curve rises so steeply in the Figure, and perhaps that the lower orbital is one of the bonding orbitals. In other words, it is very plausible to assign transition (17) to the system. The bands of the system should be of the parallel type, but a decision awaits study under higher dispersion.

According to the Figure, the following transition should lie very close to (17):

$$\cdots (a_1')(b_2')^2(a_2'')^2(b_2')^2(a_1')^2,\ {}^2A_1 \longleftarrow \cdots (a_1')^2(b_2')^2(a_2'')^2(b_2')^2(a_1'),\ {}^2A_1 \quad (18)$$

It too should lead to a considerable decrease of apex angle, though not quite as much as (17). It should also lead to an increase of N–O distance. It is an allowed transition that should be *z*-polarized. Price and Simpson (*loc. cit.*) find a second absorption system to be present in the 1600—1350-Å region. It too shows a long vibrational progression, the frequency difference being roughly the same as for the system identified as (17). It seems very plausible that this represents transition (18). When the 1600—1350-Å system is examined under high dispersion, it should be possible to test whether the bands ascribed to (18) are in fact of the expected perpendicular type.

A further transition,

$$\cdots (b_1'')(a_1')^2(b_2')^2(a_2'')^2(b_2')(a_1')^2,\ {}^2B_1 \longleftarrow \cdots (b_1'')^2(a_1')^2(b_2')^2(a_2'')^2(b_2')^2(a_1'),\ {}^2A_1 \quad (19)$$

should lie at slightly shorter wave-lengths than (17) and (18). Transition (19) is allowed, and x-polarized, and should lead to an increase of N–O distance and a small decrease of apex angle. According to Price and Simpson's photograph (*loc. cit.*), there are faint narrow bands in the 1350—1300-Å region (which may be a continuation of the 1600—1350-Å system) followed by a strong progression beginning at 1280 Å and undoubtedly representing one of the Rydberg transitions of the molecule. Just on the long-wave-length side of the latter occur weaker, diffuse bands which Price and Simpson ascribe to transitions from vibrating ground states. This seems unlikely in view of the Boltzmann factors involved. More plausibly, the bands either represent (19) or form part of the Rydberg transition, all being due to excitation from the vibrationless ground state and the (O, O) band not being the strongest member of the system.

The absorption in the 1600—1350-Å region is strong, requiring an appearance pressure hardly greater than that required to bring out the Rydberg bands at shorter wave-lengths. In part this can be attributed to the presence of two (and perhaps three) transitions all occurring in the same region. Further, a transition such as (17) can be regarded as belonging to the $V \longleftarrow N$ class. It is related to a $\sigma_g \longleftarrow \sigma_u$ transition, wherein there is no change of angle and the wave function of the upper orbital is of the same type as that of the lower except for the possession of an extra node along the chain of atoms. Both (17) and (18) have the characteristic of $V \longleftarrow N$ transitions in that an electron is partially transferred from the end atoms to the central atom. However, a fuller discussion of why transitions (17) and (18) should have high intensity (as well as an examination of the bands under high dispersion) is required before the assignments can be considered really satisfactory. Until this is done, the assignments must be regarded as tentative, though plausible.

Price and Simpson attribute the absorption to excitation of a "π" electron in the NO bonds. According to the present assignments, however, the absorption primarily involves only orbitals which are symmetrical with respect to reflection in the molecular plane.

(viii) SO_2. In the ground state of SO_2 the only low-lying vacant orbital (see the Figure) should be the uppermost b_1''. Next in order but considerably above b_1'' comes $\bar{a}_1'$. Remembering that the apex angle in the ground state is 119°, we see that the transitions of longest wave-length should fall into two groups. The first should sub-divide as follows :

$$\cdots (a_2'')^2(b_2')^2(a_1')(b_1''),\ {}^1B_1 \longleftarrow \cdots (a_2'')^2(b_2')^2(a_1')^2,\ {}^1A_1 \quad . \quad . \quad (20)$$

This is an allowed transition that should give rise to perpendicular bands (the molecule being assumed to be an approximately symmetric top in both lower and upper states). Further, it should lead to a small increase of apex angle and probably to a small increase of S–O length :

$$\cdots (a_2'')(b_2')^2(a_1')^2(b_1''),\ {}^1B_2 \longleftarrow \cdots (a_2'')^2(b_2')^2(a_1')^2,\ {}^1A_1 \quad . \quad . \quad (21)$$

This is an allowed transition that should give rise to parallel bands and lead to a moderate decrease of apex angle and increase of S–O length.

$$\cdots (a_2'')^2(b_2')(a_1')^2(b_1''),\ {}^1A_2 \longleftarrow \cdots (a_2'')^2(b_2')^2(a_1')^2,\ {}^1A_1 \quad . \quad . \quad (22)$$

This is a forbidden transition.

Experimentally, SO_2 is known to have absorption bands stretching from *ca.* 3900 to 2600 Å. Metropolis and Beutler (*Phys. Review*, 1940, **58**, 1078) showed that the regions from *ca.* 3900 to 3400 Å ($\lambda_{max.}$ *ca.* 3740 Å) and from *ca.* 3400 to 2600 Å ($\lambda_{max.}$ 2940 Å) belonged to separate transitions. The origin of the former is at 25,775 cm.$^{-1}$ (3880 Å). The latter region is the stronger, having a maximum molecular extinction coefficient of ~400 (Garrett, *J.*, 1915, 1324). The characteristics of the bands in the 3900—3400-Å region are as follows. They possess a J-structure degraded towards the red and a K-structure degraded towards the violet, the latter diverging more slowly than the former. Metropolis (*ibid.*, 1941, **60**, 283) has shown that this implies a small increase both of apex angle and of S–O length in the excited state. The conclusion is confirmed by the vibrational structure of the transition, especially by the absence of long progressions involving the deformation frequency. That the changes are small is also confirmed by the fact that the strongest band is that representing a combination of only one quantum of each of the totally symmetrical vibrations. The arrangement of the K sub-bands shows that the transition is probably of the perpendicular type. The 3900—3400-Å system thus agrees very well with the expected characteristics of transition (20). Mulliken has previously assigned (20) to the region.

From a vibrational analysis of the 3400—2600-Å region, Metropolis (*ibid.*, p. 295) showed that (1) the origin lay at 29,622 cm.$^{-1}$ (3376 Å), the transition thus covering a very extensive region, (2) both apex angle and bond distance change markedly in the transition, (3) the frequencies in the upper state are 794 (symmetrical stretching), 345 (symmetrical deformation), and 833 cm.$^{-1}$ (anti-symmetrical stretching), compared with 1152, 525, and 1361 cm.$^{-1}$ respectively in the ground state, (4) from the vibronic selection rules, because both a_1 and b_2 vibrations appear in the upper state, the electronic part of the latter is either $^1A_1$ or $^1B_2$, and (5) the values for the excited-state frequencies when substituted in theoretical formulæ for a valence force model yield an apex angle of probably *ca.* 100° compared with 120° in the ground state, *i.e.*, they show that the apex angle decreases in the transition. Metropolis also noted that the rotational structure showed none of the regularity present in, *e.g.*, the 3900—3400-Å bands. This implies that the dimensions of the molecule have so changed that the upper state is not, even approximately, a symmetric top. This in turn makes rotational analysis difficult, but at least makes it practically certain that the angle must have decreased and the length increased during the excitation. The characteristics of the 3400—2600-Å bands thus agree very well with identification of the transition as (21). Mulliken has previously assigned (21) to the system.

The second group of transitions expected from the Figure is as follows :

$$\cdots (b_2')(a_2'')^2(b_2')^2(a_1')^2(b_1''),\ ^1A_2 \longleftarrow \cdots (b_2')^2(a_2'')^2(b_2')^2(a_1')^2,\ ^1A_1 \quad . \quad (23)$$

This is a forbidden transition.

$$\cdots (a_1')(b_2')^2(a_2'')^2(b_2')^2(a_1')^2(b_1''),\ ^1B_1 \longleftarrow \cdots (a_1')^2(b_2')^2(a_2'')^2(b_2')^2(a_1')^2,\ ^1A_1 \quad . \quad (24)$$

This is an allowed transition which should lead to a moderate decrease of apex angle and a moderate increase of S–O distance. It should give rise to perpendicular bands.

$$\cdots (b_1'')(a_1')^2(b_2')^2(a_2'')^2(b_2')^2(a_1')^2(b_1''),\ ^1A_1 \longleftarrow$$
$$\cdots (b_1'')^2(a_1')^2(b_2')^2(a_2'')^2(b_2')^2(a_1')^2,\ ^1A_1 \quad . \quad (25)$$

This is also an allowed transition which should give rise to bands of the perpendicular type. It should cause a greater increase of S–O distance than transition (24) [because the (a_1') orbital is less S$\longleftrightarrow$O bonding than the (b_1'')], but comparatively little change in apex angle.

The following two allowed transitions may lie fairly close to the second group above :

$$\cdots (b_2')^2(a_1')(a_1'),\ ^1A_1 \longleftarrow \cdots (b_2')^2(a_1')^2,\ ^1A_1 \quad . \quad . \quad . \quad . \quad (26)$$
$$\cdots (b_2')(a_1')^2(\bar{a}_1'),\ ^1B_2 \longleftarrow \cdots (b_2')^2(a_1')^2,\ ^1A_1 \quad . \quad . \quad . \quad . \quad (27)$$

Transition (26) would be perpendicular, and (27) parallel. The former should cause a very marked, and the latter a small increase of apex angle.

Experimentally, SO_2 shows a region of absorption from *ca.* 2400 to 1800 Å ($\lambda_{max.}$ *ca.* 2000 Å). Its maximum molecule extinction coefficient is several times that of the 3400—2600-Å system. Duchesne and Rosen (*J. Chem. Phys.*, 1947, **15**, 631) have shown, by vibrational analysis, that at least two and probably three electronic transitions are involved in the region.* The first of these has its origin at 42,170 cm.$^{-1}$ (2371 Å), the vibrational frequencies appearing in the upper state being 963 (symmetrical stretching)

* The possible third transition is to an upper state having vibrational frequencies of 845 (symmetrical stretching) and 360 cm.$^{-1}$ (symmetrical bending). Rosen (*J. Phys. Radium*, 1948, **9**, 155) gives its origin as 45,499 cm.$^{-1}$ (2198 Å) and mentions the existence of a fourth transition whose origin is at 47,510 cm.$^{-1}$ (2105 Å) with upper-state frequencies of ~800 (ν_1) and 350 cm.$^{-1}$ (ν_2).

and 379 cm.$^{-1}$ (symmetrical bending). The high intensity and the absence of anti-symmetrical vibrations show that the transition is an allowed one. The partially resolved rotational structure of the bands of this system shows that they are of the perpendicular type, the *K*-structure being degraded towards the red. The *J*-structure remains almost completely unresolved, but seems also to be degraded towards the red. Duchesne and Rosen conclude that the transition causes a fairly marked decrease in apex angle and a small increase in bond length. These conclusions are the more reasonable when the upper-state frequencies are compared with those obtained by Metropolis for the 3400—2600-Å system. In agreement with the small increase in bond length, transitions with many quanta of the symmetrical stretching frequency are not strong. These conclusions are not as definite as one could wish, but are in fair agreement with the characteristics expected for transition (24). Mulliken has assigned (26) and (27) to the *ca.* 2000-Å region, but both these should cause a considerable increase of apex angle; in addition, (27) would be of the parallel type.

The second electronic transition involved in the 2400—1800-Å system has its origin at 44,236 cm.$^{-1}$ (2260 Å) and is stronger than the first. The vibrational frequencies appearing in the upper state are 775 (ν_1) and 375 cm.$^{-1}$ (ν_2). The intensity and the absence of anti-symmetrical vibrations make it virtually certain that the transition is electronically allowed. No rotational analysis has yet been made, except to show that the structure is not identical with that of the first transition in the region. However, the magnitude of the symmetrical stretching frequency suggests that the bond length has increased more in the second electronic transition than in the first. This conclusion is supported by the appearance of strong bands representing many quanta of the symmetrical stretching frequency. The rather scanty known characteristics of the transition are therefore not incompatible with its assignment to (25).

If the possible third and fourth transitions in the 2400—1800-Å region should be confirmed, the expectation is that they will be found to represent (26) and (27).

(ix) SeO_2. Selenium dioxide possesses the following absorption regions :

(*a*) 5000—3400 Å ($\lambda_{max.} \sim$ 4080 Å) (Duchesne and Rosen, 1947, *loc. cit.*). The origin appears to be at ~4570 Å. Prominent progressions involving the symmetrical bending frequency of the upper state (~200 cm.$^{-1}$) are present. Progressions involving the symmetrical stretching frequency are not prominent. It follows that whereas the apex angle changes appreciably in the transition, the Se–O bond length changes little. The absorption is fairly strong. This and the complete absence of anti-symmetrical vibrations show the transition to be allowed.

Duchesne and Rosen (*Nature*, 1946, **157**, 692) at first suggested that the absorption region might be analogous to that of SO_2 between 3900 and 3400 Å. Later, because this SO_2 absorption causes a more marked change of length than of angle, while the SeO_2 absorption causes a change of angle more marked than that of bond length, they abandoned this suggestion in favour of one whereby the SeO_2 transition was analogous to the SO_2 transition of origin at 2371 Å. The latter suggestion seems the more probable also because the former entails that the next absorption system of SeO_2 (3300—2300 Å) would be analogous to the SO_2 3400—2600-Å system and therefore show little or none of the expected long-wave-length shift. This implies that two electronic absorption systems of SeO_2 should be found in the infra-red region, and that the 5000—3400-Å region may be tentatively interpreted as (24).

(*b*) 3300—2300 Å ($\lambda_{max.} \sim$ 2700 Å) (Duchesne and Rosen, *Physica*, 1941, **8**, 540). The origin is at 3000 Å. The most prominent feature or this absorption is the presence of progressions involving the symmetrical stretching vibration (*ca.* 665 cm.$^{-1}$ in the upper state compared with 900 or 910 cm.$^{-1}$ in the ground state). It therefore appears that the transition causes a marked change of bond length but little change of apex angle. It is probably analogous to the SO_2 system of origin at 2260 Å, and may be tentatively identified as (25).

(x) TeO_2. Tellurium dioxide possesses an absorption region from 4500 and 3000 Å with $\lambda_{max.}$ near 3550 Å (Duchesne and Rosen, 1947, *loc. cit.*). The only upper-state frequency that appears strongly is one of 650 cm.$^{-1}$, representing the symmetrical stretching frequency (815 cm.$^{-1}$ in the ground state). This indicates that whereas the length has increased considerably during the excitation, the apex angle has changed comparatively little. There is little doubt that the system corresponds to that at 3300—2300-Å for SeO_2. It may be interpreted as (25). There should be several systems of longer wave-length to be found.

(xi) O_3. Ozone is an 18-electron molecule and has a ground-state apex angle not greatly different from that of SO_2. Its spectroscopic transitions should therefore correspond to those of SO_2. It has a weak absorption system in the visible region (7585—4380 Å; $\varepsilon_{max.} = 1{\cdot}12$), centred particularly on two strong, diffuse bands in the orange at about 5730 and 6020 Å (Colange, *J. Phys. Radium*, 1927, **8**, 254; Wulf, *Proc. Nat. Acad. Sci.*, 1930, **16**, 507). There are also very weak absorption bands in the infra-red (Wulf, *loc. cit.*). We suggest that one of these weak absorption regions represents transition (20), and that they therefore correspond to the SO_2 absorption of $\lambda_{max.}$ *ca.* 3740 Å. Ozone has a fairly strong absorption system in the region 2900—2300 Å ($\lambda_{max.}$ 2550 Å; $\varepsilon_{max.} \sim 2800$), which has about the same maximum extinction coefficient as the 2400—1800-Å system ($\lambda_{max.}$ *ca.* 2000 Å) of SO_2 (Wulf and Melvin, *Phys. Review*, 1931, **38**, 330). We suggest, therefore, that the 2900—2300-Å system is to be interpreted as involving some of the transitions (24) to (27). Transition (21) may give rise to known weaker absorption bands of O_3 between 3525 Å and the long-wave-length end of the 2900—2300-Å system (Fowler and Strutt, *Proc. Roy. Soc.*, 1917, *A*, **93**, 577; Wulf and Melvin, *loc. cit.*; Jakowlewa and Kondratjew, *Physikal. Z. Sowjetunion*, 1932, **1**, 471). If so, these absorption bands correspond to the 3300—2600-Å system for SO_2 ($\lambda_{max.}$ 2940 Å). The bands are diffuse and tend to degrade to the red, as expected for bands belonging to transition (21), but with no rotational structure resolved. Upper-state vibrational frequencies of $\sim$300 and $\sim$600 cm.$^{-1}$ appear. The former is probably ν_2 (705 cm.$^{-1}$ in the ground state), and the latter probably ν_1 (1110 cm.$^{-1}$ in the ground state). If these tentative interpretations are accepted, each transition in O_3 is shifted to long wave-lengths of its analogue in SO_2. This appears resonable, at least for those transitions [(21), (24), (25) and (27)] that involve transfer of an electron from the end atoms to the central atom.

Ozone emission bands have been observed between 4465 and 3090 Å in a mild condensed discharge through oxygen (Johnson, *Proc. Roy. Soc.*, 1924, *A*, **105**, 683). It is probable that these represent the reverse of one of the transitions involved in the 2900—2300-Å system. Transitions involving a considerable decrease in apex angle in the equilibrium form of the upper state could appear in emission (see the Figure) well to the long-wave-length side of the corresponding absorption bands.

(xii) NOCl. Nitrosyl chloride is a further 18-electron molecule, whose transitions should therefore correspond to those of SO_2 and O_3. It possesses two systems of weak absorption in the visible region (Goodeve and Katz, *Proc. Roy. Soc.*, 1939, *A*, **172**, 432). The shorter-wave-length system (maximum at 4750 Å) shows little structure. The longer-wave-length system (maximum at 6017 Å) is associated with seven more or less discrete, red-degraded bands, showing frequency differences of 1580 and 380 cm.$^{-1}$. The ground-state frequencies of nitrosyl chloride are given by Burns and Bernstein (*J. Chem. Phys.*, 1950, **18**, 1669) as 1799 cm.$^{-1}$ (a stretching vibration mainly localized in the O–N bond), 592 cm.$^{-1}$ (a stretching vibration mainly localized in the N–Cl bond), and 332 cm.$^{-1}$ (bending). The 1580-cm.$^{-1}$ upper-state frequency can only well correspond to the 1799-cm.$^{-1}$ ground-state frequency. The 380-cm.$^{-1}$ upper-state frequency may represent either the 592- or 332-cm.$^{-1}$ ground-state frequency. Both the banded absorption ($\varepsilon_{max.} \sim 1$) and the shorter-wave-length visible absorption ($\varepsilon_{max.} \sim 5$) have very low intensity. While it is possible, therefore, that one of the systems corresponds to the 3900—3400-Å region for SO_2, assignment is quite uncertain. The 3900—3400-Å region for SO_2 has been interpreted as (20). These transitions may readily be re-formulated in terms of the symbols appropriate to the C_s symmetry of NOCl.

A much stronger region of continuous absorption has a maximum at about 1900 Å (Price and Simpson, 1941, *loc. cit.*). With increasing pressure, this absorption spreads somewhat to short wave-lengths and much more to long wave-lengths. Price and Simpson suggest it is the analogue of the 1600—1350-Å system for NO_2 and the 2400—1800-Å system for SO_2. In terms of the present assignments the former analogy cannot be very close, since the analogue of the $(a_1'-\bar{\pi}_u)$ orbital is completely filled in the ground state of NOCl. Closer analogies are likely to be found with other 18-electron rather than with 17-electron molecules. We agree with Price and Simpson that the 1900-Å region of NOCl is probably analogous to the 2400—1800-Å system of SO_2. Its interpretation is therefore probably analogous to some of the transitions (24) to (27). Like the SO_2 system, the absorption is quite strong, $\varepsilon_{max.}$ being >2000 (Goodeve and Katz, *loc. cit.*).

Between the absorption in the visible region and that of the 1900-Å system, a broad maximum of absorption occurs (see Goodeve and Katz, *loc. cit.*), some of which may represent transition (21).

(xiii) CF_2. An uncondensed discharge through the vapour of a fluorocarbon gives rise to an extensive emission spectrum of many-headed bands lying between 2340 and *ca.* 5000 Å. A vibrational analysis of the shorter-wave-length end of this system has been made by Venkateswarlu (*Phys. Review*, 1950, **77**, 676) and the transition was assigned to the molecule CF_2. The analysis, and also the presence of both J- and K-rotational structure, of the bands shows, as expected from the present considerations, that the molecule is not linear. The frequencies of the sub-heads (*i.e.*, the K-structure) fit the relation expected for parallel bands.

Laird, Andrew, and Barrow (*Trans. Faraday Soc.*, 1950, **46**, 803) have observed some of the same bands in absorption between 2350 and 2650 Å. This establishes that the lower state is the ground state, so that the polarization of the transition identifies it as $^1B_2 \longleftarrow {}^1A_1$. The vibrations to appear in the emission transition are the two that are totally symmetrical (ν_1 and ν_2). It follows that $^1B_2$ is the symmetry of the electronic part of the upper-state wave function. The vibrational structure then agrees with the transition being an electronically allowed one. In absorption, the bands form a single progression with successive excitation of the bending vibration in the upper electronic state (495·5 cm.$^{-1}$ compared with 666·5 cm.$^{-1}$ in the ground state). This makes it clear that the apex angle changes considerably during the transition. Venkateswarlu found the K-structure of the bands to be shaded towards the violet, while the J-structure is shaded towards the red. This implies that the apex angle is larger in the upper than in the ground state; the question whether the bond length has increased or decreased must be left open. The existence, in the emission spectrum, of a long progression involving the symmetrical stretching frequency (1162 cm.$^{-1}$ in the ground state, 750 cm.$^{-1}$ in the upper state) but with the first and second bands the strongest indicates that the change of bond length is appreciable but not large.

Considering only the transitions so far formulated for an 18-electron molecule, we find only one—*viz.*, (27)—that is a parallel transition and leads to an increase of apex angle. It may therefore be that the transition just described is to be interpreted as (27). On the other hand, one would not expect (27) to lead to a very marked increase of apex angle. There is another possibility not so far formulated; it is (see the Figure)

$$\cdots (a_1')(\bar{b}_2'),\ ^1B_2 \longleftarrow \cdots (a_1')^2,\ ^1A_1 \quad . \quad . \quad . \quad . \quad . \quad . \quad . \quad (28)$$

where $(\bar{b}_2')$ represents the $(\bar{b}_2' - \bar{\sigma}_u)$ orbital. This should lead to a marked increase of both apex angle and bond length. Mulliken (quoted by Venkateswarlu) has earlier suggested this assignment. If correct, it means that there should be several longer-wave-length systems of CF_2. Weak bands on the long-wave-length side (3700—3300 Å) of the main part of the above emission system may represent one of these additional systems.

(xiv) ClO_2. Chlorine dioxide is a 19-electron molecule. According to the Figure its ground state should be $^2B_1$.

In absorption, the molecule possesses an extensive system of red-degraded absorption bands lying between 5225 and 2600 Å (Coon, *Phys. Review*, 1940, **58**, 926L; *J. Chem. Phys.*, 1946, **14**, 665). The (000) $\longleftarrow$ (000) band is at 21,016 cm.$^{-1}$ and $\lambda_{max.}$ is at *ca.* 3300 Å. The transition is a strong one ($\varepsilon_{max.}$ at least 2000; Goodeve and Stein, *Trans. Faraday Soc.*, 1929, **25**, 738) and is therefore to be regarded as electronically allowed. Rotational analysis of the bands (Coon, 1946, *loc. cit.*) shows them to be of parallel type. Of the longest-wave-length group of expected allowed transitions there are only two of parallel type. These may be formulated

$$\cdots (a_2'')(b_2')^2(a_1')^2(b_1'')^2,\ {}^2A_2 \longleftarrow \cdots (a_2'')^2(b_2')^2(a_1')^2(b_1''),\ {}^2B_1 \quad . \quad . \quad (29)$$

$$\cdots (b_2')(a_1')^2(b_1'')(\bar{a}_1'),\ {}^2A_2 \longleftarrow \cdots (b_2')^2(a_1')^2(b_1''),\ {}^2B_1 \quad . \quad . \quad . \quad (30)$$

Transition (29) should cause a small decrease, and (30) a small increase, of apex angle. Both should cause a considerable increase of Cl–O distance and should occupy a fairly extensive region of the spectrum. Coon (1940, *loc. cit.*) points out that twenty or more bands due to the symmetrical stretching vibration appear in the transition, but not more than two of the symmetrical bending vibration; this indicates a much more profound change of Cl–O length than of OClO angle. The magnitude of both the breathing and the bending frequency is reduced from the ground state to the upper state (945 to 708, and 447 to 290 cm.$^{-1}$, respectively). Further, the rotational analysis reveals that both K- and J-structure degrade to the red at approximately the same rate. This confirms that the change of length is more important than that of angle. Coon showed in fact that, while the Cl–O distance increased markedly, there was a small decrease in apex angle. (The figures given for the angle were $109° \pm 3°$ in the ground state and $92° \pm 6°$ in the upper state; but these were based upon an old electron-diffraction value for the Cl–O bond length and are subject to correction.) The observed characteristics of the system thus fit with assignment to (29). Mulliken earlier came to the same conclusion. Assignment to (29) makes the electron jump analogous to that supposed responsible for the SO_2 3400—2600-Å bands. It should be noted, however, that the latter bands are considerably weaker.

UNIVERSITY OF LEEDS. [*Received, May 15th*, 1952.]

---

468. *The Electronic Orbitals, Shapes, and Spectra of Polyatomic Molecules. Part III.** HAB *and* HAAH *Molecules.*

By A. D. WALSH.

The electronic orbitals possible for bent and linear HAB and HAAH molecules are correlated. Whether or not a given orbital becomes more or less weakly bound as the molecule is changed from the bent to the linear form is discussed. The results are used to interpret the shapes and spectra of HAB and HAAH molecules and radicals. HAB molecules containing 10 or less valency electrons should be linear in their ground states. Molecules with 10—14 electrons should be bent in their ground states. 16-Electron HAB molecules should be linear again.

HAAH molecules containing 10 valency electrons should be linear in their ground states. Those containing 12 electrons should be bent but planar (*cis*- and *trans*-forms). Those containing 14 electrons should be bent and non-planar. The spectra of the isoelectronic molecules HCN and C_2H_2 are particularly discussed; the first excited state of each should be non-linear.

Orbitals of HAB *Molecules.*—The lowest-energy orbitals of a linear HAB molecule may be approximately described as follows, on the assumption that they are built solely from s and p atomic orbitals: (i) An s orbital on B. It is assumed that this plays no direct part in the binding of A to B and remains largely unchanged whatever the HAB angle. (ii) A σ orbital binding the H and the A atom. It is built from an H 1s and an A sp hybrid valency. (iii) A σ orbital binding the B and the A atom. It is assumed to be built from an A sp hybrid valency and a B pure p valency. (iv) A π orbital binding the B and the A atom. It is built by the in-phase overlap of p orbitals on A and B. It is two-fold degenerate; and, since in many HAB molecules B is of greater electronegativity

* Part II, preceding paper.

than A, it is usually more localized on B than on A. (v) A $\bar{\pi}$ orbital that is anti-bonding between A and B. It is built by the out-of-phase overlap of p orbitals in A and B. It is two-fold degenerate and, by the arguments of Part II, usually more localized on A than on B. (vi) A σ orbital built by the out-of-phase overlap of an orbital on A with a B $p\sigma$ and the H 1s valency. It will usually be more localized on A than on B or H. It is analogous to the $\bar{\sigma}_g$ orbital described in Part II and therefore is assumed to be built from an s orbital of A.

The lowest-energy orbitals of a bent HAB molecule, in addition to the s lone-pair orbital on B, can be approximately described as follows: (i) An a' orbital binding the H and the A atom. When the HAB angle is 90°, the orbital is built from an H 1s and an A pure p valency. (ii) An a' orbital binding the A and the B atom. In the 90° molecule, it is built by the in-phase overlap of pure p valencies on A and B, pointed towards each other. In Fig. 1 the orbitals of the bent and the linear HAB molecules are correlated. Clearly, the two bonding a' orbitals must become the two σ orbitals; and, as explained in Parts I and II, the a'–σ curves must fall from left to right. (iii) An a'' orbital binding the A and the B atom and built by the in-phase overlap of a p orbital on A and a p orbital on B, the axes of these p orbitals being perpendicular to the molecular plane. Clearly, this orbital must become one of the bonding π orbitals in the linear molecule. There is no reason why it should change appreciably in binding energy as the HAB angle changes. It is therefore represented in Fig. 1 by a horizontal straight line. (iv) An a'' orbital that is built by the out-of-phase overlap of p orbitals on A and B, their axes being perpendicular to the molecular plane. It is anti-bonding between A and B, and becomes one of the anti-bonding $\bar{\pi}$ orbitals in the linear molecule. Like the other a'' orbital, there is no reason why its binding energy should change appreciably as the HAB angle changes. It is therefore represented in Fig. 1 by a horizontal straight line. (v) An $\bar{a}'$ orbital analogous to the $\bar{a}_1'$ orbital of Part II. By analogy with the latter, when the HAB angle is 90°, the orbital is built from a pure p_z orbital on A overlapping out-of-phase with a B pa' and the H 1s valency. The z-axis lies in the plane of the molecule and bisects the HAB angle. The orbital is usually predominantly localized on A. As the apex angle increases, the orbital tends towards the $\bar{\sigma}$ orbital of the linear molecule. Since it is largely a p orbital of A in the 90° molecule, but an s orbital of A in the linear molecule, it is represented in Fig. 1 by a line that falls steeply from left to right. (vi) An orbital which in the 90° molecule is a pure s lone pair orbital on A. It is of species a', here written $a's_A$. As the HAB angle changes towards 180° the orbital becomes increasingly built from a p orbital of A whose axis lies in the molecular plane. It is therefore represented in Fig. 1 by a line that rises steeply from left to right. Being localized largely on A, it tends to become one of the $\bar{\pi}$ (rather than one of the π) orbitals of the linear molecule. As the HAB angle approaches 180° it becomes, to some extent, A$\leftarrow\rightarrow$B anti-bonding. If the $\bar{\sigma}$ lies slightly lower than the $\bar{\pi}$ orbital, the $a's_A$ orbital curve will actually lead to the $\bar{\sigma}$ rather than to the $\bar{\pi}$ orbital. There is every expectation, however, that this is not so and that $\bar{\sigma}$ lies above $\bar{\pi}$. In the diatomic molecules isoelectronic with HAB molecules (*e.g.*, N_2 isoelectronic with HCN) the analogous $\bar{\sigma}$ orbital certainly lies above the analogous $\bar{\pi}$ orbital. Moreover, analysis (see below) of the spectrum of C_2H_2 (also isoelectronic with HCN) fits well with the assumption that $\bar{\pi}$ lies below $\bar{\sigma}$. For these reasons and also on grounds of simplicity Fig. 1 is drawn with $\bar{\sigma}$ above $\bar{\pi}$. (vii) An a' orbital which in the 90° molecule is a nearly pure p orbital on B whose axis lies in the molecular plane but perpendicular to the A–B line. It probably has a sufficiently different binding energy from the $a's_A$ orbital (or to the two bonding a' orbitals) for little interaction to occur. As the HAB angle increases towards 180°, this a' orbital interacts more and more with the orbital which at 90° is $a's_A$ because the latter becomes increasingly built from a p orbital of A. This interaction increases the binding energy of the present a' orbital. In other words, the present a' orbital becomes increasingly A$\leftarrow\rightarrow$B bonding as the apex angle increases and is represented in Fig. 1 by a line that descends from left to right. At 180° the orbital becomes one of the bonding π orbitals.

Fig. 1 shows the eight lowest-lying intra-valency-shell orbitals. A ninth, highest, is also possible but is not needed for the purposes of this paper; see also the comments on this point in Part I.

CONTRIBUTION FROM THE INSTITUTE FOR PHYSICAL CHEMISTRY
UNIVERSITY OF COPENHAGEN, DENMARK

The Electronic Structure of the Vanadyl Ion[1]

BY C. J. BALLHAUSEN AND HARRY B. GRAY[2]

Received October 2, 1961

The bonding in the molecule ion $VO(H_2O)_5^{2+}$ is described in terms of molecular orbitals. In particular, the most significant feature of the electronic structure of VO^{2+} seems to be the existence of considerable oxygen to vanadium π-bonding. A molecular orbital energy level scheme is estimated which is able to account for both the "crystal field" and the "charge transfer" spectra of $VO(H_2O)_5^{2+}$ and related vanadyl complexes. The paramagnetic resonance g factors and the magnetic susceptibilities of vanadyl complexes are discussed.

Introduction

The high oxidation states of metal ions occurring at the beginning of the transition and actinium series usually are found in complex oxycations of the types MO^{n+} and MO_2^{n+}. The remarkable stability of these complexes, along with their interesting spectral and magnetic properties, has aroused considerable theoretical speculation concerning their electronic structures. The uranyl ion, UO_2^{2+}, has been discussed most often,[3] but partly due to the lack of good wave functions for

(1) Presented at the Symposium on Ligand Field Theory, 140th National A.C.S. Meeting, Chicago, September, 1961.

(2) National Science Foundation Postdoctoral Fellow, 1960–61.

(3) For example, see R. L. Belford and G. Belford, *J. Chem. Phys.*, **34**, 1330 (1961).

the uranium atom, many features of the electronic structure of UO_2^{2+} are still uncertain.

One of the simplest oxycations of the above type is the vanadyl ion, VO^{2+}. It may be formulated as containing V^{4+}, with the electronic structure [Argon] $3d^1$, and an oxide ion. As might be expected, VO^{2+} always occurs coördinated to other groups both in the solid state and in solution, bringing the total coördination number of vanadium to five or six. Many complexes containing the VO^{2+} ion have been described, and in most cases they have a characteristic blue or purple color. For example, a number of common bidentate ligands form 2:1 complexes with VO^{2+}.[4]

The energy level scheme for vanadyl has been considered by Jørgensen[5] and by Furlani,[6] both using a simple crystal field model. Furlani's calculation considered only the $C_{\infty v}$ symmetry of VO^{2+} alone, and therefore cannot hope to account for all the observed levels. By considering VO^{2+} in aqueous solution as a tetragonal $VO(H_2O)_5^{2+}$ molecule ion, with axial destabilization, Jørgensen obtains a level scheme which qualitatively accounts for the "crystal field" part of the spectrum. However, Palma-Vittorelli, *et al.*,[7] first pointed out that the electrostatic model could not account for the observed magnetic properties of $VOSO_4 \cdot 5H_2O$ and concluded that π-bonding between vanadium and oxygen must be important.

There is now an appreciable collection of structural, spectral, and magnetic data available for the vanadyl ion and its complexes. Therefore it seems desirable to develop a theory of the electronic structure of VO^{2+} which will be consistent with its physical properties. It also might be hoped that an understanding of the principal features of the bonding in VO^{2+} will be a helpful guide in attempts to develop a general theory of the electronic structures of MO^{n+} and MO_2^{n+} complexes.

Structure of the Vanadyl Ion Complexes.— The structures of at least three different compounds containing vanadyl ion have been determined by X-ray methods. It is significant that VO_2 crystallizes in a highly distorted rutile (TiO_2) structure, in which there is one conspicuously short V–O bond (1.76 Å.) in each VO_6 unit.[8] Thus there seems to be a greater driving force to form VO^{2+} in VO_2 than there is to form TiO^{2+} in TiO_2, since the TiO_6 units in the rutile structure contain no distinguishable TiO^{2+} fragments.

Anhydrous $VO(aca)_2$ (aca is acetylacetone) appears to be only five-coördinated in the crystal, with the VO^{2+} group perpendicular to the oxygen base of a square pyramid.[9] The V–O bond length in this complex is only 1.59 Å.

The structure of $VO(SO_4) \cdot 5H_2O$ is a distorted octahedron, which clearly contains the VO^{2+} group situated perpendicular to a base containing the four water oxygens.[7] The V–O bond length for VO^{2+} is 1.67 Å., while the V–O bond lengths to the water ligands are approximately 2.3 Å. A sulfate oxygen completes the tetragonal structure by occupying the other axial position. The V atom is coplanar with the water oxygens.

The vanadyl ion in aqueous solution presumably has an analogous tetragonal structure; a $V–O^{2+}$ group with five more distantly coördinated water molecules completing the coördination sphere. There are other possibilities for the solution structure, of course. For example, potentiometric measurements confirm the existence of the doubly-charged vanadyl cation in aqueous solutions of V^{4+}, but fail to tell us whether it is VO^{2+} or $V(OH)_2^{2+}$.[10] However, the accumulated spectral and magnetic evidence, which will be discussed later, strongly support the assumption that the vanadyl ion actually retains its VO^{2+} identity in solution, and is surrounded by water molecules to complete a distorted octahedral array.

The electrostatic model for the hydrated vanadyl ion consists of V^{4+} situated in a tetragonal electric field caused by the oxide ion and five water dipoles. The crystal field energy level diagram for such a situation is given in Fig. 1.

The parameters Ds and Dt specify the degree of tetragonality present in the field.[11] If the tetragonal perturbation results in axial compression, as in $VO(H_2O)_5^{2+}$, the axial a_1 orbital is less stable than b_1, but the ordering of the e and b_2 orbitals depends on the relative values of Ds and Dt. Magnetic data on vanadyl complexes which will be discussed later indicate an orbitally non-degenerate ground state, and so the e orbitals are less stable than b_2 in this case. Thus for the ground state configuration the one d electron in

(4) M. M. Jones, *Z. Naturforsch.*, **12b**, 595 (1957).
(5) C. K. Jørgensen, *Acta Chem. Scand.*, **11**, 73 (1957).
(6) C. Furlani, *Ricerca sci.*, **27**, 1141 (1957).
(7) M. B. Palma-Vittorelli, M. U. Palma, D. Palumbo, and F. Sgarlata, *Nuovo cimento*, **3**, 718 (1956).
(8) G. Andersson, *Acta Chem. Scand.*, **10**, 623 (1956).
(9) R. P. Dodge, D. H. Templeton, and A. Zalkin, *J. Chem. Phys.*, **35**, 55 (1961).
(10) F. J. C. Rossotti and G. S. Rossotti, *Acta Chem. Scand.*, **9**, 1177 (1955).
(11) W. Moffitt and C. J. Ballhausen, *Ann. Rev. Phys. Chem.*, **7**, 107 (1956).

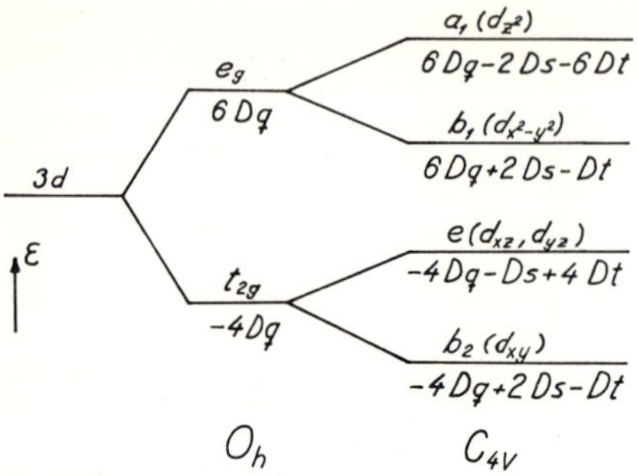

Fig. 1.—Energy levels in crystalline fields of O_h and compressed C_{4v} symmetry, with $(-3Ds + 5Dt) > 0$.

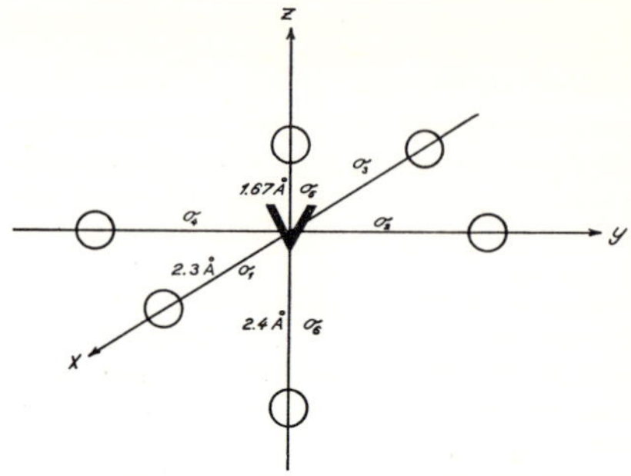

Fig. 2.—Structure of the $VO(H_2O)_5^{+2}$ molecule ion.

VO^{2+} is placed in the b_2 orbital. The predicted transitions are $b_2 \rightarrow e\ (-3Ds + 5Dt)$, $b_2 \rightarrow b_1\ (10Dq)$, and $b_2 \rightarrow a_1\ (10Dq - 4Ds - 5Dt)$. The spectrum of $VOSO_4 \cdot 5H_2O$ in aqueous solution shows two crystal field bands, at 13,000 cm.$^{-1}$ and 16,000 cm.$^{-1}$, which from Fig. 1 can be assigned to the transitions $b_2 \rightarrow e$ and $b_2 \rightarrow b_1$, respectively. The $b_2 \rightarrow a_1$ transition is expected at higher energies, but it is not observed, presumably being covered by the broad charge transfer band which sets in at about 30,000 cm.$^{-1}$.

The value of $10Dq$ is obtained directly from the $b_2 \rightarrow b_1$ transition, which gives $Dq = 1600$ cm.$^{-1}$ for V^{4+} in aqueous solution. This Dq is considerably smaller than a value of 2600 cm.$^{-1}$ which might be expected for V^{4+} by extrapolating the Dq's for $V(H_2O)_6^{2+}$ (1220 cm.$^{-1}$) and $V(H_2O)_6^{3+}$ (1900 cm.$^{-1}$).[12] The values of Ds and Dt can be calculated by making the reasonable assumption that the $b_2 \rightarrow a_1$ transition for $VOSO_4 \cdot 5H_2O$ occurs at approximately 35,000 cm.$^{-1}$ (this transition is actually observed in a number of vanadyl complexes in the neighborhood of 30,000 cm.$^{-1}$). This calculation gives $Ds = -4570$ cm.$^{-1}$ and $Dt = -143$ cm.$^{-1}$. These values may be compared with the values $Ds = -117$ cm.$^{-1}$, $Dt = -33$ cm.$^{-1}$, for tetragonal cobaltous oxide.[13]

From these results it is clear that a rather exaggerated tetragonal distortion is present in $VO(H_2O)_5^{2+}$, and that a pure crystalline field model, that is, a model which only considers σ-bonding to be present, cannot provide an adequate description of the electronic structure of VO^{2+}. It thus is evident that an accurate description of the electronic structure of the vanadyl ion and its complexes must include provisions for π-bonding, and this will be accomplished by the method of molecular orbitals.

The Molecular Orbital Description of Vanadyl Ion.—Since the crystal structure of $VOSO_4 \cdot 5H_2O$ is known, the molecule ion $VO(H_2O)_5^{2+}$ will be taken as the example. The ligand oxygens will be numbered as shown in Fig. 2. In this model the axial sulfate oxygen, present in the crystal, is replaced with a water oxygen. Such a substitution does not affect any of the energy states of interest, and has a conceptual advantage for a discussion of the VO^{2+} ion in aqueous solution.

The 3d, 4s, and 4p metal orbitals will be used for bonding, along with the 2s, $2p_\sigma$ $(2p_z)$, and $2p_\pi$-$(2p_x, 2p_y)$ orbitals of oxide oxygen, and the sp_σ hybrid orbitals for the water oxygens. In view of the longer vanadium to oxygen bond lengths (2.3 Å.), π-bonding involving the water oxygens seems unlikely, and thus will be ignored.

The transformation scheme for the metal and ligand orbitals in C_{4v} is given in Table I. In specifying the form of the molecular orbitals, use has been made of the fact that the vanadyl VO bond is undoubtedly the strongest link, the four waters in the square plane are equivalent and are attached more strongly than the axial water molecule, which is the weakest link of all. With this in mind, the bonding in $VO(H_2O)_5^{2+}$ can be pictured as follows: a strong σ bond of symmetry a_1 between the sp_σ oxygen hybrid orbital and the $(4s + 3d_{z^2})$ vanadium hybrid orbital (there is some experimental justification for sorting out a localized σ molecular orbital involving only the $3d_{z^2}$ and 4s vanadium orbitals, and the 2s and $2p_\sigma$ oxygen orbitals—this is the fact that no nitrogen

(12) L. E. Orgel, *J. Chem. Phys.*, **23**, 1819 (1955).
(13) T. S. Piper and R. L. Carlin, *ibid.*, **33**, 1208 (1960).

TABLE I

ORBITAL TRANSFORMATION SCHEME IN C_{4v} SYMMETRY

Representation	Metal Orbitals	Ligand Orbitals
a_1	$3d_{z^2} + 4s$	σ_5
	$4s - 3d_{z^2}$	$\frac{1}{2}(\sigma_1 + \sigma_2 + \sigma_3 + \sigma_4)$
	$4p_z$	σ_6
e	$3d_{xz}, 3d_{yz}$	$\pi_5(2p_x, 2p_y)$
	$4p_x, 4p_y$	$\frac{1}{\sqrt{2}}(\sigma_1 - \sigma_3), \frac{1}{\sqrt{2}}(\sigma_2 - \sigma_4)$
b_1	$3d_{x^2-y^2}$	$\frac{1}{2}(\sigma_1 - \sigma_2 + \sigma_3 - \sigma_4)$
b_2	$3d_{xy}$	

hyperfine structure is found in the paramagnetic resonance spectrum of vanadyl porphyrins)[14,15]; two π bonds of symmetry e between the oxygen $2p_x$ and $2p_y$ orbitals and the vanadium $3d_{xz}$ and $3d_{xz}$ orbitals, making a total of three vanadium to oxygen bonds in VO^{2+}; four bonds involving the sp_σ hybrid orbitals of the equivalent water oxygens and vanadium $(4s-3d_{z^2})(a_1)$, $4p_x$ and $4p_y(e)$, and $3d_{x^2-y^2}$ (b_1) orbitals; the sixth ligand, the axial water oxygen, is considered bonded to the remaining vanadium $4p_z(a_1)$ orbital; finally, the $3d_{xy}$ vanadium orbital, of symmetry b_2, is non-bonding.

The hybrid atomic orbitals used can be written in the form

$$\psi(\text{hybrid}) = (\sin\theta)\Phi(s) \pm (\cos\theta)\Phi(p \text{ or } d) \quad (1)$$

The values of θ are estimated by methods described in the Appendix; this gives $\theta = 0.455$ for the sp_σ oxygen hybrid orbital and $\theta = \pi/4$ for the sd_{z^2} vanadium hybrid orbital. The tetrahedral hybrid, $\theta = \pi/6$, is used for the water oxygen σ orbital.

In order to obtain some idea of the relative strengths of the bonds, and thus the positions of the one electron molecular orbitals, overlap integrals were evaluated, using the SCF radial functions for vanadium given by Watson,[16] and oxygen radial SCF functions taken from data given by Hartree.[17] Bond distances used were taken directly from the $VOSO_4 \cdot 5H_2O$ crystal results, and are given in Fig. 2. The overlap results are summarized in terms of the usual group overlap integrals (G_{ij}'s) for each molecular orbital, and are given in Table II. The Appendix should be consulted for details concerning any of the calculations discussed in this section.

(14) C. M. Roberts, W. S. Koski, and W. S. Caughey, *J. Chem. Phys.* **34**, 591 (1961).
(15) D. E. O'Reilly, *ibid.*, **29**, 1188 (1958).
(16) R. E. Watson, Quart. Prog. Rept., M.I.T., April, 1960.
(17) D. R. Hartree, "The Calculation of Atomic Structures," John Wiley, New York, N. Y., 1957, pp. 169–171.

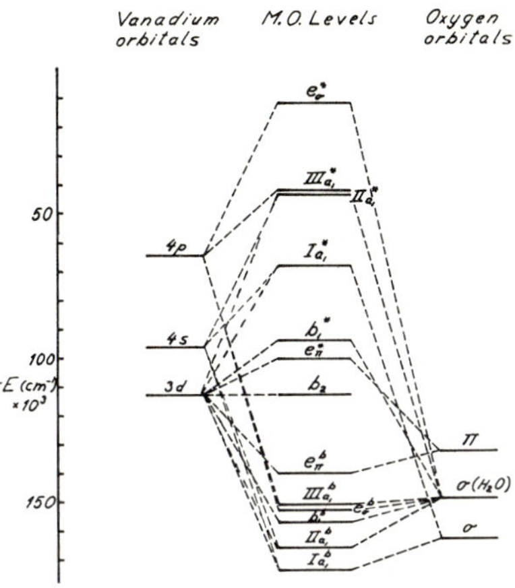

Fig. 3.—Molecular orbital scheme for $VO(H_2O)_5^{2+}$. The levels are drawn to scale.

Using the criterion that bond strengths are proportional to overlap, the ordering of the molecular orbitals is shown in Fig. 3. It is gratifying to note that the order of increasing energy of the crystal field levels b_2, e_π^*, b_1^*, and Ia_1^* is the same as given in Fig. 1 for the crystal field model.

The bonding molecular orbitals all can be written

$$\psi^b = c_1\Phi(\text{metal}) + c_2\Phi(\text{ligand}) \quad (2)$$

and similarly the antibonding levels

$$\psi^* = c_1^*\Phi(\text{metal}) + c_2^*\Phi(\text{ligand}) \quad (3)$$

with

$$c_1c_1^* + c_2c_2^* + c_1c_2^*G_{ij} + c_1^*c_2G_{ij} = 0 \quad (4)$$

where Φ (metal) and Φ (ligand) refer to the proper combination of metal and ligand orbitals for the molecular orbital in question. Approximate values of the energies of these orbitals can be obtained by solving the secular equation $|H_{ij} - G_{ij}\epsilon| = 0$. The estimation of the one electron orbital energies follows closely the procedures outlined by Mulliken[18] and Wolfsberg and Helmholz.[19] The H_{ii} and H_{jj} integrals are approximated as the VSIE's (valence state ionization energies) which are discussed in Section A of the Appendix. The H_{ij} resonance integrals are set equal to $-2G_{ij}\cdot$

(18) R. S. Mulliken, *J. Chem. Phys.*, **23**, 1841 (1955).
(19) M. Wolfsberg and L. Helmholz, *ibid.*, **20**, 837 (1952).

TABLE II

GROUP OVERLAP INTEGRALS AND ESTIMATED ORBITAL ENERGIES FOR THE $VO(H_2O)_5^{2+}$ MOLECULE ION

		Bonding levels			Antibonding levels		
Symmetry of M.O.	G_{ij}	$-\epsilon$ (cm.$^{-1}$)	c_1	c_2	$-\epsilon$ (cm.$^{-1}$)	c_1^*	c_2^*
e_π	0.139	140,026	0.446	0.834	100,422	0.907	−0.567
b_1	.194	157,126	.381	.853	94,130	.946	−0.555
Ia_1	.305	173,500	.345	.883	68,158	.996	−0.620
IIa_1	.390	165,837	.409	.765	43,395	1.006	−0.770
$IIIa_1$	.313	150,750	.154	.941	42,750	1.042	−0.474
e	.467	152,931	.201	.888	10,244	1.113	−0.698

Atomic orbital	Coulomb energy (cm.$^{-1}$)
Vanadium 3d	−112,924
4s	− 96,792
4p	− 64,528
Oxygen (2s + $2p_\sigma$)	−162,933
$2p_\pi$	−132,282
Oxygen(H_2O) (2s + $2p_\sigma$)	−148,414

TABLE III

ANALYSIS OF THE $VOSO_4 \cdot 5H_2O$ SPECTRUM

Transition	Predicted energy (polarization) (cm.$^{-1}$)	Obsd.[a] (cm.$^{-1}$) for $VOSO_4 \cdot 5H_2O$	Predicted oscillator strength $f \cdot 10^4$	Obsd.
$^2B_2 \rightarrow {}^2E(I)$	12,502($\perp$)	13,060($\perp$)	3.9	1.1
$^2B_2 \rightarrow {}^2B_1$	18,794	16,000	Vibronic	0.45
$^2B_2 \rightarrow {}^2A_1$	44,766	Covered	Vibronic	...
$^2B_2 \rightarrow {}^2E(II)$	38,800($\perp$)	41,700($\perp$)	26.4	50.3
$^2B_2 \rightarrow {}^2B_2$	44,000($\parallel$)	~50,000	44.7	150

[a] Experimentally observed energies and f's refer to aqueous solution, 0.01–0.10 M in H_2SO_4. In the $VOSO_4 \cdot 5H_2O$ crystal the $^2B_2 \rightarrow {}^2E(I)$ and the $^2B_2 \rightarrow {}^2E(II)$ bands are observed at the same energies in $\perp$ polarization only.

$\sqrt{(H_{ii})(H_{jj})}$; the geometric mean is preferable since the resonance energy is expected to decrease rapidly as the difference in the coulomb energies becomes greater. Since the VSIE's for the different atoms vary considerably with the degree of ionization, a charge distribution for $VO(H_2O)_5^{2+}$ must be assumed for the initial calculation. After completing one calculational cycle, the resulting charge distribution is calculated, using Mullikens' suggestion[20] that the overlap contribution be divided equally between two atoms. This procedure is continued until a self-consistent answer appears. The final charge distribution is $V^{+0.97}$ $O^{-0.60}$ $(5H_2O)^{+1.63}$.

The VSIE's for atoms with fractional charges are obtained by extrapolation of VSIE *vs.* degree of ionization curves. This procedure is used for all the atoms except the vanadyl oxygen. In this case the coulomb energy is raised by the excess positive charge on the neighboring atoms, and it is senseless to adjust the VSIE for any fractional charge. Therefore the coulomb energies for the vanadyl oxygen are estimated as the proper σ and π VSIE's of the neutral oxygen atom.

The results of the calculation in terms of c_1, c_2, and ϵ values are given in Table II. For purposes of comparison, Fig. 3 is drawn to this energy scale.

There are 17 electrons to place in the molecular orbitals shown in Fig. 2. The ground state is then $[(Ia_1^b)^2 (IIa_1^b)^2 (b_1^b)^2 (e_\sigma^b)^4 (IIIa_1^b)^2 (e_\pi^b)^4 (b_2)^1]$, $^2B_2$. There are several excited states with energies less than 50,000 cm.$^{-1}$ above the ground state. Table III gives the predicted energies of the transitions to these excited states, along with assignments of the solution spectrum of $VOSO_4 \cdot 5H_2O$, which is shown in Fig. 4. The oscillator strengths (f's) of the orbitally allowed transitions have been calculated using the M.O. wave functions, and also are given in Table III for purposes of comparison.

Consider first of all the so-called crystal field transitions; these involve moving the b_2 electron to the e_π^*, b_1^*, and Ia_1^* M.O.'s, which are of course essentially the 3d metal orbitals, resulting in $^2E(I)$, $^2B_1$, and $^2A_1$ excited states, respectively. To a good approximation electron repulsion

(20) R. S. Mulliken, *J. Chem. Phys.*, **23**, 1833 (1955).

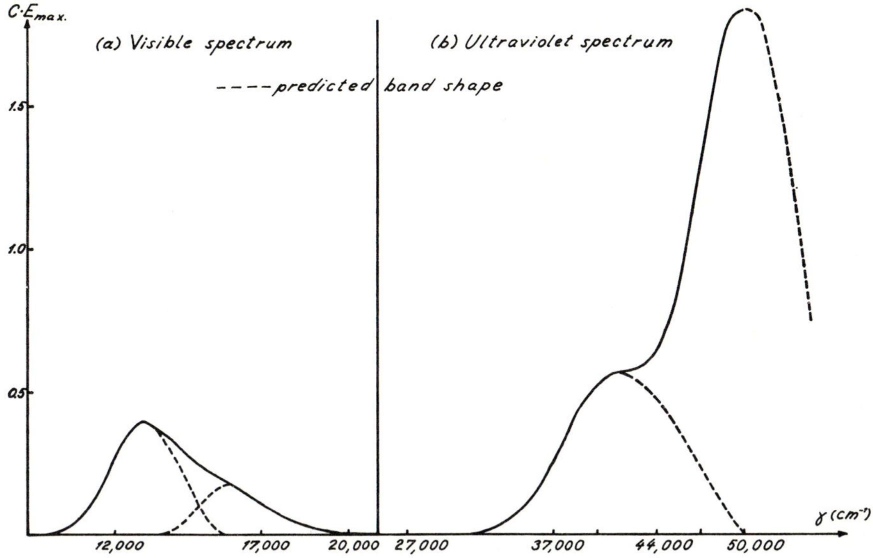

Fig. 4.—The electronic absorption spectrum of $VOSO_4{\cdot}5H_2O$ in aqueous solution: (a) complex = 0.0237 *M*, H_2SO_4 = 0.1 *M;* (b) complex = 0.00237 *M*, H_2SO_4 = 0.01 *M*.

effects can be considered to be the same for all these states. Therefore transitions are expected to occur to $^2E(I)$, $^2B_1$, and $^2A_1$ in order of increasing energy. Since the electric dipole vectors transform as A_1 and E in C_{4v}, only transitions to B_2 and E excited states are orbitally allowed; thus the $^2B_2 \rightarrow {}^2E(I)$ transition is expected to have somewhat greater intensity than the other crystal field transitions. These observations support the assignment of the more intense first band (13,000 cm.$^{-1}$) of $VOSO_4{\cdot}5H_2O$ to the transition $^2B_2 \rightarrow {}^2E(I)$. The agreement is satisfactory between the calculated and the observed *f* values.

The second transition (16,000 cm.$^{-1}$) appears as a weak shoulder and is assigned to $^2B_2 \rightarrow {}^2B_1$, it is allowed vibronically and should appear in all polarizations. The $^2B_2 \rightarrow {}^2A_1$ transition apparently is hidden under the first charge transfer band of $VOSO_4{\cdot}5H_2O$; it is observed, however, at approximately 30,000 cm.$^{-1}$ in a few vanadyl complexes.

The charge transfer spectrum below 50,000 cm.$^{-1}$ is due to the promotion of an electron from the oxygen π orbital (e_π^b) to the crystal field levels b_2 and e_π^*. Denoting the inner core σ-bonding M.O.'s as IC, the first charge transfer band (41,700 cm.$^{-1}$) is assigned

$$^2B_2[(IC)^{12}(e_\pi{}^b)^4(b_2)^1] \rightarrow {}^2E(II)[(IC)^{12}(e_\pi{}^b)^3(b_2)^2]$$

There is considerable repulsion energy involved in this transition, since an electron must be moved from a delocalized π orbital into an occupied b_2 orbital which is localized on the vanadium. This repulsion energy is estimated as *ca.* 11,700 cm.$^{-1}$ by comparing the positions of the first charge transfer band of VO^{2+} with the first band of $VOCl_3$.[21] This 11,700 cm.$^{-1}$ estimate is included in the predicted energy for the $^2B_2 \rightarrow {}^2E(II)$ transition. Both the predicted energy and the predicted *f* value agree well with the experimentally observed values.

Moving an electron from $e_\pi{}^b$ to $e_\pi{}^*$ results in the excited orbital configuration $[(IC)^{12}\ (e_\pi{}^b)^3\ (b_2)^1\ (e_\pi{}^*)^1]$. From this configuration doublet states can be constructed that transform as $^2A_1$, $^2A_2$, $^2B_1$, and $^2B_2$ in C_{4v}. Since only a transition to a $^2B_2$ state is orbitally allowed, the other states will not be considered. The energy of the first $^2B_2 \rightarrow {}^2B_2$ transition can be predicted by adding together

(21) F. A. Miller and W. B. White, *Spectrochim. Acta*, **9**, 98 (1957).

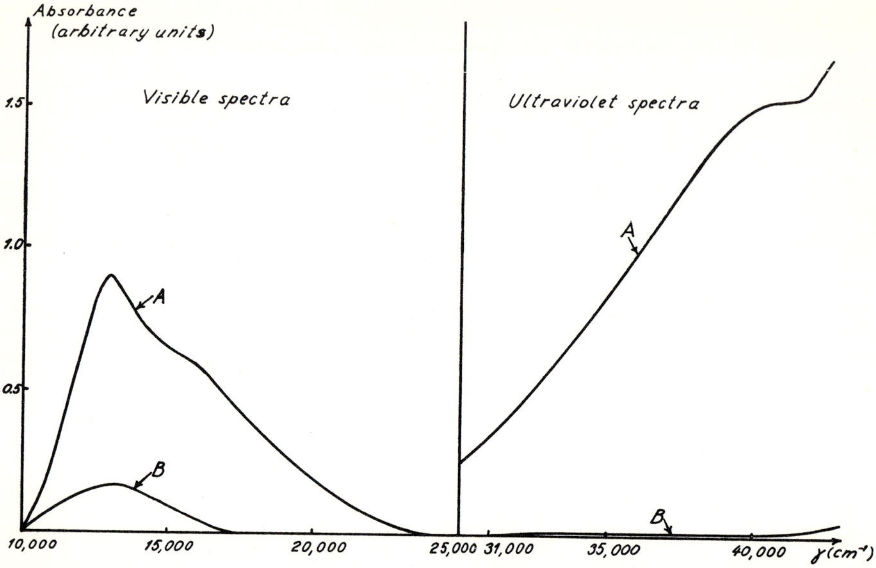

Fig. 5.—Absorption spectra of a single crystal of $VOSO_4 \cdot 5H_2O$: Curve A (cm.$^{-1}$), light polarized ⊥ to the V–O axis; Curve B, light polarized ‖ to the V–O axis.

the orbital promotional energy and the repulsion energy contributions. The repulsion energy estimate of 11,700 cm.$^{-1}$ must be corrected for the difference in repulsion energy between the $^2E(II)$ and $^2B_2$ excited states. This difference can be expressed in terms of the usual coulomb and exchange integrals, calculated from the proper determinantal wave functions for the $^2E(II)$ and the most stable $^2B_2$ excited state. This calculation predicts that the $^2B_2 \rightarrow {}^2B_2$ transition should occur at about 44,000 cm.$^{-1}$.[22] Thus it is reasonable to assign the broad band found at about 50,000 cm.$^{-1}$ to the $^2B_2 \rightarrow {}^2B_2$ transition. Furthermore, the calculated f value agrees rather well with the f value which can be estimated from the band shape.

The Crystal Spectrum of $VOSO_4 \cdot 5H_2O$ and Spectral Properties of other Vanadyl Compounds.—The electronic absorption spectra of a single crystal of $VOSO_4 \cdot 5H_2O$ have been determined for light polarized both parallel and perpendicular to the molecular V–O axis. These spectra are shown in Fig. 5. It is important to note that the positions of the absorption maxima found for the crystal are virtually the same as those found for an aqueous solution of $VOSO_4 \cdot 5H_2O$. This must mean that there are no significant structure changes in going from crystal $VOSO_4 \cdot 5H_2O$ to aqueous solution. The other important observation is that the 13,000 cm.$^{-1}$ band ($^2B_2 \rightarrow {}^2E(I)$) and the 41,700 cm.$^{-1}$ band ($^2B_2 \rightarrow {}^2E(II)$) appear principally in ⊥ polarization, providing a rather convincing confirmation of these band assignments. Since it was not possible to obtain an accurate spectrum above 45,000 cm.$^{-1}$, the predicted ‖ polarization of the second charge transfer band ($^2B_2 \rightarrow {}^2B_2$) could not be checked.

In addition to the above study, the reflectance spectrum of a powdered sample of $VOSO_4 \cdot 5H_2O$ has been determined, and the crystal field bands at 13,000 cm.$^{-1}$ and 16,000 cm.$^{-1}$ were resolved.

The reflectance spectrum of VO_2 has been measured by Rüdorff, Walter, and Stadler.[23] This measurement shows the charge transfer band at 41,700 cm.$^{-1}$ which is characteristic of vanadyl ion, but the crystal field bands were not resolved.

A survey of the spectra of common vanadyl

(22) Calculations of this type will be discussed in more detail by H. B. Gray and C. R. Hare in a forthcoming publication.

(23) W. Rüdorff, G. Walter, and J. Stadler, *Z. anorg. u. allgem. Chem.*, **297**, 1 (1958).

TABLE IV

SURVEY OF THE SPECTRA OF VANADYL COMPLEXES

Complex	Medium	Maxima ν (cm.$^{-1}$)	$f \cdot 10^4$	Assignment	Ref.
VO^{2+}	0.5 to 2 *M* $HClO_4$	13,100	1.1	$^2B_2 \rightarrow {}^2E(I)$	5
		16,000	0.48	$^2B_2 \rightarrow {}^2B_1$	
		41,700	50.8	$^2B_2 \rightarrow {}^2E(II)$	
$VOSO_4 \cdot 5H_2O$	Crystal	13,000($\perp$)	...	$^2B_2 \rightarrow {}^2E(I)$	This work
		16,000	...	$^2B_2 \rightarrow {}^2B_1$	
		41,700($\perp$)	...	$^2B_2 \rightarrow {}^2E(II)$	
VO_2	Powder	41,700	...	$^2B_2 \rightarrow {}^2E(II)$	23
$VOSO_4 \cdot 5H_2O$	Powder	12,900	...	$^2B_2 \rightarrow {}^2E(I)$	This work
		16,100	...	$^2B_2 \rightarrow {}^2B_1$	
$VO(aca)_2$	C_2H_5OH	12,800	2.6	$^2B_2 \rightarrow {}^2E(I)$	5
		17,300	1.0	$^2B_2 \rightarrow {}^2B_1$	
$VO(enta)^{2-}$	H_2O	12,800	1.5	$^2B_2 \rightarrow {}^2E(I)$	5
		17,200	1.2	$^2B_2 \rightarrow {}^2B_1$	
		29,800	2.0	$^2B_2 \rightarrow {}^2A_1$	
$VO(oxalate)_2{}^{2-}$	H_2O	12,600	2.5	$^2B_2 \rightarrow {}^2E(I)$	5
		16,500	0.70	$^2B_2 \rightarrow {}^2B_1$	
		29,400	15.6	$^2B_2 \rightarrow {}^2A_1$	
$VO(tartrate)^{2-}$	H_2O	{11,000	1.7	$^2B_2 \rightarrow {}^2E(I)$	5
		{17,000	1.4		
		18,800	2.2	$^2B_2 \rightarrow {}^2B_1$	
		25,300	4.1	$^2B_2 \rightarrow {}^2A_1$	

complexes is given in Table IV. The bands are assigned in a way consistent with the energies and intensities expected from a comparison with the $VOSO_4 \cdot 5H_2O$ spectrum. In fact, the main features of the spectra of vanadyl complexes are strikingly similar; a band at about 13,000 cm.$^{-1}$, followed by a second, less intense band at about 17,000 cm.$^{-1}$. For the enta, oxalate, and tartrate vanadyl complexes, the $^2B_2 \rightarrow {}^2A_1$ transition is observed before charge transfer spectra set in.

Magnetic Properties of Vanadyl Compounds.—The paramagnetic resonance of VO^{2+} has been investigated for a number of complexes and $< g >$ values are all very nearly 2.[7,14,15,24–29] A survey of g values for different complexes is given in Table V. In the molecular orbital description of $VO(H_2O)_5^{2+}$, the formulas for the g values become

$$g_\perp = 2\left(1 - \frac{(c_1^*)^2\xi}{\Delta E(^2B_2 \rightarrow {}^2E(Ig_\perp))}\right) \quad (5)$$

$$g_\parallel = 2\left(1 - \frac{(c_1^*)^2 4\xi}{\Delta E(^2B_2 \rightarrow {}^2B_1)}\right) \quad (6)$$

$$\langle g \rangle = \frac{1}{3}(2g_\perp + g_\parallel) \quad (7)$$

if it is assumed that $|\hat{L} + 2\hat{S}|\ \psi_2 > = 0$.

Since the approximate charge on vanadium is $+1$ in the M.O. approximation, a value of the spin orbital coupling constant $\xi = 135$ cm.$^{-1}$ is taken for V^+.[30] The calculation then gives $g_\perp = 1.983$, $g_\parallel = 1.940$, with $\langle g \rangle = 1.969$. This is in excellent agreement with the accurately known $\langle g \rangle = 1.962$ value for aqueous solutions of VO^{2+}.[25–27] The measurements on powdered samples of the vanadyl sulfates also give $\langle g \rangle$ values in reasonable agreement with this calculation.

Several measurements have been made which clearly show the predicted anisotropy of the g factor. In the case of VO(etioporphyrin II) $g_\perp = 1.988$ and $g_\parallel = 1.947$,[14] which indicates that this complex has an electronic structure which resembles $VOSO_4 \cdot 5H_2O$. The somewhat lower g values indicate that the energy levels are closer together, which is reasonable since the porphyrin nitrogens probably can π bond with the b_2 ($3d_{xy}$) metal orbital, thus raising its energy relative to e_π^* and b_1^*.

It is possible to study V^{4+} in a more symmetrical environment by substituting it into the TiO_2 lattice.[31] The symmetry is almost octahedral and presumably the dominant axial field of VO^{2+}

(24) C. A. Hutchinson, Jr., and L. S. Singer, *Phys. Rev.*, **89**, 256 (1953).

(25) R. N. Rogers and G. E. Pake, *J. Chem. Phys.*, **33**, 1107 (1960).

(26) N. S. Garif'ianov and B. M. Kozyrev, *Doklady Akad. Nauk S.S.S.R.*, **98**, 929 (1954).

(27) B. M. Kozyrev, *Discussions Faraday Soc.*, **19**, 135 (1955).

(28) R. J. Faber and M. T. Rogers, *J. Am. Chem. Soc.*, **81**, 1849 (1959).

(29) F. W. Lancaster and W. Gordy, *J. Chem. Phys.*, **19**, 1181 (1951).

(30) T. M. Dunn, *Trans. Faraday Soc.*, **57**, 1441 (1961).

(31) H. J. Gerritsen and H. R. Lewis, *Phys. Rev.*, **119**, 1010 (1960).

TABLE V
PARAMAGNETIC RESONANCE g FACTORS FOR VANADYL COMPLEXES

Compound	Details	$g_\perp$	$g_\parallel$	$\langle g \rangle$	Ref.
$VOSO_4 \cdot 5H_2O$	Powder, temp. range 4–300°K.	...	...	1.99	7
$VOSO_4 \cdot 2H_2O$	Powder	...	...	1.96	24
VO^{2+}	Aq. soln.	...	...	1.962	25–27
VO^{2+}	Alcohol or glycerol soln., temp. 90°K.	...	...	1.96	27
VO(etioporphyrin II)	Castor oil soln.	1.988	1.947	1.974	14
VO(etioporphyrin I)	Petroleum oil soln.	1.987	1.948	1.974	15
VO^{2+}	Adsorbed on (a) IR-100	1.983	1.93	1.97	28
	(b) Dowex-50	1.979	1.88	1.95	
	(c) Charcoal	1.983	...	...	
	(d) IR-4B	1.989	1.93	1.97	
$VOCl_2$	Powder	...	...	2.0	29

is eliminated. This means that an increase in $\Delta E(^2B_2 \rightarrow {}^2B_1)$ is expected, accompanied by a corresponding decrease in $\Delta E(^2B_2 \rightarrow {}^2E\ (I))$. The observed values of $g_\perp = 1.914$ and $g_\parallel = 1.957$ strongly support this interpretation. A reasonable value of $\xi = 150$ cm.$^{-1}$ gives $\Delta E(^2B_2 \rightarrow {}^2B_1) = 28{,}000$ cm.$^{-1}$ and $\Delta E(^2B_2 \rightarrow {}^2E(I)) = 3500$ cm.$^{-1}$. Recall that for the crystal field model, $\Delta E(^2B_2 \rightarrow {}^2B_1)$ is equal to 10 Dq, and the value 28,000 cm.$^{-1}$ is now in agreement with expectations for V^{4+}.

The other magnetic property of interest for VO^{2+} complexes is the susceptibility. The theoretical expression for the magnetic susceptibility is of the form[32]

$$\chi = \frac{C}{T} + \chi_{H.F.} \quad (8)$$

where C is the Curie constant and $\chi_{H.F.}$ stands for the temperature independent contributions to the susceptibility (high-frequency terms). In this case

$$C = \frac{N\beta^2(g_\parallel^2 + 2g_\perp^2)}{12k} \quad (9)$$

$$\chi_{H.F.} = 2/3\ N\beta^2 \sum_i \frac{|\langle\psi_0|\hat{L} + 2\hat{S}|\psi_i\rangle|^2}{\Delta E_{0,i}} \quad (10)$$

where β is the Bohr magneton, N is Avogadro's number, and where $\Delta E_{0,i}$ is the transition energy from ψ_0 to ψ_i. Since values of $g_\perp$ and $g_\parallel$ have been calculated above for $VO(H_2O)_5^{2+}$, they will be used to calculate C. The major contributions to $\chi_{H.F.}$ are from the $^2B_1$ and $^2E(I)$ excited states; contributions from the other excited states will be ignored. Thus eq. 8 becomes

$$\chi = \frac{N\beta^2(g_\parallel^2 + 2g_\perp^2)}{12kT} + {}^2/_3N\beta^2 \sum_i \frac{|\langle\psi_0|\hat{L} + 2\hat{S}|\psi_i\rangle|^2}{\Delta E_{0,i}} \quad (11)$$

$$\chi = \frac{N\beta^2[(1.940)^2 + 2(1.983)^2]}{12kT} + 2/3N\beta^2 \frac{(1+1)(c_1^*)^2}{\Delta E(^2B_2 \rightarrow {}^2E(I))} + \frac{(4+4)(c_1^*)^2}{\Delta E(^2B_2 \rightarrow {}^2B_1)} \quad (12)$$

so

$$\chi_{calc.} = \frac{0.363}{T} + 100 \times 10^{-6} \quad (13)$$

The magnetic susceptibility of $VOSO_4(3.5\ H_2O)$ has been measured by Perrakis[33] over a temperature range of 145°. A plot of χ *vs.* $1/T$ gives a straight line with slope 0.335 and intercept 130×10^{-6}. Thus

$$\chi_{exp.} = \frac{0.335}{T} + 130 \times 10^{-6} \quad (14)$$

in satisfactory agreement with the predicted equation (13).

Several room temperature magnetic susceptibility measurements have been made for both powdered samples and aqueous solutions of vanadyl complexes.[34,35] The effective magnetic moments arrived at are listed in Table VI. The

TABLE VI
MAGNETIC SUSCEPTIBILITY DATA FOR VANADYL COMPLEXES

Component[a]	Temp., °C.	μ_{eff} B.M.	Ref.
$VOSO_4 \cdot 5H_2O$	26	1.73	This work
VO^{2+} (aq. $HClO_4$)	20	1.72	34
$VO(aca)_2 \cdot H_2O$	17.5	1.72	35
$VO(aca)_2$	22.5	1.73	This work
VO(salicylaldehydeethylenediimine)	18.5	1.68	35
VO(salicylaldehyde-O-phenylenediimine)	17	1.68	35
$(NH_4)_2[VO(malonate)_2] \cdot 3H_2O$	17	1.70	35

[a] Powder sample unless otherwise indicated.

(32) J. H. Van Vleck, "Electric and Magnetic Susceptibilities," Oxford University Press, 1932.

(33) N. Perrakis, *J. phys. radium*, **8**, 473 (1927).

(34) S. Freed, *J. Am. Chem. Soc.*, **49**, 2456 (1927).

(35) R. W. Asmussen, "Magnetokemiske Undersøgelser over Uorganiske Kompleksforbindelser," Gjellerups Forlag, Copenhagen, 1944.

moments all are approximately equal to the spin only value of 1.73 B.M. for one unpaired spin, as is expected when the orbital contribution is completely quenched in the low symmetry field.

Discussion

It is worthwhile to summarize the main points of evidence concerning the structure of the vanadyl ion in solution. First, the position of the $^2B_2 \rightarrow {}^2E(I)$ band (13,000 cm.$^{-1}$) is the same in the crystal and for an aqueous solution of $VOSO_4 \cdot 5H_2O$. Also, the first charge transfer band (41,700 cm.$^{-1}$) is found at the same place in an aqueous solution of VO^{2+}, in the reflectance spectrum of powdered VO_2, and in the absorption spectrum of crystalline $VOSO_4 \cdot 5H_2O$. Since these transitions involve the VO^{2+} π orbitals, it is convincing evidence that the solution structure at least contains the VO^{2+} entity. Protonation of VO^{2+}, resulting in $V(OH)_2^{2+}$, would significantly affect the amount of π-bonding, and thus completely change the positions of the energy levels.

Secondly, the magnetic data are consistent with the VO^{2+} formulation for the vanadyl ion in solution. For example, the values of $\langle g \rangle$ reported for vanadyl complexes are approximately the same both for the crystal, powder, and aqueous solution measurements. A smaller $\langle g \rangle$ value would be expected for the more symmetrical $V(OH)_2^{2+}$ ion, similar to the $\langle g \rangle = 1.938$ found for V^{4+} in the TiO_2 lattice. Finally, the magnetic susceptibility of an aqueous solution of VO^{2+} is the same as the susceptibility of powdered $VOSO_4 \cdot 5H_2O$, within experimental error, and corresponds to the spin only moment of 1.73 B.M. This constancy might not be expected if the complex undergoes a structural change in going from the crystal to aqueous solution, although it must be admitted that the susceptibility difference for a structural change would hardly lie outside the realm of experimental error.

The resistance of VO^{2+} to protonation can be understood in terms of the M.O. bonding scheme. With the oxygen 2p orbitals used for π-bonding, only the non-bonding sp_σ hybrid is left for a proton; it has considerable 2s character, and is energetically unsuited for bonding purposes.

The extreme importance of ligand to metal π-bonding in the oxycations must be emphasized. In the case of VO^{2+}, this π-bonding accounts for the drastic reduction ($\sim$45%) of the free ion ξ value,[36] the resistance of VO^{2+} to protonation, and the charge transfer features of the electronic spectrum of $VOSO_4 \cdot 5H_2O$. Indeed, it is clear that any complete discussion of the electronic structures of the oxycations of the transition and actinium series must allow for substantial oxygen to metal π-bonding. Furthermore, it can be qualitatively understood why ions of this type in the first transition series usually have the formula MO^{n+}(TiO^{2+}, VO^{2+}, CrO^{3+}), while similar ions in the actinium series are invariably MO_2^{n+}(UO_2^{2+}, NpO_2^{2+}). The two $2p_\pi$ orbitals on oxygen can satisfy the π-bonding capacities of the two $3d_\pi$ orbitals of a first transition series metal ion, but it takes at least two oxygens to satisfy the combined π-bonding capacities offered by the 5f and 6d orbitals of the metal ions in the actinium series.

(36) Reference 30 gives a value of 250 cm.$^{-1}$ for the ξ of V^{4+}.

Appendix

A. Radial Functions and Atomic Orbital Energies.—Self-consistent field (SCF) radial functions for vanadium 3d and 4s orbitals were taken from Watson's report.[16] Watson gives no 4p function, so it is estimated as having approximately the same radial dependence as the 4s function. Analytic 2s and 2p oxygen SCF radial functions were obtained by fitting the numerical functions given by Hartree[17] with a linear combination of Slater functions. These radial functions are summarized

$$\text{Vanadium } R(3d) = 0.5243\phi_3(1.83) + 0.4989\phi_3(3.61) + 0.1131\phi_3(6.80) + 0.0055\phi_3(12.43) \quad (15)$$

$$(4s) = -0.02245\phi_1(23.91) - 0.01391\phi_2(20.60) + 0.06962\phi_2(10.17) + 0.06774\phi_3(9.33) - 0.09708\phi_3(5.16) - 0.2462\phi_3(3.51) + 0.04412\phi_5(3.87) + 0.3607\phi_5(1.88) + 0.6090\phi_5(1.15) + 0.1487\phi_5(0.78) \quad (16)$$

$$R(4p) = \phi_3(1.024) \quad (17)$$

$$\text{Oxygen } R(2s) = 0.5459\phi_2(1.80) + 0.4839\phi_2(2.80) \quad (18)$$

$$R(2p) = 0.6804\phi_2(1.55) + 0.4038\phi_2(3.43) \quad (19)$$

where $\phi_n(\mu) = N\mu r^{n-1}e^{-\mu r}$, and $N\mu$ is a normalization constant.

Valence state ionization energies for σ and π electrons in these orbitals are calculated for different degrees of ionization using the formulas and methods given by Moffitt[37] along with the spectroscopic data compiled by Moore.[38]

The degree of mixing in the hybrid orbitals

$$\psi(\text{oxygen } \sigma) = (\sin\theta)\Phi(2s) \pm (\cos\theta)\Phi(2p_z) \quad (20)$$

$$\psi(\text{vanadium } \sigma) = (\sin\theta)\Phi(4s) \pm (\cos\theta)\Phi(3d_{z^2}) \quad (21)$$

is estimated by requiring that the quantity

(37) W. Moffitt, *Repts. Progr. Phys.*, **17**, 173 (1954).

(38) C. E. Moore, "Atomic Energy Levels," U. S. Natl. Bur. Standards Circular 467, 1949 and 1952.

$\frac{\text{VSIE}(\theta)}{S(\theta)}$ be a minimum, where $S(\theta)$ is the overlap between the hybrid orbital and the appropriate orbital on the neighboring atom, and VSIE(θ) is the valence state ionization energy for different amounts of mixing θ. This gives $\theta = 0.455$ for the oxygen σ and $\theta = \pi/4$ for the vanadium σ. The sp$_\sigma$ tetrahedral hybrid orbital, $\theta = \pi/6$, is used for the water oxygen.

B. Overlap Integrals.—The two atom overlap integrals (S_{ij}'s) between the various atomic orbitals were found in the tabulations available in the literature.[39–42] The group overlap integrals (G_{ij}'s) of interest in the M.O. calculation are related to the S_{ij}'s as

$$G(e_\pi) = S(2p_\pi, 3d_\pi) \quad (22)$$

$$G(b_1) = \sqrt{3}S(te\sigma, 3d_\sigma) \quad (23)$$

$$G(Ia_1) = S(\sigma_5, sd_\sigma) \quad (24)$$

$$G(IIa_1) = \frac{1}{\sqrt{2}}[2S(te\sigma, 4s) + S(te\sigma, 3d_\sigma)] \quad (25)$$

$$G(IIIa_1) = S(te\sigma, 4p_\sigma) \quad (26)$$

$$G(e_\sigma) = \sqrt{2}S(te\sigma, 4p_\sigma) \quad (27)$$

where teσ is the tetrahedral sp$_\sigma$ hybrid for the water oxygens, and the oxygen orbitals are always listed first in parentheses.

C. Intensity Calculations.—The theoretical expression for the oscillator strength of a transition is given by[43]

$$f = 1.085 \times 10^{-5}(\nu_{\text{cm.}^{-1}})\ [\text{av}_\text{I}\ \Sigma_{\text{II}}\ |\int \psi_\text{I}^* |\vec{r}|\ \psi_{\text{II}}\, d\tau\ |^2] \quad (28)$$

where ψ_I and ψ_{II} refer to the initial and final states, respectively, and $\vec{r} = i_x + j_y + k_z$.

For the transition $^2B_2 \rightarrow {}^2E(II)$

$$\psi_\text{I} = [(IC)^{12}(e^b\pi_x)(e^b\pi_x)(e^b\pi_y)(e^b\pi_y)(b_2)] = B_2\ (\text{orbital}) \quad (29)$$

$$\psi_{\text{II}} = \left\{ \begin{matrix} [(IC)^{12}(e^b\pi_x)(e^b\pi_x)(e^b\pi_y)(b_2)(b_2)] \\ [(IC)^{12}(e^b\pi_x)(b_2)(e^b\pi_y)(e^b\pi_y)(b_2)] \end{matrix} \right\} = E(II)(\text{orbital}) \quad (30)$$

Simplifying, eq. 28 reduces to

$$(^2B_2 \rightarrow {}^2E(II)) = 1.085 \times 10^{-5}(\nu_{\text{cm.}^{-1}}) \left[\frac{N_{2p}c_2}{2N_{3d_{xy}}{}^{\mu_{2p}}}\right] \times 2[S(3d_\delta{}^{\mu_{2p}}, 3d_\delta)]^2 \quad (31)$$

where N_{2p} is the normalization constant for the original ligand orbitals, and $N_{3d_{xy}}{}^{\mu_{2p}}$ is the normalization constant a 3d$_{xy}$ orbital would have with an exponential factor μ_{2p} in the radial function.

Supplying the proper values of c_2 and $\nu_{\text{cm.}^{-1}}$ for the $^2B_2 \rightarrow {}^2E(II)$ transition, and obtaining the overlap integral from Jaffé's table,[41] $f = 2.64 \times 10^{-3}$ is calculated.

For the $^2B_2 \rightarrow {}^2E(I)$ transition, the same integrals are involved as in the above calculation. Only the constants $\nu_{\text{cm.}^{-1}}$ and c_2 are different. Thus eq. 31 is written

$$(^2B_2 \rightarrow {}^2E(I)) = 1.085 \times 10^{-5}(\nu_{\text{cm.}^{-1}}) \left[\frac{N_{2p}c_2^*}{2N_{3d_{xy}}{}^{\mu_{2p}}}\right]^2 \times 2[S(3d_\delta{}^{\mu_{2p}}, 3d_\delta)]^2 = 3.92 \times 10^{-4} \quad (32)$$

For the $^2B_2 \rightarrow {}^2B_2$ transition

$$\psi_{\text{II}} = \frac{1}{\sqrt{2}} \{[(IC)^{12}(e^b\pi_x)(e^*\pi_x)(e^b\pi_y)(e^b\pi_y)(b_2)] + [(IC)^{12}(e^b\pi_x)(e^b\pi_x)(e^b\pi_y)(e^*\pi_y)(b_2)]\} = B_2(\text{orbital}) \quad (33)$$

and therefore

$$f(^2B_2 \rightarrow {}^2B_2) = 1.085 \times 10^{-5}(\nu_{\text{cm.}^{-1}}) \times \left[\frac{\sqrt{2}(c_1c_2^* + c_1^*c_2)N_{2p}}{N_{3d_{xz}}{}^{\mu_{2p}}}\right]^2 [S(3d_\pi{}^{\mu_{2p}}, 3d_\pi)]^2 = 4.47 \times 10^{-3}$$

Experimental

Anhydrous VO(aca)$_2$ was prepared as described in the literature.[44] Analyzed $VOSO_4 \cdot 5H_2O$ was obtained from Struers Co., Copenhagen. Solution spectral measurements were made using a Zeiss instrument. Reflectance spectra were obtained with a Beckman DU equipped with a standard reflectance attachment. Magnesium oxide was used as a standard. The resolution of the bands in the reflectance spectrum of $VOSO_4 \cdot 5H_2O$ did not differ from the resolution achieved for the crystal absorption spectrum A shown in Fig. 5.

The crystal absorption spectra were obtained with a Zeiss instrument modified for double beam operation. It was equipped with a calcite polarizer and quartz optics which allowed accurate spectral measurements up to 45,000 cm.$^{-1}$. The optical system was designed to record spectra of very small crystals (about 2 mm.). The details of the design and operation of the crystal spectrophotometer will be reported elsewhere.[45]

The prismatic crystals of $VOSO_4 \cdot 5H_2O$ have four molecules in the monoclinic cell, with the VO groups arranged approximately parallel to the wedge of the prism. The A spectrum in Fig. 5 was recorded with the light incident on the 110 face and vibrating in a plane perpendicular to the wedge of the prism; the B spectrum was recorded with the light vibrating in a plane parallel to the wedge.

The magnetic susceptibilities of $VOSO_4 \cdot 5H_2O$ and VO-

(39) R. S. Mulliken, C. A. Rieke, D. Orloff, and H. Orloff, *J. Chem. Phys.*, **17**, 1248 (1949).

(40) H. H. Jaffé and G. O. Doak, *ibid.*, **21**, 196 (1953).

(41) H. H. Jaffé, *ibid.*, **21**, 258 (1953).

(42) D. P. Craig, A. Maccoll, R. S. Nyholm, L. E. Orgel, and L. E. Sutton, *J. Chem. Soc.*, 354 (1954).

(43) C. J. Ballhausen, *Progr. in Inorg. Chem.*, **2**, 251 (1960).

(44) *Inorganic Syntheses*, **5**, 115 (1957).

(45) A. E. Nielsen, to be published.

(aca)$_2$ were determined by the Gouy method.[46] Sample tubes were calibrated with Hg[Co(SCN)$_4$].

Acknowledgments.—The authors wish to express their gratitude to Lektor Arne E. Nielsen, who designed and built the crystal spectrophotometer and whose help was invaluable in obtaining the final crystal spectra. We also thank Dr. T. M. Dunn for allowing us to see his compilation of spin orbit coupling constants prior to publication.

(46) Chap. 6 in "Modern Coördination Chemistry," by B. N. Figgis and J. Lewis, edited by J. Lewis and R. G. Wilkins, Interscience Publishers, Inc., New York, N. Y., 1960.

From: *J. Am. Chem. Soc.*, 85, 260-265 (1963)

$\Gamma_\sigma = A_{1g} + B_{1g} + E_u$

$\Gamma_\pi(\perp) = A_{2u} + B_{2u} + E_g$

$\Gamma_\pi(\parallel) = A_{2g} + B_{2g} + E_u$

A Molecular Orbital Theory for Square Planar Metal Complexes

BY HARRY B. GRAY AND C. J. BALLHAUSEN[1]

RECEIVED JULY 25, 1962

The bonding in square planar metal complexes is described in terms of molecular orbitals. The relative single electron molecular orbital energies are estimated for the different types of π-electron systems. Both the d–d and charge transfer spectra of the halide and cyanide complexes of Ni^{2+}, Pd^{2+}, Pt^{2+} and Au^{3+} are discussed in terms of the derived molecular orbital scheme. The magnetic susceptibilities of diamagnetic d^8-metal complexes are considered, and it is concluded that a substantial "ring current" exists in $K_2Ni(CN)_4 \cdot H_2O$.

Introduction

The metal ions which form square planar complexes with simple ligands have a d^8-electronic configuration. The five degenerate d-orbitals of the uncomplexed metal ion split into four different levels in a square planar complex. Thus three orbital parameters are needed to describe the ligand field d-splittings.

Various efforts have been made to evaluate the d-orbital splittings in square planar metal complexes, and to assign the observed d-d spectral bands.[2–8] The most complete calculation has been made for Pt^{2+} complexes by Fenske, Martin and Ruedenberg,[6] and considerable confidence may be placed in their ordering of the d-orbitals. However, for other complexes the d-ordering may depend on the metal ion and the type of ligand under consideration.

It is the purpose of this paper to develop a molecular orbital theory for square planar metal complexes. Both the spectral and magnetic properties of typical square planar complexes of Ni^{2+}, Pd^{2+}, Pt^{2+} and Au^{3+} will be considered in order to arrive at consistent values for the molecular orbital energies. However, in contrast to previous workers, effort will be concentrated on the assignments of the charge transfer bands of representative square planar complexes.

Molecular Orbitals for Square Planar Complexes.—Figure 1 shows a square planar complex in a coördinate system with the central atom at the origin, and the four ligands along the x- and y-axes. The orbital transformation scheme in the D_{4h} symmetry is given in Table I.

TABLE I

ORBITAL TRANSFORMATION SCHEME IN D_{4h} SYMMETRY

Representation	Metal orbitals	Ligand orbitals
a_{1g}	d_{z^2}, s	$\frac{1}{2}(\sigma_1 + \sigma_2 + \sigma_3 + \sigma_4)$
a_{2g}	..	$\frac{1}{2}(\pi_{1h} + \pi_{2h} + \pi_{3h} + \pi_{4h})$
a_{2u}	p_z	$\frac{1}{2}(\pi_{1v} + \pi_{2v} + \pi_{3v} + \pi_{4v})$
b_{1g}	$d_{x^2-y^2}$	$\frac{1}{2}(\sigma_1 - \sigma_2 + \sigma_3 - \sigma_4)$
b_{2g}	d_{xy}	$\frac{1}{2}(\pi_{1h} - \pi_{2h} + \pi_{3h} - \pi_{4h})$
b_{2u}	..	$\frac{1}{2}(\pi_{1v} - \pi_{2v} + \pi_{3v} - \pi_{4v})$
e_g	d_{yz}, d_{xz}	$\frac{1}{\sqrt{2}}(\pi_{2v} - \pi_{4v})$, $\frac{1}{\sqrt{2}}(\pi_{1v} - \pi_{3v})$
e_u	p_x, p_y	$\frac{1}{\sqrt{2}}(\sigma_1 - \sigma_3)$, $\frac{1}{\sqrt{2}}(\sigma_2 - \sigma_4)$
		$\frac{1}{\sqrt{2}}(\pi_{2h} - \pi_{4h})$, $\frac{1}{\sqrt{2}}(\pi_{1h} - \pi_{3h})$

The single electron molecular orbitals are approximated as

$$\Psi(\text{m.o.}) = C_{M(i)}(\Phi)(\text{metal}) + C_{(L)j}\Phi(\text{ligand}) \quad (1)$$

where the C's are subject to the usual normalization and orthogonality restrictions. The pure σ-orbitals are

$$\Psi[a_{1g}(\sigma)] = C_{M(1)}\Phi[nd_{z^2}] + C_{M(2)}\Phi[(n+1)s] + C_L\Phi[\tfrac{1}{2}(\sigma_1 + \sigma_2 + \sigma_3 + \sigma_4)] \quad (2)$$

$$\Psi[b_{1g}(\sigma)] = C_M\Phi[nd_{x^2-y^2}] + C_L\Phi[\tfrac{1}{2}(\sigma_1 - \sigma_2 + \sigma_3 - \sigma_4)] \quad (3)$$

The form of the π-orbitals depends on the type of ligand under consideration. Square planar complexes in which the ligands themselves have no π-orbital system (Cl^-, Br^-, H_2O, NH_3) will be case 1; complexes in which the ligands have a π-system (CN^-), and thus both π-bonding (π^b) and π-antibonding (π^*) ligand orbitals must be considered, will be case 2.

Case 1.—The pure π-orbitals are

$$\Psi[a_{2g}(\pi)] = \Phi[\tfrac{1}{2}(\pi_{1h} + \pi_{2h} + \pi_{3h} + \pi_{4h})] \quad (4)$$

$$\Psi[a_{2u}(\pi)] = C_M\Phi[(n+1)p_z] + C_L\Phi[\tfrac{1}{2}(\pi_{1v} + \pi_{2v} + \pi_{3v} + \pi_{4v})] \quad (5)$$

$$\Psi[b_{2g}(\pi)] = C_M\Phi[nd_{xy}] + C_L\Phi[\tfrac{1}{2}(\pi_{1h} - \pi_{2h} + \pi_{3h} - \pi_{4h})] \quad (6)$$

$$\Psi[b_{2u}(\pi)] = \Phi[\tfrac{1}{2}(\pi_{1v} - \pi_{2v} + \pi_{3v} - \pi_{4v})] \quad (7)$$

$$\Psi_I[e_g(\pi)] = C_M\Phi[nd_{xz}] + C_L\Phi\left[\frac{1}{\sqrt{2}}(\pi_{1v} - \pi_{3v})\right] \quad (8)$$

$$\Psi_{II}[e_g(\pi)] = C_M\Phi[nd_{yz}] + C_L\Phi\left[\frac{1}{\sqrt{2}}(\pi_{2v} - \pi_{4v})\right] \quad (9)$$

The mixed σ- and π-orbitals are

$$\Psi_I[e_u(\sigma,\pi)] = C_M\Phi[(n+1)p_x] + C_{L(1)}\Phi\left[\frac{1}{\sqrt{2}}(\sigma_1 - \sigma_3)\right] + C_{L(2)}\Phi\left[\frac{1}{\sqrt{2}}(\pi_{2h} - \pi_{4h})\right] \quad (10)$$

$$\Psi_{II}[e_u(\sigma,\pi)] = C_M\Phi[(n+1)p_y] + C_{L(1)}\Phi\left[\frac{1}{\sqrt{2}}(\sigma_2 - \sigma_4)\right] + C_{L(2)}\Phi\left[\frac{1}{\sqrt{2}}(\pi_{1h} - \pi_{3h})\right] \quad (11)$$

Case 2.—The pure π-orbitals are

$$\Psi[a_{2g}(\pi^b)] = \Phi[\tfrac{1}{2}(\pi_{1h}^b + \pi_{2h}^b + \pi_{3h}^b + \pi_{4h}^b)] \quad (12)$$

$$\Psi[a_{2g}(\pi^*)] = \Phi[\tfrac{1}{2}(\pi_{1h}^* + \pi_{2h}^* + \pi_{3h}^* + \pi_{4h}^*)] \quad (13)$$

$$\Psi[a_{2u}(\pi)] = C_M\Phi[(n+1)p_z] + C_{L(1)}\Phi[\tfrac{1}{2}(\pi_{1v}^b + \pi_{2v}^b + \pi_{3v}^b + \pi_{4v}^b)] + C_{L(2)}\Phi[\tfrac{1}{2}(\pi_{1v}^* + \pi_{2v}^* + \pi_{3v}^* + \pi_{4v}^*)] \quad (14)$$

$$\Psi[b_{2g}(\pi)] = C_M\Phi[nd_{xy}] + C_{L(1)}\Phi[\tfrac{1}{2}(\pi_{1h}^b - \pi_{2h}^b + \pi_{3h}^b - \pi_{4h}^b)] + C_{L(2)}\Phi[\tfrac{1}{2}(\pi_{1h}^* - \pi_{2h}^* + \pi_{3h}^* - \pi_{4h}^*)] \quad (15)$$

$$\Psi[b_{2u}(\pi^b)] = \Phi[\tfrac{1}{2}(\pi_{1v}^b - \pi_{2v}^b + \pi_{3v}^b - \pi_{4v}^b)] \quad (16)$$

$$\Psi[b_{2u}(\pi^*)] = \Phi[\tfrac{1}{2}(\pi_{1v}^* - \pi_{2v}^* + \pi_{3v}^* - \pi_{4v}^*)] \quad (17)$$

$$\Psi_I[e_g(\pi)] = C_M\Phi[nd_{xz}] + C_{L(1)}\Phi\left[\frac{1}{\sqrt{2}}(\pi_{1v}^b - \pi_{3v}^b)\right] + C_{L(2)}\Phi\left[\frac{1}{\sqrt{2}}(\pi_{1v}^* - \pi_{3v}^*)\right] \quad (18)$$

$$\Psi_{II}[e_g(\pi)] = C_M\Phi[nd_{yz}] + C_{L(1)}\Phi\left[\frac{1}{\sqrt{2}}(\pi_{2v}^b - \pi_{4v}^b)\right] + C_{L(2)}\Phi\left[\frac{1}{\sqrt{2}}(\pi_{2v}^* - \pi_{4v}^*)\right] \quad (19)$$

(1) Permanent address: Institute for Physical Chemistry, University of Copenhagen, Denmark.

(2) (a) J. Chatt, G. A. Gamlen and L. E. Orgel, *J. Chem. Soc.*, 486 (1958); (b) G. Maki, *J. Chem. Phys.*, **28**, 651 (1958); **29**, 162 (1958); **29**, 1129 (1958).

(3) C. J. Ballhausen and A. D. Liehr, *J. Am. Chem. Soc.*, **71**, 538 (1959).

(4) M. I. Ban, *Acta Chim. Acad. Sci. Hung.*, **19**, 459 (1959).

(5) S. Kida, J. Fujita, K. Nakamoto and R. Tsuchida, *Bull. Chem. Soc.* (Japan), **31**, 79 (1958).

(6) R. F. Fenske, D. S. Martin and K. Ruedenberg, *Inorg. Chem.*, **1**, 441 (1962).

(7) J. R. Perumareddi, A. D. Liehr and A. W. Adamson, *J. Am. Chem. Soc.*, **85**, 249 (1963).

(8) J. Ferguson, *J. Chem. Phys.*, **34**, 611 (1961).

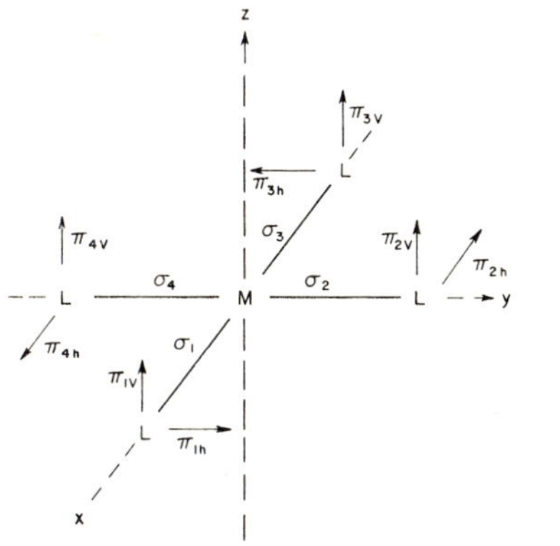

Fig. 1.—A molecular orbital coördinate system for a square planar metal complex.

The mixed σ- and π-orbitals are

$$\Psi_{\rm I}[e_u(\sigma,\pi)] = C_M\Phi[(n+1)p_x] + C_{L(1)}\Phi\left[\frac{1}{\sqrt{2}}(\sigma_1 - \sigma_3)\right] + C_{L(2)}\Phi\left[\frac{1}{\sqrt{2}}(\pi_{2h}{}^b - \pi_{4h}{}^b)\right] + C_{L(3)}\Phi\left[\frac{1}{\sqrt{2}}(\pi_{2h}{}^* - \pi_{4h}{}^*)\right] \quad (20)$$

$$\Psi_{\rm II}[e_u(\sigma,\pi)] = C_M\Phi[(n+1)p_y] + C_{L(1)}\Phi\left[\frac{1}{\sqrt{2}}(\sigma_2 - \sigma_4)\right] + C_{L(2)}\Phi\left[\frac{1}{\sqrt{2}}(\pi_{1h}{}^b - \pi_{3h}{}^b)\right] + C_{L(3)}\Phi\left[\frac{1}{\sqrt{2}}(\pi_{1h}{}^* - \pi_{3h}{}^*)\right] \quad (21)$$

The following general rules were adhered to in estimating the relative energies of the single electron molecular orbitals: (1) The order of the coulomb energies is taken to be σ (ligand), π^b(ligand), nd(metal), $(n + 1)s$ (metal), π^*(ligand), $(n + 1)p$ (metal). (2) The amount of mixing of atomic orbitals in the molecular orbitals is roughly proportional to atomic orbital overlap and inversely proportional to their coulomb energy difference. (3) Other things being approximately equal, σ-bonding molecular orbitals are more stable than π-bonding molecular orbitals, and σ-antibonding molecular orbitals correspondingly less stable than π-antibonding molecular orbitals. (4) Interactions among the ligands themselves are expressed in terms of a ligand–ligand exchange integral β: the sign of this interaction for each molecular orbital is another factor in the final relative ordering. (5) The relative molecular orbital ordering is considered final only if it is fully consistent with the available experimental results: exact differences in the single electron molecular orbital levels can only be obtained from experiment.

The general molecular orbital energy level schemes arrived at for square planar complexes are given in Fig. 2 (case 1) and Fig. 3 (case 2). Group molecular orbital overlap integrals (for $Ni(CN)_4{}^{2-}$) and ligand exchange interactions are summarized in Table II.

Almost all square planar complexes with simple ligands are diamagnetic and contain a metal ion with the d^8-electronic configuration. Thus the ground state is $^1A_{1g}$. The lowest energy excited states will be described separately for case 1 and case 2.

Case 1.—There are three spin-allowed d–d type transitions, corresponding to the one electron transitions $b_{2g}(\pi^*) \rightarrow b_{1g}(\sigma^*)$ ($^1A_{1g} \rightarrow {}^1A_{2g}$), $a_{1g}(\sigma^*) \rightarrow b_{1g}(\sigma^*)$

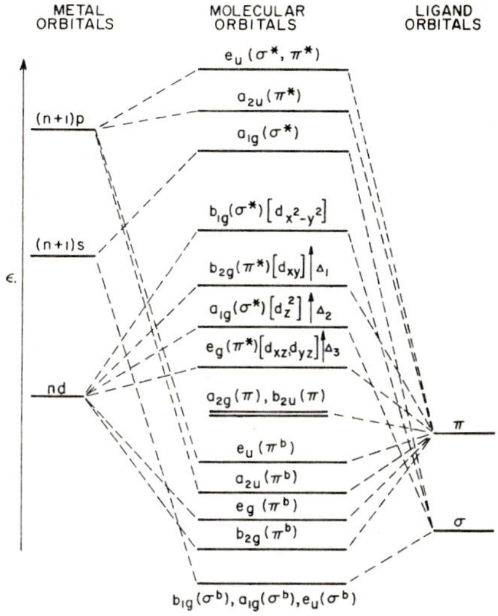

Fig. 2.—Molecular orbital energy level scheme for square planar metal complexes in which there is no intra-ligand π-orbital system (case 1).

($^1A_{1g} \rightarrow {}^1B_{1g}$) and $e_g(\pi^*) \rightarrow b_{1g}(\sigma^*)$ ($^1A_{1g} \rightarrow {}^1E_g$). All these are parity forbidden as electric dipole transitions. Two allowed charge transfer transitions may be anticipated, corresponding to the one-electron transitions $b_{2u}(\pi) \rightarrow b_{1g}(\sigma^*)$ ($^1A_{1g} \rightarrow {}^1A_{2u}$) and $e_u(\pi^b) \rightarrow b_{1g}(\sigma^*)$

TABLE II

GROUP OVERLAP INTEGRALS AND LIGAND–LIGAND INTERACTIONS FOR SQUARE PLANAR METAL COMPLEXES

Molecular orbital	Metal orbital	Ligand orbitals	$G_{ij}{}^a$	Ligand–ligand interaction
$\Psi[a_{1g}(\sigma)]$	$[nd_{z^2}]$	$[{}^1/_2(\sigma_1 + \sigma_2 + \sigma_3 + \sigma_4)]$	0.161	......
	$[(n + 1)s]$	$[{}^1/_2(\sigma_1 + \sigma_2 + \sigma_3 + \sigma_4)]$	.895	......
$\Psi[b_{1g}(\sigma)]$	$[nd_{x^2} - {}_{y^2}]$	$[{}^1/_2(\sigma_1 - \sigma_2 + \sigma_3 - \sigma_4)]$	.279	......
$\Psi[a_{2g}(\pi)]$	....	$[{}^1/_2(\pi_{1h} + \pi_{2h} + \pi_{3h} + \pi_{4h})]$	...	-2β(h,h)
$\Psi[a_{2u}(\pi)]$	$[(n + 1)p_z]$	$[{}^1/_2(\pi_{1v} + \pi_{2v} + \pi_{3v} + \pi_{4v})]$	0.464	$+2\beta$(v,v)
$\Psi[b_{2g}(\pi)]$	$[nd_{xy}]$	$[{}^1/_2(\pi_{1h} - \pi_{2h} + \pi_{3h} - \pi_{4h})]$	0.166	$+2\beta$(h,h)
$\Psi(b_{2u}(\pi)]$	....	$[{}^1/_2(\pi_{1v} - \pi_{2v} + \pi_{3v} - \pi_{4v})]$	...	-2β(v,v)
$\Psi_{\rm I}[e_g(\pi)]$	$[nd_{xz})$	$\left[\frac{1}{\sqrt{2}}(\pi_{1v} - \pi_{3v})\right]$	0.117	......
$\Psi_{\rm II}[e_g(\pi)]$	$[nd_{yz}]$	$\left[\frac{1}{\sqrt{2}}(\pi_{2v} - \pi_{4v})\right]$	.117	......
$\Psi_{\rm I}[e_u(\sigma,\pi)]$	$[(n + 1)p_x]$	$\left[\frac{1}{\sqrt{2}}(\sigma_1 - \sigma_3)\right]$	.769	......
	$[(n + 1)p_x]$	$\left[\frac{1}{\sqrt{2}}(\pi_{2h} - \pi_{4h})\right]$	.328	......
$\Psi_{\rm II}[e_u(\sigma,\pi)]$	$[(n + 1)p_y]$	$\left[\frac{1}{\sqrt{2}}(\sigma_2 - \sigma_4)\right]$	.769	......
	$[(n + 1)p_y]$	$\left[\frac{1}{\sqrt{2}}(\pi_{1h} - \pi_{3h})\right]$	.328	......

[a] The group overlap integrals are for $Ni(CN_4)^{2-}$. The atomic wave functions are: 3d and 4s radial functions for nickel from ref. 16; 2s carbon radial function from ref. 17; 2p carbon radial function from ref. 18; the 4p nickel radial function is estimated as $\Phi(4p) = R_{5p}(1.40)$, where R_{5p} (1.40) is a Slater orbital with exponent 1.40. The Ni–C distance is taken to be 1.90 Å., averaged from the data given in ref. 21.

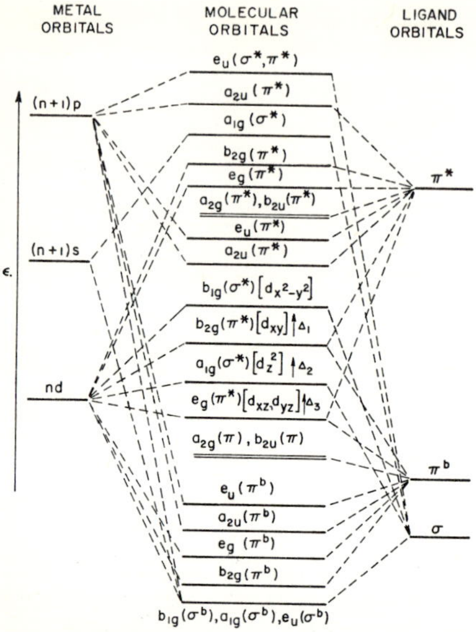

Fig. 3.—Molecular orbital energy level scheme for square planar metal complexes in which the ligands themselves have a π-orbital system (case 2).

($^1A_{1g} \rightarrow {}^1E_u$). These are transitions from molecular orbitals essentially localized on the ligands to molecular orbitals essentially localized on the metal atom. An examination of the transition moment integrals for these two transitions reveals that, for any reasonable molecular orbitals, the $^1A_{1g} \rightarrow {}^1E_u$ transition will be more intense than $^1A_{1g} \rightarrow {}^1A_{2u}$.

Case 2.—There are three *d-d* transitions; $^1A_{1g} \rightarrow {}^1A_{2g}$, $^1A_{1g} \rightarrow {}^1B_{1g}$ and $^1A_{1g} \rightarrow {}^1E_g$, as in case 1. The charge transfer bands of lowest energy are expected to involve transitions from the highest filled metal orbitals to the most stable empty "ligand" molecular orbital, the $a_{2u}(\pi^*)$. The first charge transfer transition is $b_{2g}(\pi^*) \rightarrow a_{2u}(\pi^*)$ ($^1A_{1g} \rightarrow {}^1B_{1u}$). This transition is orbitally forbidden and should have relatively little intensity. The second charge transfer band corresponds to the transition $a_{1g}(\sigma^*) \rightarrow a_{2u}(\pi^*)$ ($^1A_{1g} \rightarrow {}^1A_{2u}$). The energy of this transition is calculated to be $\Delta E(^1A_{1g} \rightarrow {}^1B_{1u}) + \Delta_2$, corrected for differences in interelectronic-repulsion energies in the $^1A_{2u}$ and $^1B_{1u}$ excited states. The third charge transfer transition is $e_g(\pi^*) \rightarrow a_{2u}(\pi^*)$ ($^1A_{1g} \rightarrow {}^1E_u$), and is calculated to be $\Delta E(^1A_{1g} \rightarrow {}^1B_{1u}) + \Delta_2 + \Delta_3$, again corrected for repulsion differences in the ^1E_u and $^1B_{1u}$ excited states. A summary of the calculated orbital and interelectronic-repulsion energies of all the excited states of interest for case 2 is given in Table III.

The $^1A_{1g} \rightarrow {}^1A_{2u}$ and $^1A_{1g} \rightarrow {}^1E_u$ transitions are allowed, with the $^1A_{1g} \rightarrow {}^1E_u$ transition expected to have considerably greater intensity. Thus a three-band charge transfer spectrum, with an intensity order $^1A_{1g} \rightarrow {}^1E_u > {}^1A_{1g} \rightarrow {}^1A_{2u} > {}^1A_{1g} \rightarrow {}^1B_{1u}$, should be typical of square planar complexes with case 2 type ligands.

Spectral Properties of Square Planar Metal Complexes. A. Halide Complexes (Case 1).—The complexes PdX_4^{2-}, PtX_4^{2-} and AuX_4^- (X^- = halide) are square planar and diamagnetic. The spectra of solid samples of K_2PdCl_4 and K_2PdBr_4 have been measured by Harris, Livingstone and Reece,[9] and by Yamada.[12] Three bands are observed which may be assigned to *d-d* transitions. In solution, $PdCl_4^{2-}$ and $PdBr_4^{2-}$ show two charge transfer bands.

TABLE III

ORBITAL AND INTERELECTRONIC-REPULSION ENERGIES OF SOME EXCITED STATES OF INTEREST FOR AN nd^8 SQUARE PLANAR COMPLEX

Term designation		Orbital energy[a]	Slater-Condon energy[a,b]
A. Singlet terms	$^1A_{1g}$	(ground state)	...
	$^1A_{2g}$	Δ_1	$-35F_4$
d-d transitions	$^1B_{1g}$	$\Delta_1 + \Delta_2$	$-4F_2 - 15F_4$
	^1E_g	$\Delta_1 + \Delta_2 + \Delta_3$	$-3F_2 - 20F_4$
Charge transfer	$^1B_{1u}$	.... $\Delta E(^1B_{1u})$	...
transitions	$^1A_{2u}$	$\Delta E(^1B_{1u}) + \Delta_2$	$-20F_2 + 100F_4$
	^1E_u	$\Delta E(^1B_{1u}) + \Delta_2 + \Delta_3$	$-15F_2 + 75F_4$
B. Triplet terms	$^3A_{2g}$	Δ_1	$-105F_4$
d-d transitions	$^3B_{1g}$	$\Delta_1 + \Delta_2$	$-12F_2 - 45F_4$
	^3E_g	$\Delta_1 + \Delta_2 + \Delta_3$	$-9F_2 - 60F_4$

[a] Referred to the ground state as zero. [b] For the approximations used to estimate the Slater–Condon energy for the charge transfer transitions, see ref. 19.

The interpretation of the spectrum of $PtCl_4^{2-}$ is complicated by the presence of one or more "spin-forbidden" bands. However, $PtCl_4^{2-}$ seems to exhibit the three spin-allowed *d-d* bands, and one charge transfer band. The spectra of the $AuCl_4^-$ and $AuBr_4^-$ complexes clearly show the two charge transfer bands.[10]

Assignments of the spectra of these halide complexes, in terms of the derived molecular orbital level scheme for case 1, are given in Table IV. The values of the single electron parameters Δ_1, Δ_2 and Δ_3 also are calculated and are given in Table V for comparison purposes.

B. Cyanide Complexes (Case 2).—The spectra of the square planar $Ni(CN)_4^{2-}$, $Pd(CN)_4^{2-}$ and $Pt(CN)_4^{2-}$ complexes are given in Table IV. Each complex exhibits the expected three charge transfer bands, but the *d-d* bands for the most part are obscured by these charge transfer bands. For $Ni(CN)_4^{2-}$, there are two shoulders on the tail of the first charge transfer band. These shoulders probably represent at least two *d-d* transitions. The charge transfer spectrum of $Au(CN)_4^-$ appears at higher energies than for $Pt(CN)_4^{2-}$. Thus only one of the expected three charge transfer bands is seen below 50,000 cm.$^{-1}$. The significance of this is discussed later.

Assignments of the spectra and the calculated values of Δ_1, Δ_2 and Δ_3 for the cyanide complexes are given in Tables IV and V, respectively.

Magnetic Properties of Square Planar Metal Complexes.—The magnetic susceptibility of the diamagnetic d^8 square planar metal complexes is given by

$$X = {}^1/_3(2X_\perp + X_{\parallel}) \quad (22)$$

where

$$X_\perp = X_A + X_{H.F.(\perp)} \quad (23)$$

$$X_{\parallel} = X_A + X_{H.F.(\parallel)} + X_{ring} \quad (24)$$

In eq. 23 and 24, X_A is the atomic contribution for the molecule in question (the sum of the Pascal constants), X_{ring} represents the expected diamagnetic contribution of the metal–ligand π-molecular orbital system (analogous to the ring diamagnetism in benzene) and

$$X_{H.F.(\perp)} = \frac{-2}{g_m} N\beta^2 \frac{|\langle\psi_0|L_x + 2S_x|\psi_i\rangle|^2}{\Delta E_{o,i}} \quad (25)$$

$$X_{H.F.(\parallel)} = \frac{-2}{g_m} N\beta^2 \frac{|\langle\psi_0|L_z + 2S_z|\psi_i\rangle|^2}{\Delta E_{o,i}} \quad (26)$$

(9) C. M. Harris, S. E. Livingstone and I. H. Reece, *J. Chem. Soc.*, 1505 (1959).

(10) A. K. Gangopadhayay and A. Chakravorty, *J. Chem. Phys.*, **35**, 2206 (1961).

Table IV

Spectral Properties of Square Planar Metal Complexes

Complex (sample)	ν_{max}, cm.$^{-1}$	ϵ_{max}	Assignment	Reference to exptl. work
	Case 1			
$PdCl_4^{2-}$ (solid K_2PdCl_4)	16,700	....	$^1A_{1g} \rightarrow {}^1A_{2g}$	9, 12, 20
	21,500	....	$^1A_{1g} \rightarrow {}^1B_{1g}$	
	23,300	....	$^1A_{1g} \rightarrow {}^1E_{1g}$	
(aq. soln., excess Cl^-)	36,000	12,000	$^1A_{1g} \rightarrow {}^1A_{2u}$	
	44,900	30,000	$^1A_{1g} \rightarrow {}^1E_u$	
$PdBr_4^{2-}$ (solid K_2PdBr_4)	16,000	....	$^1A_{1g} \rightarrow {}^1A_{2g}$	9
	20,000	....	$^1A_{1g} \rightarrow {}^1B_{1g}$	
	26,000[a]	....	$^1A_{1g} \rightarrow {}^1E_g$	
(aq. soln., excess Br^-)	30,100	10,400	$^1A_{1g} \rightarrow {}^1A_{2u}$	
	40,500	30,400	$^1A_{1g} \rightarrow {}^1E_u$	
$PtCl_4^{2-}$ (aq. soln.)	17,700	2.6	$^1A_{1g} \rightarrow {}^3A_{2g}$	2a
	21,000	15	$^1A_{1g} \rightarrow {}^1A_{2g}$	
	25,500	59	$^1A_{1g} \rightarrow {}^1B_{1g}$	
	30,200	64	$^1A_{1g} \rightarrow {}^1E_g$	
	(37,900)	250	*b*	
	46,000	9,580	$^1A_{1g} \rightarrow {}^1A_{2u}$	
$PtBr_4^{2-}$ (aq. soln., 1 MBr^-)	16,700	5	$^1A_{1g} \rightarrow {}^3A_{2g}$	This work
	19,700	15	$^1A_{1g} \rightarrow {}^1A_{2g}$	
	24,300	100	$^1A_{1g} \rightarrow {}^1B_{1g}$	
	28,200	120	$^1A_{1g} \rightarrow {}^1E_g$	
	37,300	7000	$^1A_{1g} \rightarrow {}^1A_{2u}$	
$AuCl_4^-$ (aq. soln.)	31,800	4,570	$^1A_{1g} \rightarrow {}^1A_{2u}$	10
	44,200	29,500	$^1A_{1g} \rightarrow {}^1E_u$	
$AuBr_4^-$ (aq. soln.)	26,300	5,010	$^1A_{1g} \rightarrow {}^1A_{2u}$	10
	39,400	38,900	$^1A_{1g} \rightarrow {}^1E_u$	
	Case 2			
$Ni(CN)_4^{2-}$ (aq. soln.)	(22,500)	2	$^1A_{1g} \rightarrow {}^1A_{2g}$	This work
	(30,500)	250	$^1A_{1g} \rightarrow {}^1B_{1g}$	
	32,300	700	$^1A_{1g} \rightarrow {}^1B_{1u}$	
	35,200	4,200	$^1A_{1g} \rightarrow {}^1A_{2u}$	
	37,600	10,600	$^1A_{1g} \rightarrow {}^1E_u$	
$Pd(CN)_4^{2-}$ (aq. soln.)	41,600	1,200	$^1A_{1g} \rightarrow {}^1B_{1u}$	20
	45,400	7,200	$^1A_{1g} \rightarrow {}^1A_{2u}$	
	47,200	9,000	$^1A_{1g} \rightarrow {}^1E_u$	
$Pt(CN)_4^{2-}$ (aq. soln.)	35,720	1,590	$^1A_{1g} \rightarrow {}^1B_{1u}$	
	(38,680)	26,000	$^1A_{1g} \rightarrow {}^1E_u$	
	39,180	29,500	$^1A_{1g} \rightarrow {}^1E_u$	
	41,320	1,850	$^1A_{1g} \rightarrow {}^1A_{2u}$	
$Au(CN)_4^-$ (aq. soln.)	30,960	51	$^1A_{1g} \rightarrow {}^1A_{2g}$	22
	37,880	331	$^1A_{1g} \rightarrow {}^1B_{1g}$	
	46,080	2,400	$^1A_{1g} \rightarrow {}^1B_{1u}$	

[a] This band is found at 21,700 cm.$^{-1}$ in the $K_2PdBr_4 \cdot 2H_2O$ crystal; see ref. 12. [b] Possibly a "spin-forbidden" charge transfer transition.

Table V

Metal *d*-Orbital Energies for Selected Square Planar Complexes

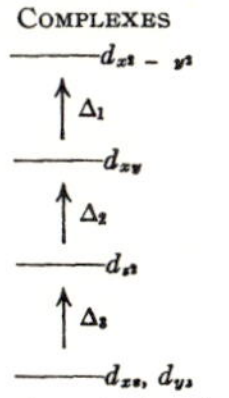

Orbital energy differences (cm.$^{-1}$) for $F_2 = 10F_4 = 700$ cm.$^{-1}$

Complex ion	Δ_1	Δ_2	Δ_3
$PdCl_4^{2-}$	19,150	6,200	1450
$PdBr_4^{2-}$	18,450	5,400	5650 (1350)[a]
$PtCl_4^{2-}$	23,450	5,900	4350
$PtBr_4^{2-}$	22,150	6,000	3550
$AuCl_4^-$	>20,000	....	....
$AuBr_4^-$	>20,000	....	....
$Ni(CN)_4^{2-}$	24,950	9,900	650
$Pd(CN)_4^{2-}$	>30,000	10,800	50
$Pt(CN)_4^{2-}$	>30,000	12,600	−4140
$Au(CN)_4^-$	33,410	10,620	....

[a] The value of Δ_3 for $K_2PdBr_4 \cdot 2H_2O$.

where g_m is the degeneracy of the ground state and $\Delta E_{o,i}$ is the energy separation between the ground state ψ_0 and excited state ψi. For the d^8 complexes under consideration, the ground state is $^1A_{1g}$, and eq. 25 and 26 reduce to

$$X_{H.F.(\perp)} = \frac{(C_M)^2 4N\beta^2}{\Delta E(^1A_{1g} \rightarrow {}^1E_g)} \quad (27)$$

$$X_{H.F.(||)} = \frac{(C_M)^2 16N\beta^2}{\Delta E(^1A_{1g} \rightarrow {}^1A_{2g})} \quad (28)$$

The magnetic susceptibility of a single crystal of $K_2Ni(CN)_4 \cdot H_2O$ has been measured by Rogers.[11] The values he obtained were

$$X_{||} = -127.8 \times 10^{-6} \text{ (c.g.s.)}$$

$$X_{\perp} = -146 \times 10^{-6} \text{ (c.g.s.)}$$

Subtracting eq. 24 and eq. 23, the unknown X_A is eliminated, yielding

$$X_{\perp} - X_{||} - (X_{H.F.(\perp)} - X_{H.F.(||)}) = -X_{ring} \quad (29)$$

Using the experimental values for the ΔE's, $X_{H.F.(\perp)} = 26 \times 10^{-6}$ and $X_{H.F.(||)} = 142 \times 10^{-6}$ are calculated; this gives $X_{ring} = -98 \times 10^{-6}$.

Discussion

The metal orbitals involved in σ-bonding in square planar complexes are the nd_{z^2}, $nd_{x^2-y^2}$, $(n+1)s$, $(n+1)p_x$ and $(n+1)p_y$. The $nd_{x^2-y^2}$, $(n+1)s$, $(n+1)p_x$ and $(n+1)p_y$ orbitals account for most of the σ-bonding, judging from the values of the overlap integrals in Table II, and the nd_{z^2} makes only a minor contribution. The most important π-molecular orbital is the a_{2u}, consisting of the $(n+1)p_z$ metal orbital and a combination of the four ligand π_v-orbitals. This gives a very stable π-bonding orbital (with only a single node in the molecular plane) which may be called the "ring" π-orbital. All the other pure π-molecular orbitals have an equal number of electrons in their bonding and antibonding levels.

The tendency of nd^8-metal ions to form square planar complexes increases in the order $Ni^{2+} < Pd^{2+} < Pt^{2+}$. Two features of the molecular orbital bonding scheme are consistent with this observation. First, the availability of the $nd_{x^2-y^2}$ metal orbital for σ-bonding undoubtedly increases in going from Ni^{2+} to Pt^{2+} ($NiCl_4^{2-}$ is tetrahedral; $PdCl_4^{2-}$ and $PtCl_4^{2-}$ are square planar). Second, the square planar configuration is stabilized by the ring π-bonding, which is expected to increase from Ni^{2+} to Pt^{2+}. The addition of a fifth group above the square plane drastically decreases the ring π-bonding, by tying up the $(n+1)p_z$ orbital in σ-bonding.

The assignments of the electronic spectra of the square planar halide and cyanide complexes of Ni^{2+}, Pd^{2+}, Pt^{2+} and Au^{3+} are given in Table IV. The halide complexes will be considered first. The spectra of K_2PdCl_4 and K_2PdBr_4 show three bands which may be assigned to *d-d* transitions. In the molecular orbital energy level scheme for case 1 ligands, the order of the single electron molecular orbitals is reasonably expected to be $e_g(\pi^*)$, $a_{1g}(\sigma^*)$, $b_{2g}(\pi^*)$ and $b_{1g}(\sigma^*)$. Thus the bands are assigned $^1A_{1g} \rightarrow {}^1A_{2g}$, $^1A_{1g} \rightarrow {}^1B_{1g}$ and $^1A_{1g} \rightarrow {}^1E_g$ in order of increasing energy. The energy order of the single electron molecular orbitals deduced here is consistent with the results of a "complete" electrostatic calculation of square planar Pt^{2+} complexes performed by Fenske, *et al.*[6]

The spectra of $PtCl_4^{2-}$ and $PtBr_4^{2-}$ are slightly more complicated; the first band is quite weak and is assigned as the first spin-forbidden transition, $^1A_{1g} \rightarrow {}^3A_{2g}$. The spin-allowed transitions are assigned in the same way as for the PdX_4^{2-} complexes. The spectra of single crystals of K_2PtCl_4 using polarized light show that the 25,500 cm.$^{-1}$ band occurs in *x,y* and the 30,200 cm.$^{-1}$ band occurs in *z* polarization.[12] These polariza-

(11) M. T. Rogers, *J. Am. Chem. Soc.*, **69**, 1506 (1947).
(12) S. Yamada, *ibid.*, **73**, 1182 (1951).

tions are expected for the assignments given here, assuming a vibronic intensity giving mechanism. Chatt, Gamlen and Orgel[2a] previously assigned the spectrum of $PtCl_4^{2-}$ somewhat differently (17,700 cm.$^{-1}$, $^1A_{1g} \rightarrow {}^3A_{2g}$; 21,000 cm.$^{-1}$, $^1A_{1g} \rightarrow {}^3E_g$; 25,500 cm.$^{-1}$, $^1A_{1g} \rightarrow {}^1A_{2g}$; 30,200 cm.$^{-1}$, $^1A_{1g} \rightarrow {}^1E_g$). However, the 25,500 cm.$^{-1}$ and the 30,200 cm.$^{-1}$ bands are both quite sensitive to axial perturbations (both bands shift to the red on changing the solvent from 12 *M* HCl to water).[13] This is evidence against the assignments of Chatt, *et al.*,[2a] and is consistent with the assignments proposed here.

The $AuCl_4^-$ and $AuBr_4^-$ complexes do not exhibit any bands which can unambiguously be assigned to *d-d* transitions. In these cases the two charge transfer bands appear at lower energies (than for PtX_4^{2-} complexes) and mask the weaker *d-d* bands.

The only bands which can clearly be identified in the cyanide complexes are due to charge transfer transitions. A rough calculation may be performed to show that the charge transfer bands in $Ni(CN)_4^{2-}$ appear at about the energy expected for transitions from metal *d*-orbitals to the ligand $a_{2u}(\pi^*)$ orbital. The ionization potential of HCN is about 14 e.v.[14] The calculated separation of the single electron π^b and π^* levels in HCN is 9 e.v.[15] Thus the coulomb energy of the π^* level of complexed CN^- may be estimated at -5 e.v. The coulomb energy of the Ni^{2+} *d*-orbitals in $Ni(CN)_4^{2-}$ may be estimated at about -10 e.v., using the spectroscopic data compiled by Moore,[24] and assuming that the net charge on the complexed Ni^{2+} is actually about $+1$. This leaves 5 e.v., or about 40,000 cm.$^{-1}$, as the difference in energy of the single electron *d* (Ni^{2+}) and $\pi^*(CN^-)$ orbitals. Since the $a_{2u}(\pi^*)$ level is stabilized by the empty $4p_z$ metal orbital (also by $+2\beta$(v,v,)) and since the $3d_{xy}$, $3d_{xz}$, $3d_{yz}$ and $3d_z{}^2$ orbitals are actually weakly antibonding, the appearance of the three charge transfer bands in $Ni(CN)_4^{2-}$ between 32,900 cm.$^{-1}$ and 37,500 cm.$^{-1}$ is in accord with theory.

The separation Δ_1 for the cyanide complexes is by far the largest of the three orbital parameters, and illustrates that the $d_{x^2-y^2}$ is much more strongly antibonding than any of the other *d*-orbitals. The separation Δ_3 decreases regularly in going from Ni^{2+} to Pt^{2+} (for $Ni(CN)_4^{2-}$ and $Pd(CN)_4^{2-}$, the d_{z^2} is more unstable than d_{xz}, d_{yz}; for $Pt(CN)_4^{2-}$, the d_{z^2} is more stable than d_{xz}, d_{yz}). This is compatible with the idea that the axial interaction of the water molecules in water solution decreases in going from Ni^{2+} to Pt^{2+}. The two components observed for the intense charge transfer band in $Pt(CN)_4^{2-}$ are consistent with its assignment as $^1A_{1g} \rightarrow {}^1E_u$; the ^1E_u excited state is expected to be split by second-order spin-orbit effects.

The assignments given in this paper for the cyanide complexes differ considerably with assignments given by other investigators.[4,5,7] The reason for this disagreement is found in our "band assignment philosophy"—*assign only the bands which can clearly be identified as to energy and intensity*. In the case of the cyanide complexes these bands are almost certainly due to charge transfer transitions, as judged by their large intensities.

A careful comparison of the charge transfer bands for the $PtCl_4^{2-}$, $AuCl_4^-$, $Pt(CN)_4^{2-}$ and $Au(CN)_4^-$ complexes offers further proof of the assignments given here. Thus the first charge transfer band appears at lower energy for $AuCl_4^-$ than for $PtCl_4^{2-}$, as expected for a ligand $\rightarrow$ metal type electronic transition, since Au^{3+} is a better electron acceptor than Pt^{2+}. On the other hand, the first charge transfer band for $Au(CN_4^-$ occurs at *higher energy* than for $Pt(CN)_4^{2-}$, expected only if the charge transfer for these cyanide complexes is of the metal $\rightarrow$ ligand type.

The charge transfer band systems observed for the halide (case 1) and cyanide (case 2) complexes should serve as a guide in assigning the charge transfer in other square planar complexes as ligand (π) $\rightarrow$ metal (case 1) or metal $\rightarrow$ ligand $[a_{2u}(\pi^*)]$ (case 2). In the halide complexes the two charge transfer bands are separated by 10,000 to 13,000 cm.$^{-1}$. Thus it can be estimated that the $b_{2u}(\pi)$ and $e_u(\pi^b)$ molecular orbitals are separated by at least 10,000 cm.$^{-1}$ in a typical square planar complex. This is a reasonable value since the b_{2u} orbital is non-bonding with a ligand interaction of -2β(v,v), and the e_u orbital is π-bonding.[25]

The three charge transfer bands due to metal $\rightarrow$ ligand $[a_{2u}[\pi^*)]$ transitions and exhibited by the cyanide complexes are much more closely spaced. The two orbitally allowed metal $\rightarrow$ ligand transitions ($a_{1g}(\sigma^*) \rightarrow a_{2u}(\pi^*)$, $e_g(\pi^*) \rightarrow a_{2u}(\pi^*)$) are only 2,000–3,000 cm.$^{-1}$ apart in all the cyanide complexes, since the $a_{1g}(\sigma^*)$ and $e_g(\pi^*)$ metal orbitals are virtually non-bonding and thus very nearly equal in energy in these complexes.

Thus these two types of charge transfer show quite different properties and should be distinguishable in other nd^8 square planar complexes. The ligand (π) to metal type will have two bands, spaced by about 10,000 cm.$^{-1}$, the second more intense than the first. The metal $\rightarrow$ ligand $[a_{2u}(\pi^*)]$ type will have three closely spaced bands with intensities $b_{2g}(\pi^*) \rightarrow a_{2u}(\pi^*)$ [small], $a_{1g}(\sigma^*) \rightarrow a_{2u}(\pi^*)$ [intermediate] and $e_g(\pi^*) \rightarrow a_{2u}(\pi^*)$ [large]. The exact positions of these bands give of course the best clues to the relative ordering of the *d*-orbital energy levels. In all probability there will be cases in which the $b_{2g}(\pi^*)$ level is fairly close to $a_{1g}(\sigma^*)$ and $e_g(\pi^*)$ (Δ_2 small), and thus only the two orbitally allowed bands will be observed. In any event the spacing of the two orbitally allowed charge transfer bands will easily distinguish between ligand $\rightarrow$ metal and metal $\rightarrow$ ligand type transitions for the nd^8 square planar complexes.

The "ring" diamagnetism apparently observed for $K_2Ni(CN)_4 \cdot H_2O$ may be considered as experimental evidence of electron delocalization through the cyanide system. The electrons in the $a_{2u}(\pi^b)$ orbital, in particular, would be expected to generate considerable "ring current."

Experimental.—The complex $K_2Ni(CN)_4 \cdot H_2O$ was prepared as described in the literature.[23] The electronic spectral measurements were made using a Cary 14 spectrophotometer.

Acknowledgments.—The authors thank Dr. Andrew D. Liehr for making available a preprint of his paper with Dr. Adamson and Dr. Perumareddi, and for stimu-

(13) R. G. Pearson, H. B. Gray and F. Basolo, *J. Am. Chem. Soc.*, **82**, 787 (1960).

(14) F. H. Field and J. L. Franklin, "Electron Impact Phenomena," Academic Press, Inc., New York, N. Y., 1957, p. 273.

(15) K. Iguchi, *J. Chem. Phys.*, **23**, 1983 (1955).

(16) J. W. Richardson, W. C. Nieuwpoort, R. R. Powell and W. F. Edgell, *ibid.*, **36**, 1057 (1962).

(17) D. R. Hartree, "The Calculation of Atomic Structures," John Wiley and Sons, Inc., New York, N. Y., 1957, pp. 169–171.

(18) C. Zener, *Phys. Rev.*, **36**, 51 (1930).

(19) H. B. Gray and C. R. Hare, *Inorg. Chem.*, **1**, 363 (1962).

(20) C. K. Jørgensen, "Absorption Spectra of Complexes of Heavy Metals," Report to U. S. Army, Frankfurt am Main, October, 1958.

(21) "Interatomic Distances," The Chemical Society, London, 1958, p. M 178.

(22) A. Kiss, J. Csaszar and L. Lehotai, *Acta Chim. Acad. Sci. Hung.*, **14**, 225 (1958).

(23) W. C. Fernelius and J. J. Burbage, *Inorg. Syntheses*, **2**, p. 227.

(24) C. E. Moore, "Atomic Energy Levels," U. S. Natl. Bur. Standards Circular 467, 1949 and 1952.

(25) There is another assignment possible for the two charge transfer bands in the square planar halide complexes. The first band could contain both the $b_{2u}(\pi) \rightarrow b_{1g}(\sigma^*)$ and $e_u(\pi^b) \rightarrow b_{1g}(\sigma^*)$ transitions; this would leave the allowed $e_u(\sigma^b) \rightarrow b_{1g}(\sigma^*)$ transition responsible for the second intense band.

lating discussions. Acknowledgment is made to the donors of the Petroleum Research Fund of the American Chemical Society for partial support of this research.

[Contribution from the Department of Chemistry, Columbia University, New York 27, N. Y.]

The Electronic Structures of Octahedral Metal Complexes. I. Metal Hexacarbonyls and Hexacyanides

By Harry B. Gray and N. A. Beach

Received April 20, 1963

The bonding in metal hexacarbonyls and hexacyanides is described in terms of molecular orbitals. Vapor phase electronic spectra for the metal hexacarbonyls are reported in the range 3500–1700 Å. A molecular orbital energy level scheme is presented which is able to account for the observed d–d and charge-transfer absorption bands in the d^6 metal complexes. The charge-transfer transitions all are assigned as metal (d) to ligand (π^*).

Introduction

Transition metal carbonyls and cyanides are among the simplest complexes in which both metal-to-ligand and ligand-to-metal types of π-bonding are of importance. Thus a knowledge of the electronic energy levels for typical metal carbonyls and cyanides is very desirable.

Recently progress has been made in assigning the electronic spectra of several metal cyanide complexes: Perumareddi, Liehr, and Adamson have treated the d^1, d^2, d^8, d^9, and d^{10} metal complexes[1]; the spectrum of $Fe(CN)_6^{3-}$ has received considerable attention[2]; Robin has provided a theoretical interpretation of the charge-transfer processes in Prussian blue[3]; an interpretation of the spectra of $M(CN)_8^{n-}$ complexes has been given[4]; and the d–d spectrum of the $Cr(CN)_6^{3-}$ complex is well understood.[5]

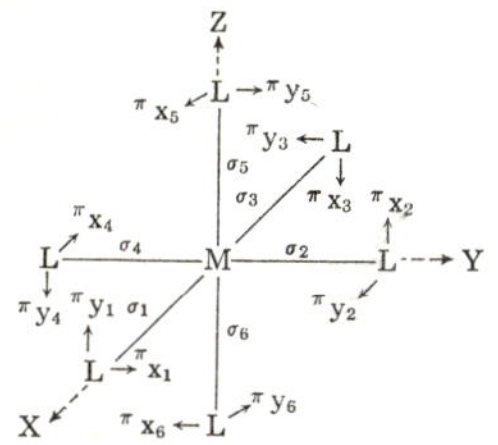

Fig. 1.—Coordinate system for the σ- and π-molecular orbitals of an ML_6 complex.

One of the present authors and colleagues have dealt with the square-planar metal cyanides[6] and the metal pentacyanonitrosyl complexes.[7–9] Of the metal carbonyls, complete spectral assignments have been given only for the $Mn(CO)_5X$ type complexes.[10]

It is important to establish the main similarities and differences in the electronic spectra of isoelectronic metal carbonyls and cyanides and to relate these spectral comparisons to the nature of the M–CN and M–CO bonds. In this paper the electronic spectra of d^6 metal carbonyls and cyanides are assigned on the basis of a derived molecular orbital energy level scheme. The differences in the energies of the single electron molecular orbitals for representative metal hexacarbonyls and hexacyanides are obtained and a general discussion of electronic structure is presented.

Molecular Orbitals for Octahedral Metal Carbonyls and Cyanides.—The coordinate system adopted for the case of full σ- and π-bonding in the O_h symmetry is given in Fig. 1. The metal valence orbitals are nd, $(n+1)$s, and $(n+1)$p. The carbon 2s and $2p_z$ orbitals will be used for σ-bonding; for π-bonding, both the ligand π-bonding (π^b) and π-antibonding (π^*) molecular orbitals will be combined with the d_π and p_π metal orbitals to form the π-molecular orbital system for the complex. The single electron molecular orbitals all are assumed to have the form

$$\psi(\text{m.o.}) = C_M\Phi(M) + C_L\Phi(L) \qquad (1)$$

where the C's are subject to conditions of normalization and orthogonality and the Φ's are proper metal and ligand orbital combinations for the molecular orbital in question. These proper combinations for $\Phi(M)$ and $\Phi(L)$ are given in Table I. Group overlap integrals are given for $Cr(CO)_6$.

The σ-molecular orbital system utilizes the $nd_{x^2-y^2}$ and nd_{z^2} (e_g), $(n+1)$s (a_{1g}), and $(n+1)p_x$, $(n+1)p_y$, and $(n+1)p_z$ (t_{1u}) metal orbitals with the proper linear combinations of ligand σ-orbitals. Thus there are formed six bonding and six antibonding σ-molecular orbitals.

The pure π-molecular orbitals are composed of the nd_{xy}, nd_{xz}, and nd_{yz} (t_{2g}) metal orbitals and the t_{2g} combinations of π^b and π^* ligand molecular orbitals. The instability order of the combining π-orbitals is always $\pi^b(L) < nd_\pi < \pi^*(L)$. Thus there are formed three (degenerate) strongly bonding π-molecular orbitals mainly localized on the CN^-'s or CO's; three virtually nonbonding (which may be either weakly bonding or antibonding, depending on the stabilities of $\pi^b(L)$ and $\pi^*(L)$ relative to nd_π) molecular orbitals mainly localized on the metal; and three strongly antibonding molecular orbitals mainly localized on the CN^-'s or CO's.

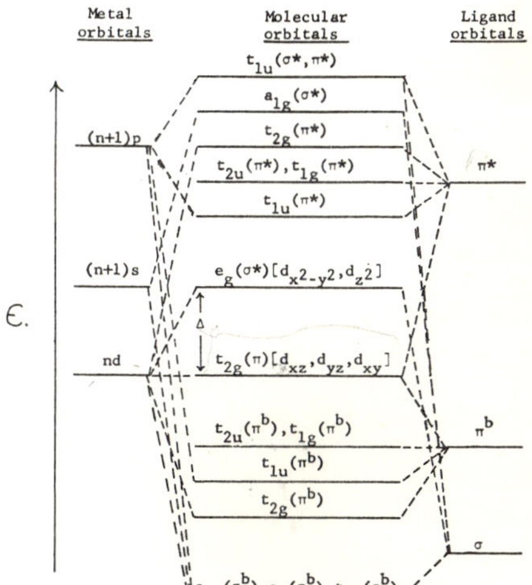

Fig. 2.—Molecular orbital energy level diagram for metal hexacyanides and hexacarbonyls.

Although the $(n+1)$p metal orbitals (t_{1u}) are expected to be primarily involved in σ-bonding, they may interact with the t_{1u} ligand π^b and π^* orbital combinations to yield three bonding and three antibonding molecular orbitals mostly localized on the ligands and three strongly antibonding molecular orbitals, with both σ- (*vide supra*) and π-character, mainly localized on the metal. The instability order $\pi^*(L) < (n+1)$p is assumed.

Finally, there are t_{1g} and t_{2u} π^b and π^* ligand orbital combinations which do not interact with metal orbitals. The molecular orbital energy level scheme expected for the bonding situation described above is shown in Fig. 2.

(1) J. R. Perumareddi, A. D. Liehr, and A. W. Adamson, *J. Am. Chem. Soc.*, **85**, 249 (1963).

(2) (a) C. S. Naiman, *J. Chem. Phys.*, **35**, 323 (1961); (b) G. Basu and R. L. Belford, *ibid.*, **37**, 1933 (1962); (c) C. S. Naiman, *ibid.*, in press; (d) the spectrum of $Cr(CN)_5NO^{3-}$ has been related to the d^5 metal hexacyanide level diagram by C. S. Naiman, *ibid.*, **35**, 1503 (1961).

(3) M. B. Robin, *Inorg. Chem.*, **1**, 337 (1962).

(4) (a) G. Ghemann, *Theoret. chim. Acta*, **1**, 14 (1962); (b) E. König, *ibid.*, **1**, 23 (1962).

(5) (a) C. E. Sháffer, private communication; (b) also quoted in C. K. Jørgensen, "Absorption Spectra and Chemical Bonding in Complexes," Pergamon Press, New York, N. Y., 1962, p. 291.

(6) H. B. Gray and C. J. Ballhausen, *J. Am. Chem. Soc.*, **85**, 260 (1963).

(7) C. J. Ballhausen and H. B. Gray, *Inorg. Chem.*, **2**, 426 (1963).

(8) H. B. Gray and C. J. Ballhausen, *J. Chem. Phys.*, **36**, 1151 (1962).

(9) H. B. Gray, I. Bernal, and E. Billig, *J. Am. Chem. Soc.*, **84**, 3404 (1962).

(10) H. B. Gray, E. Billig, A. Wojcicki, and M. Farona, *Can. J. Chem.*, **41**, 1281 (1963).

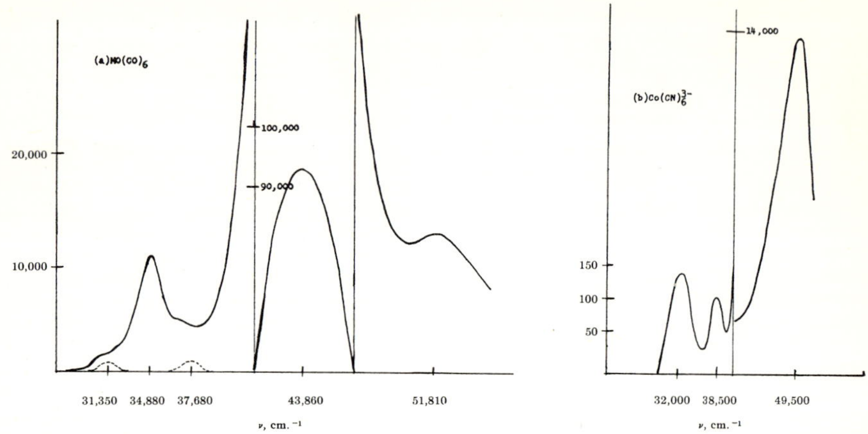

Fig. 3.—Electronic spectra of (a) $Mo(CO)_6$ (left curves) in the vapor phase and (b) $Co(CN)_6^{3-}$ (right curves) in water solution.

There will always be 36 electrons (6 from each ligand) to place in the molecular orbitals for the complex. In addition, the metal may furnish as many as 6 valence electrons for an octahedral metal carbonyl or cyanide complex. The ground state for a d^6 metal hexacarbonyl or hexacyanide is therefore

$$[t_{1u}(\sigma^b)]^6[a_{1g}(\sigma^b)]^2[e_g(\sigma^b)]^4[t_{2g}(\pi^b)]^6[t_{1u}(\pi^b)]^6\ [t_{2u}(\pi^b)]^6$$
$$[t_{1g}(\pi^b)]^6[t_{2g}(\pi)]^6 = {}^1A_{1g}$$

The most stable excited states for the d^6 complexes result from the following electronic transitions: (1) d–d transitions—these are the transitions that involve the molecular orbitals which are mainly composed of the central metal d-orbitals. The $t_{2g}(\pi) \rightarrow e_g(\sigma^*)$ promotion gives spin-allowed excited states ${}^1T_{1g}$ and ${}^1T_{2g}$. The Slater–Condon energies of ${}^1T_{1g}$ and ${}^1T_{2g}$ are $-35F_4$ and $16F_2 - 115F_4$, respectively, referred to the ground state ${}^1A_{1g}$ as zero. Electronic absorption bands for the transitions ${}^1A_{1g} \rightarrow {}^1T_{1g}$ and ${}^1A_{1g} \rightarrow {}^1T_{2g}$ are expected to be weak, since these transitions are forbidden for electric dipole radiation. (2) Charge transfer from metal to ligand—the lowest energy transitions of this type are $t_{2g}(\pi) \rightarrow t_{1u}(\pi^*)$, $t_{2g}(\pi) \rightarrow t_{2u}(\pi^*)$, $t_{1g}(\pi^*)$, and $t_{2g}(\pi) \rightarrow t_{2g}(\pi^*)$. The $t_{2u}(\pi^*)$ and $t_{1g}(\pi^*)$ energies may be estimated at about $-35{,}000$ cm.$^{-1}$ from ionization potential[11] and spectroscopic[12] data for CO and HCN. The $t_{1u}(\pi^*)$ level should be more stable, and the $t_{2g}(\pi^*)$

Table I

Group Overlap Integrals for $Cr(CO)_6$ and Ligand–Ligand Interactions for Octahedral Metal Complexes

Molecular orbital	Metal orbital	Ligand orbitals	G_{ij}[a]	Ligand–ligand interaction[b]
$\Psi[a_{1g}(\sigma)]$	$[(n+1)s]$	$[(1/\sqrt{6})(\sigma_1 + \sigma_2 + \sigma_3 + \sigma_4 + \sigma_5 + \sigma_6)]$	0.990	..
$\Psi_I[e_g(\sigma)]$	$[nd_{z^2}]$	$[(1/2\sqrt{3})(2\sigma_5 + 2\sigma_6 - \sigma_1 - \sigma_2 - \sigma_3 - \sigma_4)]$	.493	..
$\Psi_{II}[e_g(\sigma)]$	$[nd_{x^2-y^2}]$	$[(1/2)(\sigma_1 - \sigma_2 + \sigma_3 - \sigma_4)]$	.493	..
$\Psi_I[t_{1u}(\sigma,\pi)]$	$[(n+1)p_x]$	$[(1/\sqrt{2})(\sigma_1 - \sigma_3)]$	.710	..
	$[(n+1)p_x]$	$[(1/2)(\pi_{y2} + \pi_{x5} - \pi_{x4} - \pi_{y6})]$	.480	$+2\beta_\parallel$
$\Psi_{II}[t_{1u}(\sigma,\pi)]$	$[(n+1)p_y]$	$[(1/\sqrt{2})(\sigma_2 - \sigma_4)]$	.710	..
	$[(n+1)p_y]$	$[(1/2)(\pi_{x1} + \pi_{y5} - \pi_{y3} - \pi_{x6})]$	.480	$+2\beta_\parallel$
$\Psi_{III}[t_{1u}(\sigma,\pi)]$	$[(n+1)p_z]$	$[(1/\sqrt{2})(\sigma_5 - \sigma_6)]$	.710	..
	$[(n+1)p_z]$	$[(1/2)(\pi_{y1} + \pi_{x2} - \pi_{x3} - \pi_{y4})]$	.480	$+2\beta_\parallel$
$\Psi_I[t_{2g}(\pi)]$	$[nd_{xz}]$	$[(1/2)(\pi_{y1} + \pi_{x5} + \pi_{x3} + \pi_{y6})]$	.352	$+2\beta_\perp$
$\Psi_{II}[t_{2g}(\pi)]$	$[nd_{yz}]$	$[(1/2)(\pi_{x2} + \pi_{y5} + \pi_{y4} + \pi_{x6})]$	.352	$+2\beta_\perp$
$\psi_{III}[t_{2g}(\pi)]$	$[nd_{xy}]$	$[(1/2)(\pi_{x1} + \pi_{y2} + \pi_{y3} + \pi_{x4})]$	.352	$+2\beta_\perp$
$\psi_I[t_{2u}(\pi)]$	...	$[(1/2)(\pi_{y2} - \pi_{x5} - \pi_{x4} + \pi_{y6})]$	...	$-2\beta_\parallel$
$\psi_{II}[t_{2u}(\pi)]$	...	$[(1/2)(\pi_{x1} - \pi_{y5} - \pi_{y3} + \pi_{x6})]$	...	$-2\beta_\parallel$
$\psi_{III}[t_{2u}(\pi)]$	...	$[(1/2)(\pi_{y1} - \pi_{x2} - \pi_{x3} + \pi_{y4})]$	...	$-2\beta_\parallel$
$\psi_I[t_{1g}(\pi)]$	...	$[(1/2)(\pi_{y1} - \pi_{x5} + \pi_{x3} - \pi_{y6})]$	...	$-2\beta_\perp$
$\psi_{II}[t_{1g}(\pi)]$	...	$[(1/2)(\pi_{x2} - \pi_{y5} + \pi_{y4} - \pi_{x6})]$	...	$-2\beta_\perp$
$\psi_{III}[t_{1g}(\pi)]$	...	$[(1/2)(\pi_{x1} - \pi_{y2} + \pi_{y3} - \pi_{x4})]$	...	$-2\beta_\perp$

[a] The group overlap integrals (G_{ij}'s) are for $Cr(CO)_6$. The radial wave functions are: 3d and 4s for chromium from ref. 13; 2s for carbon from ref. 14; 2p for carbon from ref. 15; the 4p chromium radial function is taken to be $\Phi(4p) = R_{5p}(1.20)$ where $R_{5p}(1.20)$ is an STO with exponent 1.20. The Cr–C distance in $Cr(CO)_6$ is 1.92 Å., from ref. 16. The σ-valence orbital of CO and CN^- is assumed to be an sp-hybrid. [b] For the π-molecular orbitals only, where, for example: $\beta_\parallel = \int(\pi_{y1})\mathcal{H}(\pi_{x2})d\tau$; $\beta_\perp = \int(\pi_{y1})\mathcal{H}(\pi_{x5})d\tau$.

(11) F. H. Field and J. L. Franklin, "Electron Impact Phenomena," Academic Press, Inc., New York, N. Y., 1957, pp. 273 and 280.

(12) G. Herzberg, "Spectra of Diatomic Molecules," D. Van Nostrand and Co., Inc., New York, N. Y., 1950, p. 452; for HCN see K. Iguchi, *J. Chem. Phys.*, **23**, 1983 (1955).

(13) J. W. Richardson, W. C. Nieuwpoort, R. R. Powell, and W. F. Edgell, *J. Chem. Phys.*, **36**, 1057 (1962).

Table II

Electronic Spectra of d^6 Metal Hexacarbonyls and Hexacyanides

Complex (sample)	λ_{max}, Å.	γ_{max}, cm.$^{-1}$	$f \times 10^2$	Assignment
$Cr(CO)_6$ (vapor)	3190	31,350	0.69	${}^1A_{1g} \rightarrow {}^1T_{1g}$
	2795	35,780	19.1	${}^1A_{1g} \rightarrow {}^1T_{1u}{}^{(1)}$
	2248	44,480	145	${}^1A_{1g} \rightarrow {}^1T_{1u}{}^{(2)}$
	1950	51,280	2.5	$t_{2g}(\pi) \rightarrow t_{2g}(\pi^*)$
$Mo(CO)_6$ (vapor)	3190	31,350	0.53	${}^1A_{1g} \rightarrow {}^1T_{1g}$
	2867	34,880	13.9	${}^1A_{1g} \rightarrow {}^1T_{1u}{}^{(1)}$
	2654	37,680	0.41	${}^1A_{1g} \rightarrow {}^1T_{2g}$
	2280	43,860	178	${}^1A_{1g} \rightarrow {}^1T_{1u}{}^{(2)}$
	1930	51,810	22.4	$t_{2g}(\pi) \rightarrow t_{2g}(\pi^*)$
$W(CO)_6$ (vapor)	3070	32,570	0.36	${}^1A_{1g} \rightarrow {}^1T_{1g}$
	2865	34,900	12.6	${}^1A_{1g} \rightarrow {}^1T_{1u}{}^{(1)}$
	2690	37,170	0.30	${}^1A_{1g} \rightarrow {}^1T_{2g}$
	2240	44,640	208	${}^1A_{1g} \rightarrow {}^1T_{1u}{}^{(2)}$
	1980	50,510	4.6	$t_{2g}(\pi) \rightarrow t_{2g}(\pi^*)$
	1900	52,630	4.0	$t_{2g}(\pi) \rightarrow t_{2g}(\pi^*)$
$Fe(CN)_6^{4-}$ (aq. soln.)	4220	23,700	0.002	${}^1A_{1g} \rightarrow {}^3T_{1g}$
	3225	31,000	0.84	${}^1A_{1g} \rightarrow {}^1T_{1g}$
	2700	37,040	0.47	${}^1A_{1g} \rightarrow {}^1T_{2g}$
	2180	45,870	53.5	${}^1A_{1g} \rightarrow {}^1T_{1u}{}^{(1)}$
	2000	50,000	2.3	${}^1A_{1g} \rightarrow {}^1T_{1u}{}^{(2)}$
$Ru(CN)_6^{4-}$ (aq. soln.)[a]	3225	31,000	. . .	${}^1A_{1g} \rightarrow {}^1T_{1g}$
	2060	48,500	85	${}^1A_{1g} \rightarrow {}^1T_{1u}{}^{(1)}$
	1920	52,000	45	${}^1A_{1g} \rightarrow {}^1T_{1u}{}^{(2)}$
$Os(CN)_6^{4-}$ (aq. soln.)[a]	2130	47,000	110	${}^1A_{1g} \rightarrow {}^1T_{1u}{}^{(1)}$
	1950	51,300	95	${}^1A_{1g} \rightarrow {}^1T_{1u}{}^{(2)}$
$Co(CN)_6^{3-}$ (aq. soln.)	3120	32,050	0.30	${}^1A_{1g} \rightarrow {}^1T_{1g}$
	2600	38,460	0.28	${}^1A_{1g} \rightarrow {}^1T_{2g}$
	2020	49,500	31.8	${}^1A_{1g} \rightarrow {}^1T_{1u}{}^{(1)}$

[a] Estimated from Fig. 2 in ref. 3.

less stable than the $-35,000$ cm.$^{-1}$. The $t_{2g}(\pi)$ orbital energy depends strongly on the metal under consideration, but is likely to be in the $-70,000$ to $-85,000$ cm.$^{-1}$ range. Thus the metal-to-ligand charge transfer transitions should appear in the 30,000–55,000 cm.$^{-1}$ region of the spectrum.

Only ${}^1A_{1g} \rightarrow {}^1T_{1u}$ transitions are allowed for d^6 metal hexacarbonyls and hexacyanides. Since both $t_{2g}(\pi) \rightarrow t_{1u}(\pi^*)$ and $t_{2g}(\pi) \rightarrow t_{2u}(\pi^*)$ promotions give ${}^1T_{1u}$ excited states, there should be two intense charge-transfer bands of the metal-to-ligand type in the 30,000–50,000 cm.$^{-1}$ range. (3) Charge transfer from ligand to metal—the lowest energy allowed transition of this type is $t_{2u}(\pi^b) \rightarrow e_g(\pi^*)$, ${}^1A_{1g} \rightarrow {}^1T_{1u}$. The energy of the $t_{2u}(\pi^b)$ orbital can be estimated at about $-110,000$ cm.$^{-1}$ from the ionization potentials of CO and HCN.[10,17] The $e_g(\sigma^*)$ energy is obtained from the previous estimate of the $t_{2g}(\pi)$ orbital energy and adding 35,000 cm.$^{-1}$ for Δ (*vide infra*). Thus the ligand-to-metal charge transfer for the d^6 complexes is not anticipated below 60,000 cm.$^{-1}$.

Electronic Spectra of d^6 Metal Hexacarbonyls and Hexacyanides. A. Metal Hexacarbonyls.—The electronic spectrum of $Cr(CO)_6$ in the vapor phase shows two intense absorption bands at 35,780 and 44,480 cm.$^{-1}$, which are assigned as the allowed metal-to-ligand charge transfer transitions, ${}^1A_{1g} \rightarrow {}^1T_{1u}{}^{(1)}$ and ${}^1A_{1g} \rightarrow {}^1T_{1u}{}^{(2)}$, respectively. In addition, shoulders indicating maxima at 31,350 and 51,280 cm.$^{-1}$ are in evidence. The low energy weak band is assigned as the first d–d transition, ${}^1A_{1g} \rightarrow {}^1T_{1g}$. The weak band at 51,280 cm.$^{-1}$ probably results from the $t_{2g}(\pi) \rightarrow t_{2g}(\pi^*)$ transition.

The spectra of $Mo(CO)_6$ and $W(CO)_6$ are very similar to the spectrum of $Cr(CO)_6$. Figure 3 shows the spectrum of $Mo(CO)_6$. The intense bands at about 35,000 and about 43,000 cm.$^{-1}$ are present. Both the ${}^1A_{1g} \rightarrow {}^1T_{1g}$ and the ${}^1A_{1g} \rightarrow {}^1T_{2g}$ d–d bands apparently are resolved, the former on the low energy side, the latter on the high energy side of the first intense band. The separation of the two d–d bands is 6330 cm.$^{-1}$ for $Mo(CO)_6$; it is only 4600 cm.$^{-1}$ for $W(CO)_6$. The predicted difference between ${}^1T_{1g}$ and ${}^1T_{2g}$ is $16F_2 - 80F_4$. Thus we obtain the reasonable values $F_2 = 10F_4 = 790$ cm.$^{-1}$ for $Mo(CO)_6$ and $F_2 = 10F_4 = 575$ cm.$^{-1}$ for $W(CO)_6$. The bands observed at higher energy than ${}^1A_{1g} \rightarrow {}^1T_{1u}{}^{(2)}$ are assigned to states arising from the transition $t_{2g}(\pi) \rightarrow t_{2g}(\pi^*)$. Complete spectral assignments for the $M(CO)_6$ complexes are summarized in Table II.

B. Metal Hexacyanides.—The electronic spectrum of $Fe(CN)_6^{4-}$ shows relatively weak bands at 31,000 and 37,040 cm.$^{-1}$. These bands are assigned as the expected d–d transitions, ${}^1A_{1g} \rightarrow {}^1T_{1g}$ and ${}^1A_{1g} \rightarrow {}^1T_{2g}$, respectively. There is a very weak band at 23,700 cm.$^{-1}$, which is assigned as the first spin-forbidden transition, ${}^1A_{1g} \rightarrow {}^3T_{1g}$. This assignment agrees with the $-70F_4$ Slater–Condon energy of ${}^3T_{1g}$ relative to ${}^1T_{1g}$. The two charge-transfer bands are at 45,870 (${}^1A_{1g} \rightarrow {}^1T_{1u}{}^{(1)}$) and 50,000 cm.$^{-1}$ (${}^1A_{1g} \rightarrow {}^1T_{1u}{}^{(2)}$). The spectra of $Ru(CN)_6^{4-}$ and $Os(CN)_6^{4-}$ are similar, except the weaker bands for $Os(CN)_6^{4-}$ apparently are obscured by the more intense charge transfer.

The absorption spectrum of $K_3Co(CN)_6$ in aqueous solution is shown in Fig. 3. The two d–d bands, ${}^1A_{1g} \rightarrow {}^1T_{1g}$ and ${}^1A_{1g} \rightarrow {}^1T_{2g}$, are displayed as sharp maxima at 32,050 and 38,460 cm.$^{-1}$. The first charge-transfer band is located at 49,500 cm.$^{-1}$ (${}^1A_{1g} \rightarrow {}^1T_{1u}{}^{(1)}$). The ${}^1A_{1g} \rightarrow {}^1T_{1u}{}^{(2)}$ transition should appear at about 53,000 cm.$^{-1}$. However, since it is difficult to obtain high resolution solution spectra in this region, the exact position of the ${}^1A_{1g} \rightarrow {}^1T_{1u}{}^{(2)}$ band was not located. The spectral assignments for the $M(CN)_6^{n-}$ complexes are given in Table II.

Discussion

The d–d transitions in most of the d^6 metal carbonyls and cyanides are hard to locate with certainty, since the large Δ-values mean that metal-to-ligand charge transfer may occur at about the same energy as the two bands due to the $t_{2g}(\pi) \rightarrow e_g(\sigma^*)$ transition. For the complexes in which the metal furnishes fewer than 6 valence electrons (such as the d^5 $Fe(CN)_6^{3-}$), additional charge-transfer bands due to ligand-to-metal transitions further obscure the weak d–d bands.[2] Nevertheless, in many cases reliable values for Δ are available for $M(CO)_6$ and $M(CN)_6^{n-}$ complexes. These Δ-values are summarized in Table III.

With the exception of the d^1–d^3 complexes, the Δ-values in Table III all are approximately 35,000 cm.$^{-1}$. It is somewhat surprising to find similar Δ-values for complexes with widely varying charges on the central metal. One of the good examples for comparison purposes is the negligible change in Δ in going from $Fe(CN)_6^{4-}$ to $Fe(CN)_6^{3-}$. Recall that in halide, ammine, and aquo metal complexes, an increase in the

(14) D. R. Hartree, "The Calculation of Atomic Structures," John Wiley and Sons, Inc., New York, N. Y., 1957, pp. 169–171.

(15) C. Zener, *Phys. Rev.*, **36**, 51 (1930).

(16) "Interatomic Distances," The Chemical Society, London, 1958, p. M188.

(17) The π^b-level cannot be less stable than $-110,000$ cm.$^{-1}$ in either CO or HCN. The ionization potentials of CO and HCN are both approximately 14 e.v. The CO π^b-level is very likely to be considerably more stable than $-110,000$ cm.$^{-1}$, since in this case there is evidence[18] that the ionization potential refers to the removal of one of the carbon 2s electrons.

(18) H. H. Jaffé and M. Orchin, *Tetrahedron*, **10**, 212 (1960).

Table III

Magnitude of the d–d Splitting Parameter Δ for Different Metal Hexacarbonyls and Hexacyanides

Complex	Δ, cm.$^{-1}$	Complex	Δ, cm.$^{-1}$
$Cr(CO)_6$	34,150[a]	$Fe(CN)_6^{3-}$	35,000[e]
$Mo(CO)_6$	34,150[a]	$Co(CN)_6^{3-}$	34,800[a]
$W(CO)_6$	34,570[b]	$Fe(CN)_6^{4-}$	33,800[a]
$Ti(CN)_6^{3-}$	22,300[c]	$Ru(CN)_6^{4-}$	33,800[a]
$V(CN)_6^{3-}$	23,390[c]	$Os(CN)_6^{4-}$	>34,000
$Cr(CN)_6^{3-}$	26,600[d]		

[a] For $35F_4$ = 2800 cm.$^{-1}$. [b] For $35F_4$ = 2000 cm.$^{-1}$. [c] From ref. 1. [d] From ref. 5. [e] From ref. 2a.

positive charge on the central metal ion means a substantial increase in Δ. For example, for $Fe(H_2O)_6^{2+}$, Δ = 10,400 cm.$^{-1}$; for $Fe(H_2O)_6^{3+}$, Δ = 14,300 cm.$^{-1}$.[19] However, halide, ammonia, and water ligands do not engage the metal in metal (d_π)-to-ligand (π^*) π-bonding, and since Δ is the difference in the orbital energies of $e_g(\sigma^*)$ and $t_{2g}(\pi)$, the expected increase in Δ due to an increase in σ-bonding in going from Fe^{2+} to Fe^{3+} (in the hexacyanides) is canceled by the accompanying decrease in Δ due to a decrease in $d_\pi \rightarrow (CN^-)$ back donation.[20] This observed constancy of Δ with a change in the charge on the central metal ion may be expected in other octahedral complexes in which there is extensive back bonding,[20] provided systems with approximately the same number of valence electrons are compared. Indeed, the Δ-values for the metal hexacyanides decrease rapidly as the number of d-electrons decreases in the M^{3+} complexes. Thus we have the Δ-order: $Co(CN)_6^{3-} \approx Fe(CN)_6^{3-} >> Cr(CN)_6^{3-} > V(CN)_6^{3-} > Ti(CN)_6^{3-}$. This of course indicates that ligand (π^b)-to-metal π-bonding is relatively more important in the d^1–d^3 cases, and metal-to-ligand (π^*) π-bonding is more important in the d^5–d^6 cases.

Maximum back bonding in the cases here under investigation occurs in the metal hexacarbonyls, since the $t_{2g}(\pi)$ level must be assumed quite stable to account for the large Δ-values observed. It is interesting that Δ in the $M(CO)_6$ series increases only slightly in the order $Cr(CO)_6$, $Mo(CO)_6$, and $W(CO)_6$. This follows the usual pattern of Δ(3d) < Δ(4d) < Δ(5d) in a consistent series of complexes, but the effect here is very small, indicating that the σ- and π-changes almost cancel in the 3d, 4d, and 5d valence orbitals in this case. The constancy of the total M–CO bond order for M = Cr, Mo, or W in an analogous series of complexes is also indicated from vibrational spectral results.[21]

The metal hexacarbonyls and hexacyanides all appear to exhibit the two allowed charge transfer bands predicted for metal-to-ligand (π^*) type transitions. There is considerable evidence in favor of these assignments. First of all, the band positions are in agreement with the results of a simple calculation (*vide supra*). Also, recall that for the other type of charge transfer (ligand (π^b) → metal), the predicted energies are far larger than the energies observed. The most convincing evidence, however, is the comparison of band positions in the series $Cr(CO)_6$, $Fe(CN)_6^{4-}$, and $Co(CN)_6^{3-}$. The first band shifts to *higher* energies in this series in the order: $Cr(CO)_6 < Fe(CN)_6^{4-} < Co(CN)_6^{3-}$. This is the order expected for metal-to-ligand (π^*) type charge transfer, since the ionization potential of electrons in metal d_π-type orbitals must increase in the order $Cr^0 < Fe^{2+} < Co^{3+}$. This *increase* in the energy of a particular charge-transfer transition on increasing the positive charge of the central metal ion is characteristic of metal-to-ligand (π^*) transitions and has been used as evidence for this same transition type in square-planar cyanide complexes. The ligand-to-metal type charge transfer, which occurs in metal halide complexes, shifts to *lower* energy, as expected, on increasing the positive charge of the central metal ion. This is demonstrated nicely in the $PtCl_4^{2-}$ and $AuCl_4^-$ complexes.[6]

The difference in the energies of the $^1A_{1g} \rightarrow {}^1T_{1u}^{(1)}$ and $^1A_{1g} \rightarrow {}^1T_{1u}^{(2)}$ transitions is due to two factors. First, ligand–ligand interactions result in a stabilization of the $t_{1u}(\pi^*)$ molecular orbitals; this stabilization is $4\beta_\parallel$ relative to the $t_{2u}(\pi^*)$ molecular orbitals. This part of the total energy difference is probably between 1000 and 2000 cm.$^{-1}$ for all the $M(CO)_6$ and $M(CN)_6^{n-}$ complexes. The remainder of the difference in the $^1A_{1g} \rightarrow {}^1T_{1u}^{(1)}$ and $^1A_{1g} \rightarrow {}^1T_{1u}^{(2)}$ transition energies is due to the stabilization of the $t_{1u}(\pi^*)$ ligand orbital combination by π-interaction with the metal $(n + 1)$p-orbitals. Thus the magnitude of the total splitting gives an indication of the degree of involvement of the $(n + 1)$-p-metal orbitals in π-bonding.

An examination of Table II reveals that the splitting of the two charge-transfer bands is 3500–4300 cm.$^{-1}$ for the $M(CN)_6^{4-}$ complexes and 8700–9740 cm.$^{-1}$ for $M(CO)_6$ complexes. Thus the involvement of the $(n + 1)$p-metal orbitals in π-interaction increases considerably in going from M^{2+} hexacyanides to M^0 hexacarbonyls. This is reasonable since the $(n + 1)$p-orbitals are undoubtedly more expanded for the neutral metal atoms.

The intensity of the 45,000 cm.$^{-1}$ band in the $M(CO)_6$ complexes is unusually large. The allowed transition is $t_{2g}(\pi) \rightarrow t_{2u}(\pi^*)$ ($^1A_{1g} \rightarrow {}^1T_{1u}^{(2)}$), which involves a molecular orbital ($t_{2g}(\pi)$) substantially delocalized out to the CO's and a molecular orbital ($t_{2u}(\pi^*)$) essentially localized on the CO's. *Thus a large contribution to the intensity from ligand–ligand integrals is suggested.* This particular contribution to the intensity would increase with an increase in back donation and, indeed, the *f*-value of the $^1A_{1g} \rightarrow {}^1T_{1u}^{(2)}$ band may be related to the amount of back donation. The order of increasing *f* of the $^1A_{1g} \rightarrow {}^1T_{1u}^{(2)}$ band is $Co(CN)_6^{3-} < Fe(CN)_6^{4-} < Ru(CN)_6^{4-} < Os(CN)_6^{4-} < Cr(CO)_6 < Mo(CO)_6 < W(CO)_6$. The $Co(CN)_6^{3-}$ complex is assumed to have the smallest $^1A_{1g} \rightarrow {}^1T_{1u}^{(2)}$ band intensity, since no maximum or shoulder separate from the $^1A_{1g} \rightarrow {}^1T_{1u}^{(1)}$ band could be resolved. The above order is certainly a reasonable prediction of the order of increasing back donation in these complexes. We are now investigating this matter in some detail.

Experimental

The $K_3[Co(CN)_6]$ complex was prepared as described in the literature.[22] A sample of $Cr(Co)_6$ was donated by the Ethyl Corporation. Samples of $Mo(CO)_6$ and $W(CO)_6$ were provided by the Climax Molybdenum Co.

Spectra in the 3500–1900 Å. region were taken with a Cary 14 spectrophotometer purged with nitrogen. The vapor phase spectra were obtained using quartz gas cells, with path lengths from 1 to 10 cm. The vapor phase spectral measurements were extended to 1700 Å. using a nitrogen-purged Perkin–Elmer Model 350 spectrophotometer.

Acknowledgments.—Acknowledgment is made to the donors of the Petroleum Research Fund, admin-

(19) See ref. 5b, p. 110.

(20) In molecular orbital language, back donation (or back bonding) means considerable participation of empty ligand orbitals (usually π^*) in the occupied π-molecular orbitals mostly localized on the metal. We are fortunate to have the simple words "back donation" to describe such a state of affairs.

(21) F. A. Cotton and C. S. Kraihanzel, *J. Am. Chem. Soc.*, **84**, 4432 (1962).

(22) J. H. Bigelow and J. C. Bailar, *Inorg. Syntheses*, **2**, 225 (1946).

istered by the American Chemical Society, for support of this research. The authors are grateful to the Ethyl Corporation and the Climax Molybdenum Co. for providing samples of the metal hexacarbonyls, and to Dr. Thomas Porro of the Perkin–Elmer Co. for several spectral measurements.

We thank Dr. C. S. Naiman for allowing us to see his theoretical work on $Fe(CN)_6^{3-}$ prior to publication.

[CONTRIBUTION FROM THE DEPARTMENT OF CHEMISTRY, HARVARD UNIVERSITY]

The Electronic Structure of Bis-cyclopentadienyl Compounds

BY W. MOFFITT

RECEIVED JANUARY 14, 1954

The electronic structure of bis-cyclopentadienyl compounds is discussed in terms of molecular orbital theory. The way in which symmetry arguments may be used to facilitate such an analysis is described in some detail.

The purpose of this note is twofold. On the one hand, it is hoped to present a plausible and useful account of the electronic structures attributable to bis-cyclopentadienyl compounds of the transition metals. The current interest in these molecules[1] and their continuous proliferation prompts a rather more detailed examination of their structure than was put forward by either Dunitz and Orgel[2] or by Jaffé.[3] And on the other hand, these systems are beautifully symmetrical. They therefore also offer an opportunity to illustrate in a simple manner, the principles by means of which symmetry arguments are used to elucidate electronic properties.

Since the appearance of Pauling's book,[4] several others have been published, of which that by Coulson[5] is perhaps the most recent. They have served to acquaint primarily experimental chemists with the work that is being done in valence theory. There is, however, a considerable gap that remains to be covered before the current theoretical literature is comprehensible to the student of such books. In particular, the use of group theory in the resolution of problems with high symmetry has only been treated in the later chapters of considerably more exacting texts on quantum chemistry.[6] Whereas it is not, of course, intended to bridge this gap in the present note, it is hoped that the principles underlying the use of symmetry may be illustrated in a straightforward fashion, and that this may aid the experimentalist in deciding for himself the relative merits of proposed electronic structures.

The electronic structure, particularly of ferrocene, has been discussed by several authors.[2,3,7] All are open to some criticism.[1] The approach to be followed here is that of molecular orbital theory, and therefore similar to that of Dunitz and Orgel[2] and of Jaffé.[3] It differs in several respects from these treatments, however, and has been useful in the correlation of more recent information about these bis-cyclopentadienyl compounds. Throughout the analysis, a working rule of least disturbance will be invoked. The local structure of, for example, a cyclopentadienyl group will be disturbed as little as possible so that its considerable resonance energy is not lost. If, as will be shown to be the case, strong primary sources of binding can be found without violating this rule, the structures to which one is led may be regarded as sufficient to describe the molecules, at least to a first approximation. It is then quite possible to discuss second-order terms but in the present instance this is hardly necessary.

(1) G. Wilkinson, P. L. Pauson and F. A. Cotton, THIS JOURNAL, **76**, 1970 (1954), where an exhaustive series of references may be found.

(2) J. D. Dunitz and L. E. Orgel, *Nature*, **171**, 121 (1953).

(3) H. H. Jaffé, *J. Chem. Phys.*, **21**, 156 (1953).

(4) L. Pauling, "The Nature of the Chemical Bond," Cornell Univ. Press, Ithaca, N. Y., 1939.

(5) C. A. Coulson, "Valence," Oxford Univ. Press, Oxford, 1952.

(6) For example, H. Eyring, J. Walter and G. E. Kimball, "Quantum Chemistry," John Wiley and Sons, Inc., New York, N. Y., 1944.

(7) G. Wilkinson, M. Rosenblum, M. C. Whiting and R. B. Woodward, THIS JOURNAL, **74**, 2125 (1952); E. O. Fischer and W. Pfab, *Z. Naturforschung*, **7B**, 377 (1952).

(a) The Orbitals of the Metallic Atom or Ion.—For simplicity, we shall confine our attention to the transition metals of the first long period. In their ground states, the atoms of these metals have electronic structures of a common type, namely, $(K)(L)(3s)^2(3p)^6(3d)^m(4s)^n$. Whereas there is only one 4s orbital, there are five linearly independent 3d orbitals. These may be distinguished by the components of angular momentum $m_l(h/2\pi)$ which they have about some given axis, 0z say, where $m_l = 0, \pm 1, \pm 2$. Explicitly, their angular dependence is given by the formulas[8]

$$m_l = \pm 2: (15/32\pi)^{1/2}\sin^2\theta \cdot e^{\pm 2i\phi}R(r) = de^{\pm}_{2g}$$
$$m_l = \pm 1: (15/8\pi)^{1/2}\sin\theta\cos\theta \cdot e^{\pm i\phi}R(r) = de^{\pm}_{1g} \quad (1)$$
$$m_l = 0: (5/16\pi)^{1/2}(3\cos^2\theta - 1)R(r) = da_g$$

The last form of writing will be explained below. In order to show the nature of these orbitals, their contours on the zx-plane (*i.e.*, the two half-planes $\phi = 0, \pi$) have been plotted in Figs. 1–4. The specific example chosen is neutral iron, and the radial functions $R(r)$ were taken from the calculations of Manning and Goldberg.[9] A three-dimensional picture of the moduli of these orbitals (one-electron wave functions) is obtained by ignoring the minus signs wherever they appear and rotating the figures about the z-axis; the contours, on which the absolute values of the orbitals are constant, then appear as surfaces of revolution. For example, the 3d orbitals with $|m_l| = 1$ both have moduli which may be represented diagrammatically by the paraboloidal surfaces of Fig. 5. The squares of these moduli, of course, give the corresponding electron densities.

Now these 3d orbitals have certain interesting properties. In the first place, suppose we invert them in the origin, namely, the atomic nucleus. That is, let us construct a new set of functions whose values at the point (x, y, z) are the same as those of the original set, evaluated at the point $(-x, -y, -z)$. Then an examination of Figs. 1–4, or an inspection of equation (1) shows that the new orbitals are identical in every way with the old ones: the orbitals are simply reproduced, without change of sign, under the inversion. For this reason, they are said to be "even" (or "gerade") and we use subscripts g in their specification. This is an example of what are known as transformation properties. It is easy to see that the 4s orbital is also "even" in this sense.

(8) C. A. Coulson, reference 5, p. 27.

(9) M. F. Manning and L. Goldberg, *Phys. Rev.*, **53**, 662 (1938).

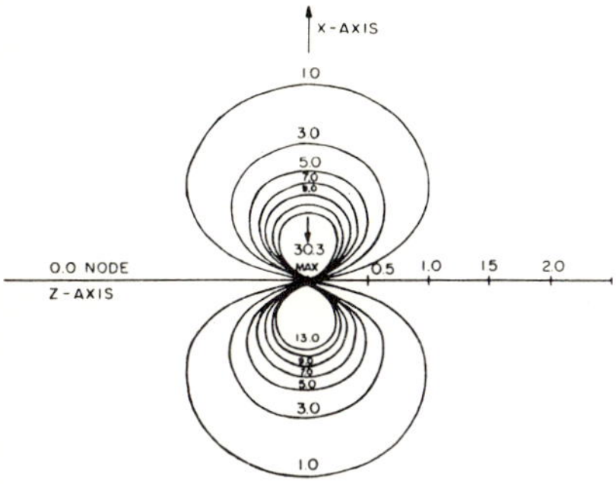

Fig. 1.—Contours on the *zx*-plane of the iron 3*d* orbital with $m_l = \pm 2$ ($de^{\pm}_{2g}$) (distances along the *z*-axis in Å. units).

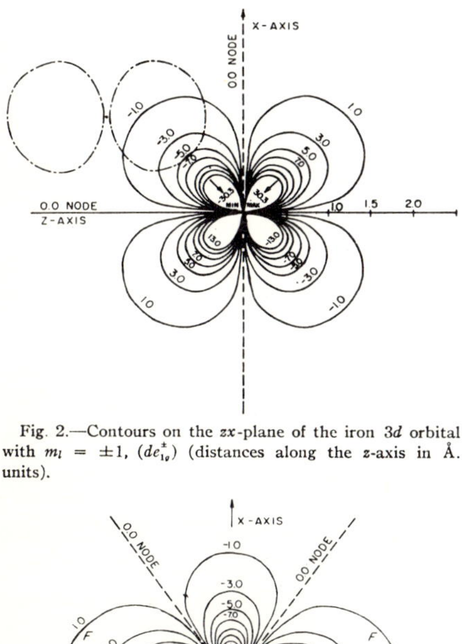

Fig. 2.—Contours on the *zx*-plane of the iron 3*d* orbital with $m_l = \pm 1$, ($de^{\pm}_{1g}$) (distances along the *z*-axis in Å. units).

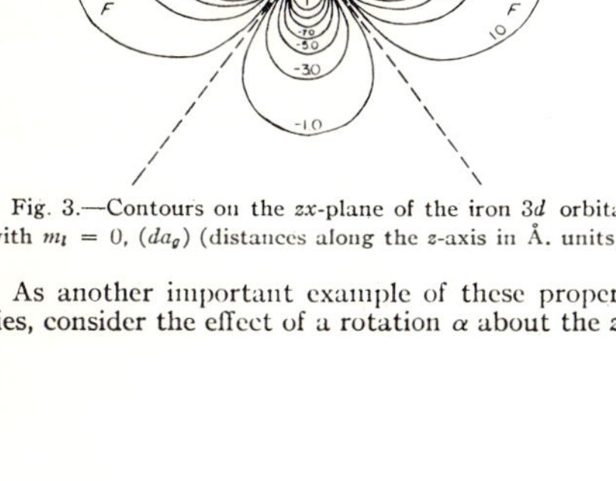

Fig. 3.—Contours on the *zx*-plane of the iron 3*d* orbital with $m_l = 0$, (da_g) (distances along the *z*-axis in Å. units).

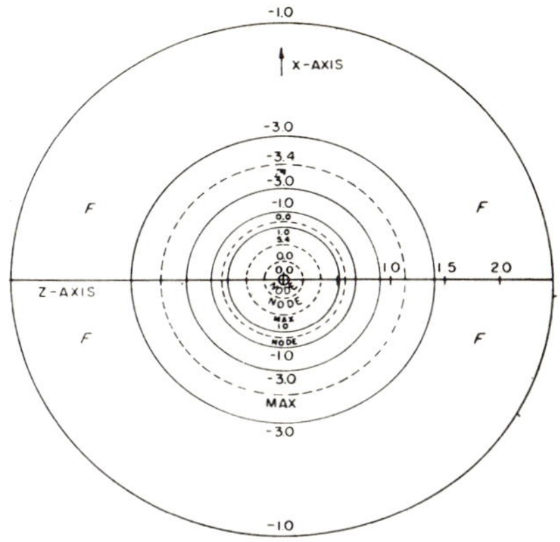

Fig. 4.—Contours on the *zx*-plane of the iron 4*s* orbital, (sa_g), (distances along the *z*-axis in Å. units).

As another important example of these properties, consider the effect of a rotation α about the *z*-axis. That is, construct a new set of orbitals whose values at the point (r, θ, ϕ) are the same as those of the original set at the point (r, θ, $\phi + \alpha$). Then it is seen from equation (1) that the new 3*d* orbital with $m_l = \pm 2$ is the same as the old one, apart from a factor $e^{\pm 2i\alpha}$. Similarly, the $m_l = 0$ orbital remains completely unchanged, but the $m_l = \pm 1$ orbitals which arise from the rotation contain the additional factors $e^{\pm i\alpha}$. The $m_l = \pm 2$ orbitals are multiplied by factors which are the complex conjugates, one of another, and are called $de^{\pm}_{2g}$; *g* because they are "even," *e* because they are conjugate pairs, and 2 because of the appearance of this term in the appropriate rotation factors $e^{\pm 2i\alpha}$. The $m_l = \pm 1$ orbitals are called $de^{\pm}_{1g}$ for analogous reasons, but the $m_l = 0$ orbital, which remains invariant and, of the 3*d* orbitals, is unique in this respect, is called da_g.[10] The spherically symmetrical 4*s* orbital, which is also "even," remains invariant under the rotation and therefore qualifies for the symbol a_g as well; we call it sa_g. An understanding of these simple transformation properties is of great value in handling problems of high symmetry.

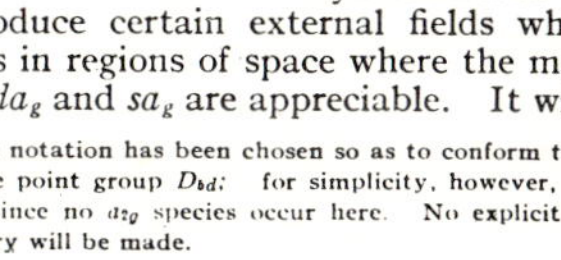

Fig. 5.—Diagrammatic representation of $|de^{\pm}_{1g}|$ contours in space.

Lastly, it is profitable to consider the possibility of a particular kind of hybridization. Suppose we introduce certain external fields which repel electrons in regions of space where the magnitudes of both da_g and sa_g are appreciable. It will be pos-

(10) The notation has been chosen so as to conform to representations of the point group D_{5d}; for simplicity, however, a_{1g} has been called a_g, since no a_{2g} species occur here. No explicit reference to group theory will be made.

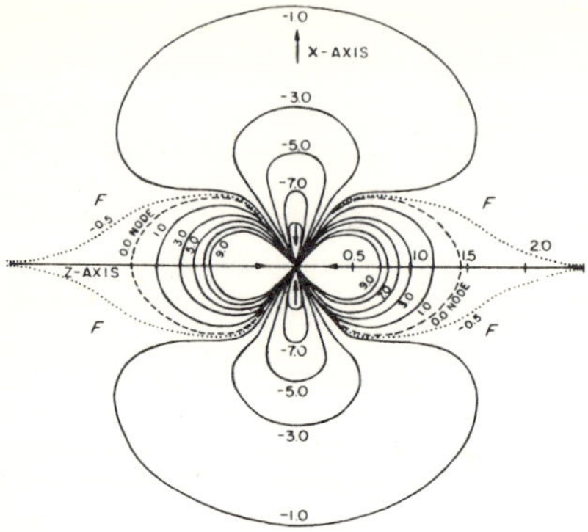

Fig. 6.—Contours on the zx-plane of the stable hybrid orbital ha_g in the field F.

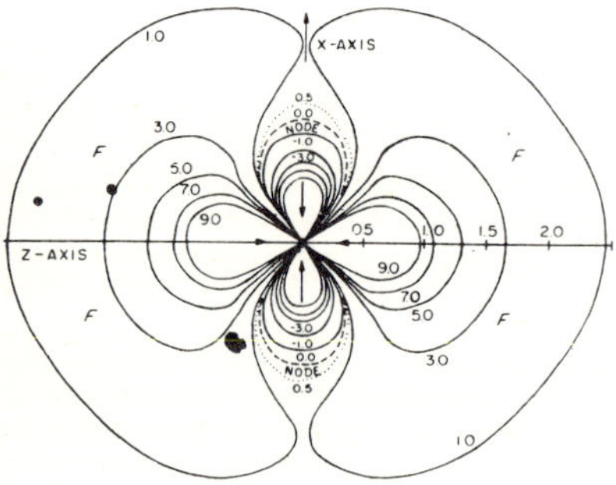

Fig. 7.—Contours on the zx-plane of the unstable hybrid orbital ka_g in the field F.

sible to construct new metal orbitals of a_g type by linearly combining da_g and sa_g. The interference of these wave functions will increase the values of the hybrid orbitals (and therefore the densities of any electrons they may contain) in some places and diminish them in others. For example, in Figs. 6 and 7 we illustrate, respectively, the two orthogonal hybrid orbitals

$$\begin{aligned} ha_g &= (1/\sqrt{2})(da_g + sa_g) \\ ka_g &= (1/\sqrt{2})(da_g - sa_g) \end{aligned} \qquad (2)$$

It is clear that if the repulsive fields are operative in those regions where ha_g is small, namely, in the areas labeled F in the diagrams, then ha_g may be considerably more stable than ka_g and, indeed, more stable than either of the original orbitals, sa_g and da_g. Regarding the field as a perturbation, we see that the effect of the field on the da_g and sa_g orbitals is to mix them in such a way as to produce one orbital, ha_g, which is more stable than either and another, ka_g, which is more highly excited. If there are two electrons of a_g type, therefore, these will be most firmly bound in the hybrid orbital ha_g, and not in either sa_g or da_g. (The precise values of the coefficients determining the amount of hybridization are not in general those given in equation (2)—these were chosen for convenience in illustrating fields like F—but must be calculated.) The effect is important whenever the fields F are of the same order of magnitude as the difference between the energies of the unperturbed $4s$ and $3d$ orbitals; an inspection of spectroscopic data on, for example, the iron atom, shows that this energy difference is, in fact, quite small.[11]

Other metal orbitals which are reasonably close to, but more highly excited than the $3d$ and $4s$ levels arise from the $4p$ set. It may be shown that these are "odd" (or "ungerade") with respect to the operation of inversion—that is, they change sign. Moreover, that $4p$ orbital with $m_l = 0$ is invariant under a rotation α, whereas the $m_l = \pm 1$ orbitals are multiplied by factors $e^{\pm i\alpha}$. These orbitals are therefore called pa_u, $pe_{1u}^{\pm}$, respectively.

(b) The Cyclopentadienyl Orbitals.—We shall now consider the description of an isolated cyclopentadienyl radical. This will be a regular planar system each of whose carbon atoms is joined by σ bonds to its carbon neighbors and to the hydrogen atom with which it is associated.[12] The five electrons which are not yet accounted for are the radical's unsaturation or π electrons. These are assigned to molecular orbitals encompassing all five nuclei and containing the ring system as a nodal plane: the orbitals change sign on reflection in the molecular plane. They are prescribed as linear combinations of the atomic $2p\pi$ orbitals[13]

$$\begin{aligned} \psi_0 &= \nu_0(\phi_1 + \phi_2 + \phi_3 + \phi_4 + \phi_5) = cpa; \quad \omega = e^{2\pi i/5} \\ \psi_{+1} &= \nu_1(\phi_1 + \omega\phi_2 + \omega^2\phi_3 + \omega^3\phi_4 + \omega^4\phi_5) = cpe_1^+ \\ \psi_{-1} &= \nu_1(\phi_1 + \omega^{-1}\phi_2 + \omega^{-2}\phi_3 + \omega^{-3}\phi_4 + \omega^{-4}\phi_5) = cpe_1^- \\ \psi_{+2} &= \nu_2(\phi_1 + \omega^2\phi_2 + \omega^4\phi_3 + \omega^6\phi_4 + \omega^8\phi_5) = cpe_2^+ \\ \psi_{-2} &= \nu_2(\phi_1 + \omega^{-2}\phi_2 + \omega^{-4}\phi_3 + \omega^{-6}\phi_4 + \omega^{-8}\phi_5) = cpe_2^- \end{aligned} \qquad (3)$$

where the ν are normalizing factors, ϕ_r is the $2p\pi$ orbital of atom r, and the last form of writing will be explained presently. That the coefficients with which the ϕ_r appear in the molecular orbitals ψ have been well chosen may be verified by solving the requisite secular equation, or otherwise. Since ψ_{-1} is the complex conjugate of ψ_{+1}, and ψ_{-2} the conjugate of ψ_{+2}, the energies of these orbitals are equal in pairs—that is, they fall into two doubly-degenerate levels. More specifically, it may be demonstrated that the energies of the orbitals ψ_0, $\psi_{\pm 1}$, $\psi_{\pm 2}$ are $\alpha + 2\beta$, $\alpha + 2\beta \cos(2\pi/5)$, $\alpha - 2\beta \cos(\pi/5)$, respectively, where α is a constant and β, which is negative, is the usual resonance integral. It appears that ψ_0 is a very strongly bonding orbital, that the $\psi_{\pm 1}$ are less strongly bonding and the $\psi_{\pm 2}$ are quite powerfully anti-bonding. In the ground state of the cyclopentadienyl radical, therefore, the electrons are allotted first to ψ_0 and then to $\psi_{\pm 1}$. There are two possible assignments, $(\psi_0)^2(\psi_{+1})^2\cdot(\psi_{-1})$ and $(\psi_0)^2(\psi_{-1})^2(\psi_{+1})$; both have the same energy, so that the ground state of the radical is doubly-degenerate. (Additional degeneracy also

(11) R. F. Bacher and S. Goudsmit, "Atomic Energy States," McGraw-Hill Book Co., Inc., New York, N. Y., 1932.

(12) This statement, although perfectly adequate for the purposes of this note, is not strictly true; it will be amplified in a later communication.

(13) C. A. Coulson, reference 5, pp. 238–240.

arises from the two possible orientations of the spin of the unpaired electron in each of these assignments.) The resonance energy of the radical is the difference between the energy of a single structure containing two localized double bonds, $(5\alpha + 4\beta)$, and that of the above allocations, namely, $[5\alpha + 4\beta + 6\beta \cos(2\pi/5)]$. Taking $\beta = -20$ kcal.,[13] this leads to the value of 37 kcal./mole.[14,15]

Now the molecular orbitals in equation (3) also have interesting transformation properties. Consider the effect of a rotation through $2\pi/5$ radians about the fivefold axis of symmetry, perpendicular to the plane of the ring. For example, in place of ψ_{+1} take that function which is obtained from it by replacing atom 5 (and thus ϕ_5) by atom 4 (and thus ϕ_4), 4 by 3, and so on, namely

$$\psi'_{+1} = \nu_1(\phi_5 + \omega\phi_1 + \omega^2\phi_2 + \omega^3\phi_3 + \omega^4\phi_4) = \omega\psi_{+1}$$

since $\omega^5 = 1$. Clearly the effect of the rotation, through an angle $2\pi/5 = \alpha$, say, is to multiply $\psi_{\pm1}$ by $\omega = e^{i\alpha}$. Similarly ψ_{-1} is simply multiplied by $\omega^{-1} = e^{-i\alpha}$. By the same criterion that we used in discussing the rotational properties of the orbitals of the metals, this pair of orbitals may be called $e_1^{\pm}$; since they are on the cyclopentadienyl ring, we use the more specific notation $cpe_1^{\pm}$.[16] Similarly the molecular orbitals $\psi_{\pm2}$ acquire factors $e^{\pm 2i\alpha}$ under the rotation, and are therefore referred to as $cpe_2^{\pm}$. The strongly bonding ψ_0 orbital is invariant under the transformation and is called *cpa* on this account.

Let us now consider a system of two cyclopentadienyl radicals. We suppose that these are sufficiently far apart that the $2p\pi$ orbitals of the one do not overlap the $2p\pi$ orbitals of the other at all appreciably. Moreover, we arrange that the fivefold axes of the two rings coincide; this is to be called the *z*-axis. In order to fix the signs of the $2p\pi$ orbitals on the one ring (see Fig. 8) relative to those of the other, we choose the point midway between the two rings as the origin and adopt the convention that the negative lobes of the two sets of $2p\pi$ orbitals are directed toward this origin, and therefore toward each other. Finally, we arrange for the two cyclopentadienyl radicals to be *skew*—that is, they will form a pentagonal antiprism. Labeling the carbon atoms of the first ring serially, from 1 to 5, we label those of the second ring 1′ to 5′, such that carbon 1 goes over into carbon 1′, 2 into 2′ and so on, under an inversion in the origin, which is therefore a center of symmetry (see also Fig. 9).

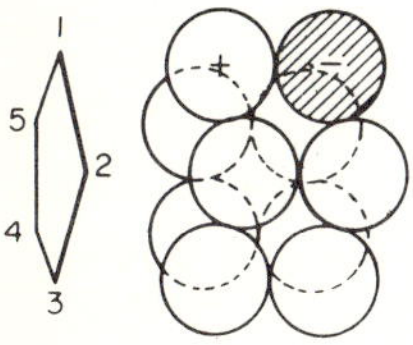

Fig. 8.—A cyclopentadienyl radical.

It is profitable to discuss what happens when the molecular orbitals of the two rings are subjected to this operation of inversion. The molecular orbitals of the first ring Cp_A[16] are called cp_Aa, $cp_Ae_1^{\pm}$, $cp_Ae_2^{\pm}$

(14) F. A. Cotton and G. Wilkinson, This Journal, **74**, 5764 (1952).

(15) J. L. Franklin and F. H. Field, *ibid.*, **75**, 2819 (1953).

(16) We use the abbreviations Cp for a cyclopentadienyl radical and *cp* to designate orbitals of this radical.

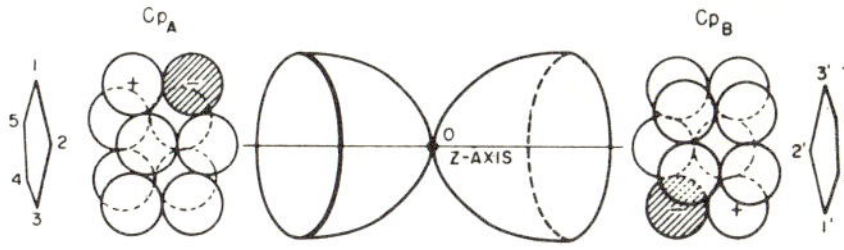

Fig. 9.—The arrangement of the two cyclopentadienyl radicals with respect to the bonding de_{1g} orbitals of the metal.

and those of the second ring Cp_B are called cp_Ba, $cp_Be_1^{\pm}$, $cp_Be^{\pm}$. Under the inversion, atom 1 and therefore, with our convention of signs, also the $2p\pi$ orbital ϕ_1, go over into atom 1′ and orbital $\phi_{1'}$, respectively. Thus the local molecular orbitals of the first ring, namely, the cp_A's, are transformed into the local molecular orbitals of the second ring by the inversion

$$cp_Aa \rightarrow cp_Ba,\ cp_Ae_1^{\pm} \rightarrow cp_Be_1^{\pm},\ cp_Ae_2^{\pm} \rightarrow cp_Be_2^{\pm} \quad (4)$$

And, *mutatis mutandis*, those of the second ring go over into those of the first.

Now it is easy, and convenient, to form linear combinations of the cp_A's and cp_B's which transform into themselves, and not into each other as in (4), under the inversion. The new orbitals, so formed, encompass both rings in this case

$$cpa_g = (1/\sqrt{2})(cp_Aa + cp_Ba),\ cpa_u = (1/\sqrt{2})(cp_Aa - cp_Ba)$$
$$cpe_{1g}^{\pm} = (1/\sqrt{2})(cp_Ae_1^{\pm} + cp_Be_1^{\pm}),\ cpe_{1u}^{\pm} = (1/\sqrt{2})(cp_Ae_1^{\pm} - cp_Be_1^{\pm}) \quad (5)$$
$$cpe_{2g}^{\pm} = (1/\sqrt{2})(cp_Ae_2^{\pm} + cp_Be_2^{\pm}),\ cpe_{2u}^{\pm} = (1/\sqrt{2})(cp_Ae_2^{\pm} - cp_Be_2^{\pm})$$

Using relations (4), it may be shown that cpa_g is indeed "even" with respect to the operation—as the subscript *g* implies. Moreover, since this orbital contains equal admixtures from cp_Aa and cp_Ba, an electron in the cpa_g orbital spends half its time on any one particular ring, and the remainder on the other. And similarly for the remaining orbitals of (5). The ground state of the system of two non-interacting cyclopentadienyl rings is normally described by the two separate assignments of unsaturation electrons, namely

$$(cp_Aa)^2(cp_Ae_1)^3 \text{ and } (cp_Ba)^2(cp_Be_1)^3 \quad (6)$$

where we have not troubled to distinguish between the e_1^+ and e_1^- orbitals, but tacitly agreed that the doubly-degenerate cpe_1 level may accommodate as many as four electrons, and therefore the three which are indicated. However, if we use relations (5), it may be shown that the three electron assignments each describe states of the composite system

$$(cpa_g)^2(cpa_u)^2(cpe_{1g})^2(cpe_{1u})^4 \quad (7a)$$
$$(cpa_g)^2(cpa_u)^2(cpe_{1g})^3(cpe_{1u})^3 \quad (7b)$$
$$(cpa_g)^2(cpa_u)^2(cpe_{1g})^4(cpe_{1u})^2 \quad (7c)$$

of two rings which, in these simple molecular orbital terms, have the same energy (and, in particular, the same resonance energy) as the specification (6). For, on the average, there are two electrons in cp_Aa and two in cp_Ba, as well as three electrons in each of the levels cp_Ae_1, cp_Be_1. Any one of these assignments, such as (7c) which may contain two unpaired electrons, may therefore be used in order to link the two cyclopentadienyl radicals to the metal atom, without destroying the resonance energy of either conjugated ring.

(c) **The Use of Symmetry Arguments.**—In the analysis which we have so far undertaken, namely, the orbital description of the fragments which together form the bis-cyclopentadienyl compounds, we have stressed the transformation properties of these orbitals. Before going on to discuss these molecules themselves, we shall explain the reason for this emphasis.

If it is desired to form an electron-pair bond between two atoms or groups by means of electrons in orbital ϕ_w of the one moiety and orbital ϕ_v of the other, then a good criterion of the bonding is given by the overlap of these two orbitals. Now this overlap is gauged by the integral

$$S_{wv} = \int \phi_w^* \cdot \phi_v dv$$

Whenever S_{wv} vanishes, it follows that the orbitals do not overlap, and that they are not suited to binding between the two atoms or groups. On the other hand, if S_{wv} is appreciably different from zero, the overlap is correspondingly large and the incipient binding strong. We shall apply this idea to binding between the orbitals of a metallic atom—which were considered in a—and the group orbitals (5) of the two cyclopentadienyl rings, which were defined and described in b.

The importance of the symmetry properties comes in when we note that the overlap integral S between two orbitals often vanishes identically, because of the way in which the latter transform. They therefore give us an exceedingly powerful method of analyzing the possible sources of binding, or of eliminating the impossible sources. For simplicity, we shall describe a one-dimensional example to illustrate the arguments which are to be used. Let $\phi_w(x)$ be some real function of a single variable x which is "even," that is, $\phi_w(-x) = \phi_w(x) = \phi_w^*(x)$. Also, let $\phi_v(x)$ be some "odd" function of x: $\phi_v(-x) = -\phi_v(x) = -\phi_v^*(x)$. These are shown diagrammatically in Fig. 10. Now let us evaluate the overlap integral

$$S_{wv} = \int_{-\infty}^{+\infty} \phi_w(x)\phi_v(x)dx$$

Consider the contribution to this integral of increments Δx around the two points $x = -b$, $x = b$. At the former, we have

$$\Delta_- S_{wv} = \phi_w(-b)\phi_v(-b)\Delta x = -\phi_w(b)\phi_v(b)\Delta x$$

since ϕ_w is "even" and ϕ_v is "odd." At the latter

$$\Delta_+ S_{wv} = \phi_w(b)\phi_v(b)\Delta x$$

The sum of these increments therefore vanishes. Moreover, since the whole domain of integration, from $-\infty$ to $+\infty$, may be broken up into pairs of increments of this type, it follows that the value of the complete integral vanishes also. That is, since ϕ_w and ϕ_v transform differently under the inversion in the origin, the overlap between them vanishes identically. When ϕ_w and ϕ_v are both either "even" or "odd," this will no longer be true in general, and these orbitals "overlap."

More generally, it may be shown for the three-dimensional case that whenever ϕ_w and ϕ_v differ in any one transformation property, then the overlap between them vanishes. Thus if ϕ_w acquires the factor $e^{i\alpha}$ under a rotation through $\alpha = 2\pi/5$ radians and is "even" with respect to inversion (*i.e.*, it is of e_{1g}^+ type), and if ϕ_v is multiplied by $e^{-2i\alpha}$ under the rotation but may also be "even" (of species e_{2g}^-), then those two orbitals do not overlap at all. The proof of this theorem requires a more exacting notation, but essentially no different arguments from those which we used in the one-dimensional case above. As an important corollary to this result, it follows that unless the transformation properties of two orbitals are identical, these are not suitable for the formation of electron-pair bonds. This limits the possibilities of binding very considerably and eases the subsequent discussion.

(d) **The Structure of Bis-cyclopentadienyliron (II) (Ferrocene).**—In order to fix our ideas, let us consider the electronic structure of ferrocene. On the extreme right of Fig. 11, we list the orbitals of the isolated iron atom, indicating the symmetries and approximate locations of the $3d$, $4s$ and $4p$ levels. On the left, we show the orbitals of the group of two cyclopentadienyl radicals. Some preliminary calculations, which will be published in full at a later date, show that the cpe_1 orbitals are at about the same level as the iron $3d$ orbitals—that is, their ionization potentials are about equal. As was demonstrated in b, any one of the three assignments (7) describes a state of minimum energy of the two cyclopentadienyl rings (each of whose ground states is degenerate). We choose a particular one of

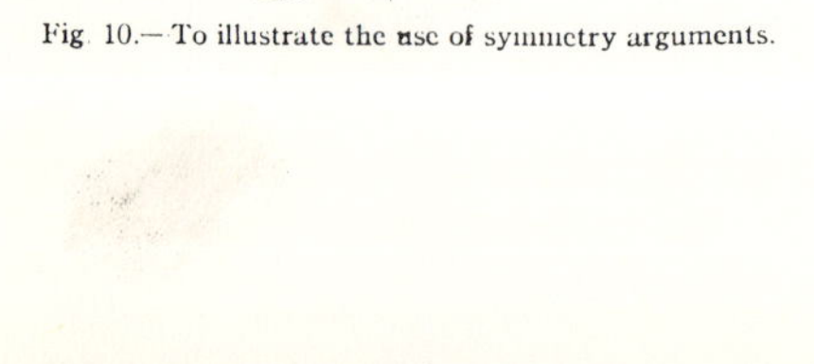

Fig. 10.—To illustrate the use of symmetry arguments.

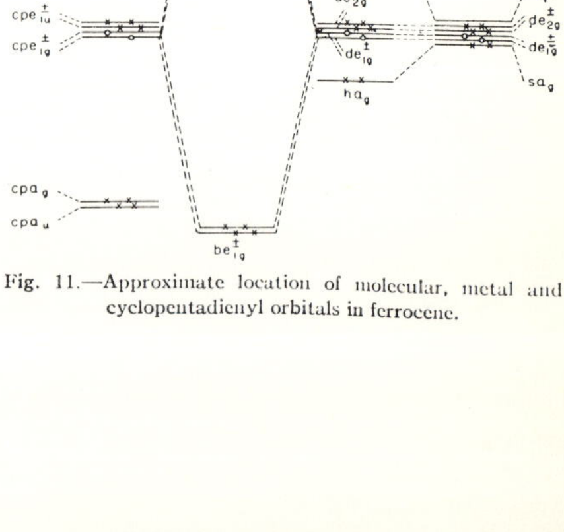

Fig. 11.—Approximate location of molecular, metal and cyclopentadienyl orbitals in ferrocene.

these, namely (7a), illustrating electrons in filled shells by crosses, and unpaired electrons by circles on the diagram. It will easily be seen that (7b) and (7c) are not suitable for binding. There are also several different ways of allotting iron's eight outer electrons to the stable 4*s* and 3*d* orbitals. Only one of these is useful, however, so that we choose to put four electrons into the de_{2g} orbitals, two into the sa_g orbital and one each into de_{1g}^{+} and de_{1g}^{-}.

Now let the two cyclopentadienyl radicals approach the central metal atom in the manner illustrated in Fig. 9. (The only iron orbitals shown in this diagram are rough representations of the de_{1g} orbitals, for reasons which will appear below.)

Before considering the binding possibilities, there is an important effect of Coulomb repulsions which must be considered. When the rings have assumed the positions they occupy in the stable ferrocene molecule, their electrons, which are very strongly bound, have high electron densities in the neighborhood of regions labeled *F* in Figs. 3, 4, 6 and 7. They will powerfully repel electrons in these regions. Accordingly, as was explained in a, the sa_g and da_g orbitals will hybridize to form stable ha_g and unstable ka_g. We may say that the effect of the σ electrons of the two rings is to "polarize" the sa_g and da_g orbitals. The two electrons which we assigned to sa_g for the isolated iron atom, will therefore drop into the favorable hybrid ha_g when the two rings are brought up. The complementary unfavorable hybrid ka_g will probably now lie in the region of the 4*p* orbitals of the iron atom. This reorganization is illustrated, diagrammatically, in the third column of Fig. 11. (The precise location of ha_g and ka_g is difficult to determine. It is only important for our argument that ka_g should not be appreciably more stable than the 4*p* orbitals.)

Finally, we come to consider the nature of the iron–carbon bonding. By the symmetry arguments of c, we need only consider binding between orbitals with the same transformation properties. Moreover, these should contain unpaired electrons. The primary source of the binding must therefore lie between the cpe_{1g} and de_{1g} orbitals. The overlap between these orbitals is not only allowed by symmetry, but also favored by superposition: in Fig. 2, we show a dotted representation of one $2p\pi$ orbital of Cp_A. (More specifically, it is a contour chosen so as just to touch the similar contour on a neighboring carbon atom.) Thus it appears that the ring orbitals cpe_{1g}, which encompass all ten carbon atoms, appreciably overlap the paraboloidal contours of the de_{1g} orbitals (see also Fig. 10). All the prerequisites for strong bonding between them are thus satisfied. Since the energies of the de_{1g} and cpe_{1g} orbitals are also at about the same level, this bonding should not involve charge separation. The net charges on the iron and cyclopentadienyl rings should vanish, in first approximation. Four electrons are therefore involved, two for each Cp-Fe bond.

The bonding orbitals be_{1g}, which in molecular orbital theory will be

$$be_{1g}^{\pm} \approx (1/\sqrt{2})(cpe_{1g}^{\pm} + de_{1g}^{\pm}),$$

are approximately located in the second column of Fig. 11. The complementary antibonding or repulsive orbital

$$re_{1g}^{\pm} \approx (1/\sqrt{2})(cpe_{1g}^{\pm} - de_{1g}^{\pm})$$

is also shown.

There are various secondary types of bonding, which may be considered. For example, the filled de_{2g} orbitals may donate electrons into the vacant cpe_{2g} orbitals. The latter are strongly antibonding in the carbon–carbon sense, and therefore are not inclined to accept electrons—that is, the electronegativity exhibited by the cyclopentadienyl rings is small. Similarly, the filled cpe_{1u} orbitals may formally donate electrons into the vacant pe_{1u} orbitals. However, the cpe_{1u} orbitals are bonding in the carbon–carbon regions, and the electron affinity of iron for 4*p* electrons is low, so this effect will also be small. There may be a little of both occurring simultaneously, leaving no net charge on the rings or on the metal. But this would have to be at the expense of the resonance energy of the cyclopentadienyl rings. Since very favorable binding conditions have already been found, it seems unnecessary to invoke these secondary forces.

This description of the molecule provides a very satisfactory rationalization of the organic reactions of ferrocene, which behaves in many respects like an aromatic system.[17] In our view, each ring has a high local resonance energy, discussed in b and remains uncharged. This is in agreement with the reactivity of the molecule and with the acid dissociation constants of its carboxylic acid derivatives.[17] Since the densities of the bonding iron orbitals are non-directional rotationally but only axially, as shown in Figs. 5 and 9, the two unsubstituted rings should be freely rotating to good approximation. The absence of any nearby unfilled orbitals accounts for the molecule's diamagnetism.

Other Bis-cyclopentadienyl Compounds.—It is interesting to give similar accounts of other neutral molecules like ferrocene, where the central iron atom is replaced by Co, Ni, Cr or other atoms of the transition elements. The energy level diagram will not be very different from Fig. 11 in all these cases, and we may treat the molecules by adding, or subtracting the requisite number of electrons.

The simplest example is bis-cyclopentadienylcobalt(II), where we add one electron. This may go either into one of the 4*p* orbitals or into the relatively unstable ka_g orbital. In either of these cases, of course, it has one unpaired electron (*i.e.*, it is in a doublet state). For bis-cyclopentadienylnickel-(II),[1] on the other hand, two electrons must go into these orbitals. Whether they go, one into ka_g and the other into some 4*p* level, or both into the 4*p* levels, is in many ways immaterial. The proximity of the ka_g orbital energy to that of the 4*p* orbitals will ensure that we shall be left with two singly occupied orbitals whose electrons are in a triplet state with their spins parallel (Hund's principle of maximum multiplicity). In bis-cyclopentadienylchromium-(II) two electrons are removed. If the ha_g orbital is appreciably more stable than the de_{2g} orbitals, both will come from the latter level, leaving one

(17) R. B. Woodward, M. Rosenblum and M. C. Whiting, This Journal, **74**, 3458 (1952).

electron in de_{2g}^{+} and the other in de_{2g}^{-}, so that their spins become parallel, and we have a triplet once again. If the ha_g level is close to the de_{2g} levels, however, one electron will be taken from each and we get another triplet condition. (The ha_g level is certainly not appreciably less stable than de_{2g}, for then it would lose both electrons and leave a diamagnetic molecule—contrary to experience.[18]) More detailed magnetic studies may well be able to settle some of these ambiguities, which are due to our ignorance of the precise location of the ha_g and ka_g levels. However, since these are only concerned with the details of the electronic structure of the metal atom, and do not in any way affect the nature of the metal–carbon bonding, the uncertainties are unimportant in the above instances.

Bis-cyclopentadienylmanganese(II) would be isoelectronic with the ferricinium ion, and arise from the loss of a de_{2g} electron. It is interesting to speculate on the magnetic properties of bis-cyclopentadienyltitanium(II) and -vanadium(II). If, as we may suspect, the ha_g orbital is sufficiently more stable than de_{2g}, in these cases, then $(C_5H_5)_2Ti$, with only two electrons not involved in metal–carbon bonding, would be diamagnetic, both being assigned to the ha_g orbital. However, it is also quite possible that the extra Hund stabilization, which occurs when two electrons can orient their spins parallel to one another, is sufficiently large that one electron goes into ha_g and the other into de_{2g}. The molecule would then be a paramagnetic triplet. If $(C_5H_5)_2Ti$ is diamagnetic, then $(C_5H_5)_2V$ with one additional electron (in de_{2g}) would most probably have only one unpaired electron (doublet state). On the other hand, if $(C_5H_5)_2Ti$ is paramagnetic, it would be supposed that $(C_5H_5)_2V$ has three electrons with their spins all parallel, one each in ha_g, de_{2g}^{+} and de_{2g}^{-} (quartet state).

Throughout, we have regarded, for example, ferrocene as being made up of two neutral cyclopentadienyl radicals and of a neutral iron atom. Moreover, we have shown that this tacit nomenclature reflects the actual charge distribution in the stable molecule. However, since many of the reactions leading to synthesis of bis-cyclopentadienyl compounds occur under ionic conditions, and in view of the oxidation potentials of the compounds, it is in some ways chemically appropriate to regard the metal atom in $(C_5H_5)_2M$ as being in the II oxidation state. Whereas it is not, of course, intended to suggest that the metal has a formal charge of $+2$, this notation is very convenient also in other cases. For example, the name bis-cyclopentadienyltitanium(IV) dibromide refers to the molecule $Ti^{+4}(C_5H_5^-)_2(Br^-)_2$. However, it is better, for structural purposes, to regard this molecule as $[Ti(C_5H_5)_2]^{++}Br_2^-$, so that the titanium has two de_{1g} electrons at hand for the metal–carbon bonding—which we consider to be the essential feature of all these bis-cyclopentadienyl molecules. The existence of this particular compound[19] is, in fact, a strong piece of evidence in favor of our picture of the metal–carbon bonding; it can have no more than two electrons available for this purpose.

As a final point, there is no reason why the pentagonal pyramid formed by any one cyclopentadienyl ring and a metal should be a structure unique to bis-cyclopentadienyl compounds. It is quite possible that compounds of a different type, *e.g.*, $[(C_5H_5)M^nA_y]X_{n-1}$ where A is a neutral group and X a singly charged anion, can exist. The first examples of such cyclopentadienyl compounds have in fact been made[18]; these are the molybdenum and tungsten cyclopentadienyl carbonyls, (C_5H_5)-$Mo(CO)_6Mo(C_5H_5)$ and $(C_5H_5)W(CO)_6W(C_5H_5)$.

In conclusion, it is hoped that an understandable account has been given of the way in which symmetry arguments may be used to simplify the analysis of the electronic structures which may be ascribed to bis-cyclopentadienyl compounds of the transition elements. More particularly, it has been shown that the essential binding is between the de_{1g} orbitals of the metal and the cpe_{1g} orbitals of the cyclopentadienyl rings, forming two electron-pair bonds. In order to understand the magnetic properties of the compounds, the effect of the electrostatic fields on the da_g and sa_g orbitals of the metal, to produce hybridization, must be considered. Dunitz and Orgel appreciated the predominant role of the e_{1g} orbitals but neglected to consider the s or p electrons of the metal. On the other hand, Jaffé, in considering all possible combinations of ring orbitals and metal s, d and p orbitals, gave structures whose physical significance it is difficult to understand. Both treatments give erroneous predictions of one sort or another.[1,20]

The author would like to express his cordial thanks to Professors G. Wilkinson and R. B. Woodward for arousing his interest in these molecules.

CAMBRIDGE, MASSACHUSETTS

(18) G. Wilkinson, THIS JOURNAL, **76**, 209 (1954).

(19) G. Wilkinson, P. L. Pauson, J. M. Birmingham and F. A. Cotton, *ibid.*, **75**, 1011 (1953).

(20) NOTE ADDED IN PROOF.—More recently, in a private communication to the author, Dr. L. Orgel has expressed views which are in substantial agreement with those described in the present note.

The Group Relation Between the Mulliken and Slater-Pauling Theories of Valence

J. H. VAN VLECK, *Harvard University*

(Received October 7, 1935)

By means of the group theory of characters, it is shown that there is an intimate relation between Mulliken's molecular orbitals and the Slater-Pauling directed wave functions. One can pass from the former to the latter by making a simple transformation from an irreducible to a reducible representation. Consequently the same formal valence rules are usually given by either method, and one can understand generally why wave functions of the central atom which are nonbonding in Mulliken's procedure are likewise never employed in constructing Pauling's "hybridized" linear combinations.

TWO distinct viewpoints have been particularly developed in applying quantum mechanics to problems of valence,[1] namely, the Heitler-London-Pauling-Slater method of the electron pair bond, and the method of molecular orbitals used by Hund, Lennard-Jones, Mulliken, and others. The two procedures represent different approximations to the solution of a complicated secular equation. The method of molecular orbitals permits factorization into one-electron problems, but at the expense of adequate cognizance of the terms due to electron repulsion, which are too fully recognized in the H-L-P-S procedure. A characteristic feature of the latter is the "hybridization," whereby linear combinations of states of different azimuthal quantum number for the central atom are necessary in fields of, for example, tetrahedral symmetry. It was shown by the writer that in the case of carbon compounds the two theories, though superficially different, predicted similar results on geometrical arrangement.[2] It is the purpose of the present paper to show that the equivalence is general in the sense that one formulation will give the same formal stereochemical valence principles as the other. Thus it is futile to discuss whether the Mulliken or Pauling theory will give better working rules in compounds formally amenable to electron pair treatment. Both, for instance, suggest the now classic Pauling square configuration for $Ni(CN)_4^{--}$ in view of the diamagnetism of this ion. (A tetrahedral model would give paramagnetism with either method.[3]) One can only inquire which procedure involves the more reasonable hypotheses. It seems to us that in the case of the transition elements, one must probably decide in favor of the Mulliken formation as a simple qualitative description, though perhaps a poor quantitative approximation. It is difficult to believe, for instance, that the $Fe(CN)_6^{4-}$ radical has the Pauling structure $Fe^{4-}(CN)_6$, since the Fe ion certainly is unwilling to swallow four extra electrons. The conventional ionic model $Fe^{++}(CN^-)_6$, on the other hand, probably goes too far in the other direction. The Mulliken viewpoint has here the advantage of allowing an arbitrary distribution of charge between Fe and $(CN)_6$, depending on how one weights the various atomic orbitals in forming a molecular orbital as a linear combination of them. Only a limited significance should, however, be given to any purported preference between the two methods, as each represents a solution of the secular equation only under certain extreme conditions. The true wave function is in reality a combination of H-LJ-M and H-L-P-S functions, along with many ingredients intermediate between these two extremes, and so either theory is bound to have some semblance of truth. The latter functions, for instance, can be amplified

[1] L. Pauling, J. Am. Chem. Soc. **53**, 1367 (1931); J. C. Slater, Phys. Rev. **38**, 1109 (1931); R. S. Mulliken, Phys. Rev. **40**, 55; **41**, 49, 751; **43**, 279 (1932–3); J. Chem. Phys. **1**, 492 (1933); **3**, 375, 506 (1935). For other references, or more detailed introduction on the methods which we compare, see J. H. Van Vleck and A. Sherman, Rev. Mod. Phys. **7**, 167 (1935). Other workers besides Mulliken have contributed to the molecular orbital procedure, but we sometimes refer to the latter as Mulliken's method, since we are concerned with the application to polyatomic molecules in the light of symmetry groups, an aspect considered primarily by Mulliken.

[2] J. H. Van Vleck, J. Chem. Phys. **1**, 219 (1933).

[3] Cf. Pauling, reference 1, and J. H. Van Vleck and A. Sherman, Rev. Mod. Phys. **7**, 206, 221 (1935).

by taking linear combinations with terms representing different stages of charge transfer until finally just the right polarity is obtained.

Molecular Orbitals

We shall confine our discussion to the case where a central atom attaches n atoms arranged in some symmetrical fashion (tetrahedron, square, etc.) characteristic of a crystallographic point group.[4] Let $\psi(\Gamma)$ be a wave function of the central atom which has the proper symmetry, i.e., whose transformation scheme under the covering operations of the group is that characteristic of some irreducible representation Γ. Let ψ_i be a wave function of attached atom i. We shall assume that only one orbital state need be considered for each attached atom, and that this state is either an s state or else is symmetric (as in a $2p\sigma$ bond) about the line joining the attached to the central atom. The method of molecular orbitals in its simplest form[5] seeks to construct solutions of the form

$$\psi=\psi(\Gamma)+\Sigma_i a_i\psi_i. \quad (1)$$

The coefficients a_i must be so chosen that $\Sigma_i a_i\psi_i$ transforms in the fashion appropriate to the irreducible representation Γ. Now the important point is that bases for only certain irreducible representations can be constructed out of linear combinations of the ψ_i. To determine which, one ascertains the group characters associated with the transformation scheme, usually reducible, of the original attached wave functions ψ_i before linear combinations are taken. This step is easy, as the character χ_D for a covering operation D is simply equal to q, where q is the number of atoms left invariant by D. This result is true inasmuch as D leaves q of the atoms alone, and completely rearranges the others, so that the diagonal sum involved in the character will contain unity q times, and will have zeros for the other entries. The scheme for evaluating the characters is reminiscent of that in the group theory of molecular vibrations employed by Wigner and by E. B. Wilson, Jr.,[6] but is simpler since we are not interested in displacements of atoms from equilibrium, and in consequence all nonvanishing entries are unity rather than some root of unity. After the characters have been found, the determination of the constituent irreducible representations proceeds in the usual way by means of the theorem that the unresolved characters must equal the sum of the primitive characters contained therein.

As an illustration, we may consider a complex containing six atoms octahedrally arranged, i.e., located at the centers of the six cube faces. Then the symmetry group is the cubic one O_h, and the characters associated with the arrangement of attached atoms are

	E	C_2	C_4	C_2'	C_3	I	IC_2	IC_4	IC_2'	IC_3	
$\chi=$	6	2	2	0	0	0	4	0	2	0.	(2)

Here $\chi(C_4)$, for instance, means the character for the covering operation consisting of rotation about one of the fourfold or principal cubic axes (normals to cube faces) by $2\pi/4$. Any rotation about such an axis leaves two atoms invariant, and hence $\chi(C_2)=\chi(C_4)=2$. On the other hand, $\chi(C_2')=\chi(C_3)=0$ since no atoms are left invariant under rotations about the twofold or secondary cubic axes (surface diagonals) or about the threefold axes (body diagonals). Inversion in the center of symmetry is denoted by I. By using tables of characters for the group O_h, one finds that the irreducible representations contained in the character scheme (2) are, in Mulliken's notation,[4]

$$A_{1g},\ E_g,\ T_{1u}. \quad (3)$$

The irreducible representations corresponding to various kinds of central orbitals are shown below:

orbit	s	p	$d\gamma$	$d\epsilon$	f	
rep.	A_{1g}	T_{1u}	E_g	T_{2u}	A_{2u}, T_{1u}, T_{2u}.	(4)

The notation for the various kinds of d wave functions is that of Bethe,[7] *viz.*,

---

[4] See R. S. Mulliken, Phys. Rev. **43**, 279 (1933) or references 6 and 7 if further background is desired on the aspects of group theory and crystallographic symmetry which we use.

[5] Called by Mulliken the LCAO ("linear combination of atomic orbitals") form. For a critique of this type of approximation see R. S. Mulliken, J. Chem. Phys. **3**, 375 (1935). It appears to have been first suggested by Lennard-Jones.

[6] E. Wigner, Gött. Nachr., p. 133 (1930); E. B. Wilson, Jr., J. Chem. Phys. **2**, 432 (1934); Phys. Rev. **45**, 706 (1934).

[7] H. Bethe, Ann. d. Physik **3**, 165 (1929).

TABLE I.

Symmetry	Central orbitals s	p	d	f	Attached orbitals No.	Representations
Tetrahedral (T_d)	A_1	T_2	$E(d\gamma)$, $T_2(d\epsilon)$	A_1, T_1, T_2	4	A_1, T_2
Trigonal (D_{3h})	A_1'	A_2'', E'	A_1', E', E''	A_1', A_2', A_2'', E', E''	3	A_1', E'
					6	A_1', A_2'', E', E''
Tetragonal (D_{4h})	A_{1g}	A_{2u}, E_u	A_{1g}, B_{1g}, B_{2g}, E_g	A_{2u}, B_{1u}, B_{2u}, $2E_u$	4	A_{1g}, B_{1g}, E_u
					8	A_{1g}, A_{2u}, B_{1g}, B_{2u}, E_u, E_g

$$\psi(d\gamma_1)=(1/12)^{\frac{1}{2}}f(r)(3z^2-r^2),\quad \psi(d\gamma_2)=\tfrac{1}{2}f(r)(x^2-y^2), \tag{5}$$

$$\psi(d\epsilon_1)=f(r)xy,\quad \psi(d\epsilon_2)=f(r)xz,\quad \psi(d\epsilon_3)=f(r)yz. \tag{6}$$

By comparison of (3) and (4) we see that is it impossible for $d\epsilon$ orbitals to form any partnerships with attached orbitals, in agreement with Mulliken's conclusion[8] that $d\gamma$ rather than $d\epsilon$ is particularly adapted to forming octahedral bonds. Mulliken's arguments were mainly of a qualitative nature. The preceding considerations enable us to formulate the situation more succinctly, as they show that the $d\epsilon$ orbitals are entirely nonbonding.

In Table I we give the irreducible representations, in Mulliken's notation, contained in the central and attached orbitals for compounds of other types of symmetry. When 3 or 4 atoms are trigonally or tetragonally attached, we have supposed that the plane of these atoms is a plane of symmetry, as in $(NO_3)^-$ or $Ni(CN)_4^{--}$. When there is no such symmetry plane, as in NH_3, the distinctions between u and g, or between primes and double primes, are to be abolished,[9] and the symmetries degenerate to C_{3v}, C_{4v} instead of D_{3h}, D_{4h}. When 6 atoms are attached in the scheme D_{3h}, or 8 in D_{4h}, they are arranged respectively at the corners of a trigonal and a square prism.

The results given in Table I are obtained by the same method as in the octahedral example. The explicit forms of the linear combinations of the attached orbitals which transform irreducibly, or in other words the values of the coefficients a_i in (1) have been tabulated by Van Vleck and Sherman[10] in many instances, and so need not be repeated here. The a_i for the octahedral case are also given in Eqs. (2)–(7) of the following paper.

Note particularly that in the tetrahedral complexes, the $d\epsilon$ orbitals of the central atom are bonding, as there are attached orbitals of similar group properties with which they can combine, while the $d\gamma$ orbitals are nonbonding. The reverse was true of octahedral compounds—a result at first a little surprising in view of the isomorphism of the groups T_d and O_h. This reversal was also deduced by Mulliken[8] from the geometrical study of the way the central wave functions "overlap."

It will be observed that if eight atoms are attached, their full bonding power is not utilized unless one includes f wave functions for the central atom, since the representation B_{2u} is not included in s, p, or d. Now f wave functions usually have too high energy to be normally available, or else are so sequestered in the interior of the atom as to be of no value for bonding because of small overlapping. Even if eight atoms are attached at the corners of a cube, central f wave functions must be included in order to realize all possible bonding partnerships, for results always true of tetragonal symmetry surely apply to cubic symmetry, which is a special case of the latter. On the other hand, no f functions are needed for six atoms attached either octahedrally or at the corners of a trigonal prism. We thus have an indication of why it is that coordination numbers of six are common in nature, while those of eight are rare.

METHOD OF DIRECTED ELECTRON PAIRS

We now turn to the method of Pauling and Slater. Here the procedure is to use hybridized

[8] R. S. Mulliken, Phys. Rev. **40**, 55 (1932).

[9] In adapting Table I to the case C_{3v}, the following irregularity, however, is to be noted: one must replace A_1', A_1'', A_2', A_2'', respectively, by A_1, A_2, A_2, A_1 rather than by A_1, A_1, A_2, A_2 as one would guess.

[10] J. H. Van Vleck and A. Sherman, Rev. Mod. Phys. **7**, 219 (1935).

central orbitals, i.e., linear combinations of orbitals of different azimuthal quantum number, in such a way that the resulting central wave function projects out especially in some one direction in space, and so is adapted to form an electron pair with one particular attached atom. Thus for tetrahedral compounds Pauling and Slater[1] use wave functions which are linear combinations of s and p, or alternatively as Pauling[1] shows, of s and $d\epsilon$ wave functions. For octahedral compounds, Pauling finds that $sp^3d\gamma^2$ combinations are appropriate, and $sp^2d\gamma$ for tetragonal. Hultgren[11] proves that eight atoms cannot be attached (at least symmetrically) by means of unidirectional electron pair bonds formed from s, p, and d wave functions. Incidentally, the present paper shows that sp^3d^3f functions are needed to hold eight atoms.[12] It will be noted that the wave functions involved in the Pauling unidirectional linear combinations are precisely those which are bonding in the method of molecular orbitals. For example, Pauling, like Mulliken, makes no use of $d\gamma$ orbitals for tetrahedral compounds, or of $d\epsilon$ for octahedral. Such coincidences have hitherto appeared something of a mystery, but as immediate explanation, as follows, is furnished by group theory.

In the Pauling-Slater theory, one desires the central wave functions to possess unilateral directional properties so as to be correlated with one particular attached atom. Hence the P-S central functions must have the same transformation properties as do those ψ_i of the attached atoms before linear combinations of the latter are taken. Thus the problem of finding the linear combinations of the central orbitals which exhibit the proper directional properties is simply the reverse of finding the proper linear combinations of the attached orbitals in the Mulliken procedure. The difference is only that in the P-S theory, the linear combinations are in the central rather than attached portion, and their construction corresponds to transformation from an irreducible representation to a reducible one, of structure similar to that belonging to the original ψ_i's, rather than to the inverse transformation.[13] Clearly, the same irreducible representations are needed in the construction of a given reducible representation as those contained in the resolution of the latter into its irreducible parts. Pauling has obviously shown considerable ingenuity in constructing his wave functions without using the status in terms of group theory.

[11] R. Hultgren, Phys. Rev. **40**, 891 (1932).

[12] Similar conclusions on the type of bonds necessary to attach eight atoms have also been obtained in unpublished work of R. S. Mulliken.

Generality of the Results

The argument underlying Table I, etc., ostensibly assumed that the molecular orbital be expressible as a linear combination of atomic orbitals, but is readily seen to be still applicable provided only that the charge cloud of any attached orbital be symmetric about the line joining the given attached atom to the central one. Hence the atomic orbitals can be of what James calls the flexible type, i.e., contain parameters which can be varied in the Ritz method, and which allow for the fact that chemical combination distorts the atomic orbitals from what they would be in the free condition. This admission of flexibility is fortunate, for it is well known that it is a bad quantitative approximation[5] to express a molecular orbital as a linear combination of undistorted atomic orbitals. The Ritz variational problem is, of course, to be of the 1 rather than n electron type, so that the generality in our analysis by means of molecular orbitals is roughly comparable with that in the Hartree method.

One thing which the preceding analysis does not do is to tell us what is the best arrangement of atoms in case the symmetry group does not uniquely determine this arrangement. For instance, by examining the overlapping of wave functions, Hultgren[11] finds that when six atoms are attached at the corners of a trigonal prism, the binding is firmest if the sides of the prism are square. This fact cannot, however, be inferred from our group theory considerations, as there are no additional elements of symmetry when the sides are square rather than rectangular.

[13] This transformation has been explicitly given by the writer in the case of methane (J. Chem. Phys. **1**, 177 (1933)), but he did not discuss its group-theoretical significance.

The Spectra and Electronic Structure of the Tetrahedral Ions MnO_4^-, CrO_4^{--}, and ClO_4^- *†

MAX WOLFSBERG‡ AND LINDSAY HELMHOLZ
Department of Chemistry, Washington University, St. Louis, Missouri

(Received January 14, 1952)

We have made use of a semiempirical treatment to calculate the energies of the molecular orbitals for the ground state and the first few excited states of permanganate, chromate, and perchlorate ions. The calculation of the excitation energies is in agreement with the qualitative features of the observed spectra, i.e., absorption in the far ultraviolet for ClO_4^-, two strong maxima in the visible or near ultraviolet for MnO_4^- and $CrO_4^=$ with the chromate spectrum displaced toward higher energies. An approximate calculation of the relative f-values for the first transitions in $CrO_4^=$ and MnO_4^- is also in agreement with experiment.

The data on the absorption spectra of permanganate ion in different crystalline fields is interpreted in terms of the symmetries of the excited states predicted by our calculations.

INTRODUCTION

In the past twenty years the electronic structures of many organic molecules, particularly benzene and related compounds, have been discussed in terms of the molecular orbital and valence bond methods.[1] During the same period the structures of inorganic ions have been inferred from the bond distances;[2] a bond distance shorter than the sum of the conventional radii has been attributed to the resonance of double bonded structures with the single bonded or Lewis structure.

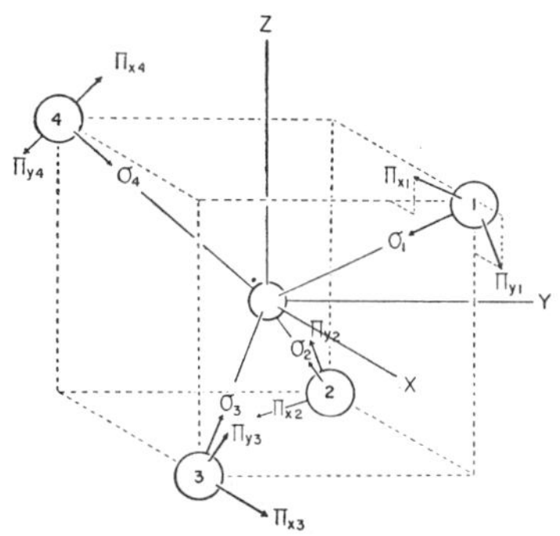

FIG. 1. Orientation of the orbitals. The orientation of the oxygen orbitals relative to the axes on the central atom is shown in this figure. The direction cosines of the O_1 orbitals are σ_1: $-1/3^{\frac{1}{2}}$, $-1/3^{\frac{1}{2}}$, $-1/3^{\frac{1}{2}}$; πx_1: $1/6^{\frac{1}{2}}$, $-2^{\frac{1}{2}}/3^{\frac{1}{2}}$, $1/6^{\frac{1}{2}}$; πy_1: $1/2^{\frac{1}{2}}$, 0, $-1/2^{\frac{1}{2}}$. The direction cosines of the other oxygen orbitals may be obtained from these by the action of the twofold axes of T_d, which coincide with the x, y, and z axes shown. The central atom orbitals are defined with respect to the axes on the central atom.

It has been shown[3], however, that this shortening need not necessarily be attributed to double bond formation. We hope to come to some conclusion concerning the bonding in inorganic molecules from molecular orbital calculations which can be checked against the observed spectra.

We have applied to inorganic complex ions a semiempirical method similar to that which has yielded significant results in the case of organic molecules and have chosen to treat permanganate, chromate, and perchlorate ions because the existing data[4] seem to afford the greatest number of tests of the correctness of the calculations.

The tetrahedral ions, $PO_4^{\equiv}$ $SO_4^=$ and ClO_4^- show no near ultraviolet absorption whereas the ions, XO_4^-, of the fourth-row transition elements, which have the same number of valence electrons, show characteristic visible and near ultraviolet absorption, two strong maxima with the corresponding peaks displaced toward shorter wavelengths with decreasing atomic number of the central atom. This trend is also observed for the fifth- and sixth-row transition element compounds. A satisfactory theory must account for these differences and regularities.

The relative intensities of absorption calculated may be compared with those observed as an additional check of the method.

Data on the absorption of permanganate ion in different crystalline fields[4] indicate the existence of several excited states to which dipole transitions are forbidden in the absence of perturbing effects. The correlation of these data with the results of our calculations gives important evidence as to their correctness.

PROCEDURE

The molecular orbital method in its LCAO form is employed. We assume that the nonvalence shell electrons of X and O are unaffected by the bonding and that the X and O nuclei plus these inner shell electrons form an effective core into the field of which the mo-

* Presented at the Symposium on Molecular Structure and Spectroscopy, Columbus, Ohio, June, 1951.

† Abstracted from the thesis submitted by Max Wolfsberg in partial fulfillment of the requirements for the Degree of Doctor of Philosophy at Washington University.

‡ Atomic Energy Commission Predoctoral Fellow. Present address: Brookhaven National Laboratory, Upton, Long Island, New York.

[1] For reference on both methods see R. S. Mulliken, J. chim. phys. **46**, 497 (1949).

[2] L. Pauling, *The Nature of the Chemical Bond* (Cornell University Press, Ithaca, 1940).

[3] K. S. Pitzer, J. Am. Chem. Soc. **90**, 2140 (1948).

[4] J. Teltow, Z. Phys. Chem. **B43**, 198 (1939); Z. Phys. Chem. **B40**, 397 (1938).

TABLE I. Molecular orbital combinations—XO_4, symmetry T_d.

X orbitals	O orbitals		Irreducible representations
s	$(\frac{1}{2})[\sigma_1+\sigma_2+\sigma_3+\sigma_4]$		A_1
d_{z^2}[a]		$(\frac{1}{4})[\pi_{x1}+\pi_{x2}+\pi_{x3}+\pi_{x4}-3^{\frac{1}{2}}(\pi_{y1}+\pi_{y2}+\pi_{y3}+\pi_{y4})]$	E
$d_{x^2-y^2}$		$(\frac{1}{4})[\pi_{y1}+\pi_{y2}+\pi_{y3}+\pi_{y4}+3^{\frac{1}{2}}(\pi_{x1}+\pi_{x2}+\pi_{x3}+\pi_{x4})]$	
p_x, d_{yz}	$(\frac{1}{2})[\sigma_1+\sigma_3-\sigma_2-\sigma_4]$	$(\frac{1}{4})[\pi_{x4}+\pi_{x2}-\pi_{x1}-\pi_{x3}+3^{\frac{1}{2}}(\pi_{y4}+\pi_{y2}-\pi_{y1}-\pi_{y3})]$	
p_y, d_{xz}	$(\frac{1}{2})[\sigma_1+\sigma_2-\sigma_3-\sigma_4]$	$(\frac{1}{2})[\pi_{x1}+\pi_{x2}-\pi_{x3}-\pi_{x4}]$	T_2
p_z, d_{xy}	$(\frac{1}{2})[\sigma_1+\sigma_4-\sigma_2-\sigma_3]$	$(\frac{1}{4})[\pi_{x3}+\pi_{x2}-\pi_{x1}-\pi_{x4}+3^{\frac{1}{2}}(\pi_{y4}+\pi_{y1}-\pi_{y2}-\pi_{y3})]$	
		$(\frac{1}{4})[\pi_{y2}+\pi_{y4}-\pi_{y3}-\pi_{y1}+3^{\frac{1}{2}}(\pi_{x1}+\pi_{x3}-\pi_{x2}-\pi_{x4})]$	
		$\frac{1}{2}[\pi_{y1}+\pi_{y2}-\pi_{y3}-\pi_{y4}]$	T_1
		$(\frac{1}{4})[\pi_{y2}+\pi_{y3}-\pi_{y1}-\pi_{y4}+3^{\frac{1}{2}}(\pi_{x2}+\pi_{x3}-\pi_{x1}-\pi_{x4})]$	

[a] The designation of the d-orbitals is that given by Eyring, Walter, and Kimball, *Quantum Chemistry* (John Wiley and Sons, Inc., New York, 1944). We have also used throughout the group theoretical symbolism which is essentially that of these same authors.

lecular electrons are to be placed. Thus we consider only the $3d$, $4s$, and $4p$ atomic orbitals of the central atoms ($3s$ and $3p$ in the case of Cl) and only the $2p$ atomic orbitals of the oxygen atoms in constructing the molecular orbitals. In the case of the molecule ions here discussed twenty-four electrons are to be placed in the molecular orbitals, the ground state being constructed by placing twenty-four electrons in the lowest-lying molecular orbitals.

In setting up molecular orbitals for a system possessing equivalent atomic orbitals on equivalent atoms one finds that the requirement that the individual molecular orbitals belong to irreducible representations of the point symmetry group of the system determines immediately the relative coefficients with which these equivalent orbitals enter into the final molecular orbitals. With the oxygen orbitals and X orbitals oriented as in Fig. 1, one finds by standard group theoretical methods[5] the irreducible representations to which the central atom orbitals belong and determines the appropriate linear combinations of oxygen orbitals transforming according to the same rows of the same irreducible representations as the central atom orbitals. The final results are shown in Table I. The sets of combinations of oxygen orbitals are considered normalized, the overlap of oxygen wave functions being neglected for the time being in view of the qualitative nature of our calculations.

The calculations of the molecular orbital energies ϵ (which we employ in the same manner as is usual in the semiempirical methods for organic molecules) and the evaluation of the coefficients of the atomic orbitals and sets of atomic orbitals in Table I in the final molecular orbital requires the solution of secular determinants (one for each irreducible representation) of the form $|H_{ij}-G_{ij}\epsilon|=0$, where H_{ij} has its usual meaning and G_{ij} is the group overlap integral to be further defined later. It is seen from Table I that this involves the solution of a quartic for the orbitals with symmetry T_2 and quadratics for the orbitals of symmetries E and A_1. The orbital of symmetry T_1 will be nonbonding and is fully determined by symmetry considerations.

The values of H_{ii} (the energy of an electron in the ith orbital in the field of the nuclear skeleton and the remaining valence electrons) have been used essentially as parameters in the solution of the equations for the case of chromate ion. The values were chosen not greatly different from the valence state ionization potentials[6] for Cr^0 and estimated for oxygen from the ionization energies of some inorganic compounds in which the formal charge on oxygen has been estimated to be in the neighborhood of minus one-half. The values thus obtained were altered by trial and error to satisfy the following requirements: (1) the first and second excitation energies should be at least qualitatively correct and (2) the resulting charge distribution should lead to formal charges approximately those assumed in setting up the atomic orbital wave functions. The values of H_{ii} for permanganate were then obtained from the chromate values by varying them in accord with what seems reasonable chemically; the oxygen and central atom should be somewhat less negative and the values of H_{ii} should correspondingly decrease. The values assumed are shown in Fig. 2. For oxygen we have assigned $H_{\sigma\sigma}$ smaller values than $H_{\pi\pi}$ to take into account the fact that the σ-orbitals penetrate farther into the region of positive charge on the central atom. That this is not unreasonable is indicated by the results of a more exact calculation for the CO_2 molecule[7] in which the calculated $H_{\sigma\sigma}$ for the oxygen atoms was found to be considerably lower than for the $H_{\pi\pi}$.

The values of H_{ij} we have approximated by setting them proportional to the corresponding overlap integrals according to the following relation:

$$H_{ij}=F_xG(i,j)(H_{ii}+H_{jj})/2.$$

[5] E. Wigner, *Gruppen Theorie und ihre Anwendung auf die Quantenmechanik der Atomspektren* (Friedr. Vieweg und Sohn, Braunschweig, 1931).

[6] R. S. Mulliken, J. Chem. Phys. 2, 782 (1934).

[7] J. F. Mulligan, J. Chem. Phys. 19, 347 (1951).

The $G(i, j)$ are the "group overlap" integrals, for example:

$$\int 4s(X)\cdot\tfrac{1}{2}(\sigma_1+\sigma_2+\sigma_3+\sigma_4)d\tau = G_{A_1}(s, \sigma).$$

These may be obtained in terms of the simple σ and π diatomic overlaps[8] (S_{ij}) by expressing the central atom orbitals in terms of a linear combination of equivalent orbitals oriented relative to the axes on the oxygen atoms.

The relationship between the G_{ij} and S_{ij} is given in Table II and the values of G_{ij} for the three cases under discussion are given in Table III. The group overlap integrals calculated using wave functions derived by

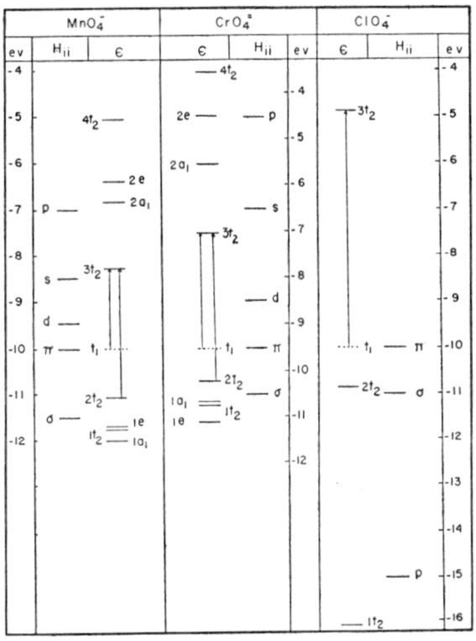

Fig. 2. The calculated molecular orbital energies for MnO_4^-, $CrO_4^=$, and ClO_4^-.

the Slater formulas[9] were found to give results which could not be reconciled with experiment. In particular, the $G_{A_1}(s, \sigma)$'s were so small that the first excited orbital was one of symmetry A_1. As will be seen later, this is in disagreement with the spectra. The Slater $2p$ and $3d$ atomic wave functions differ radically from the self-consistent field functions. We have therefore derived a general set of analytic atomic wave functions[10] which more closely approximates the existing self-consistent field functions and, with the exception of the

Table II.

$$G_{T_2}(d, \sigma) = (2/\sqrt{3})S(d\sigma, p\sigma)$$
$$G_{T_2}(d, \pi) = (2\sqrt{2}/\sqrt{3})S(d\pi, p\pi)$$
$$G_{T_2}(p, \sigma) = (2/\sqrt{3})S(p\sigma, p\sigma)$$
$$G_{T_2}(p, \pi) = -(2\sqrt{2}/\sqrt{3})S(p\pi, p\pi)$$
$$G_E(d, \pi) = (2\sqrt{2}/\sqrt{3})S(d\pi, p\pi)$$
$$G_{A_1}(s, \sigma) = 2S(s, p\sigma)$$

($4p$, x) overlaps, these functions have been used to calculate the values in Table III. The values of $G_{T_2}(4p, \sigma)$ and $G_{T_2}(4p, \pi)$ we have estimated from the calculations using Slater orbitals by varying the values in accordance with the trends observed in the $G_{T_2}(3d, x)$ from $CrO_4^=$ to MnO_4^-. This is certainly a doubtful procedure but, since the omission of the $4p$ orbitals from the calculation entirely produces no qualitative change in the energy levels§ or the relative intensities, we do not feel that this approximation is serious for our present purposes. The values of F_x employed were $F_\sigma = 1.67$ for σ overlaps and $F_\pi = 2.00$ for π-overlaps. These values are in moderately good agreement with those found by us to give satisfactory results in making calculations of the energy levels of homonuclear diatomic molecules of first and second row elements. These values are $F_\sigma = 1.6$, $F_\pi = 1.87$.

With the simplifications outlined above the solution of the secular equations is straightforward and leads to the energies of the molecular orbitals illustrated in Fig. 2. The twenty-four valence electrons are to be placed in the orbitals of lowest energy. It is seen that there exist four t_2 orbitals (threefold degenerate) each capable of holding six electrons. The orbital $1t_2$‖ is strongly bonding and $2t_2$ slightly bonding while $3t_2$ and $4t_2$ are antibonding. In addition there are two e orbitals (twofold degenerate), $1e$ strongly bonding and $2e$ antibonding. The strong bonding of the $1e$ electrons is π-oxygen bonding with the central atom and may be considered as similar to the effect of double bonding in the simple valence bond treatment. It is to be remarked, however, that the stability of the $1e$ orbital as given by our calculations is much greater than is ordinarily attributed to the π-electrons in a double

Table III.

G	MnO_4^-	$CrO_4^=$	ClO_4^-
G_{T_2} (p, σ)	0.10	0.00	0.33
G_{T_2} (p, π)	−0.25	−0.20	−0.26
G_{T_2} (d, σ)	0.12	0.11	...
G_{T_2} (d, π)	0.15	0.18	...
G_E (d, π)	0.26	0.32	...
G_{A_1} (s, σ)	0.28	0.27	0.56

[8] For the system of coordinates used and general method see Mulliken, Rieke, Orloff, and Orloff, J. Chem. Phys. **17**, 1248 (1949).

[9] J. C. Slater, Phys. Rev. **36**, 57 (1930).

[10] The general form of these functions and a discussion of the method of obtaining them will be submitted for publication in the near future.

§ The fact that the neglect of the $4p$ orbitals in the $CrO_4^=$ and MnO_4^- cases lead to no qualitative change in the calculated orbital energies has led us to neglect also the $3d$ orbital in the ClO_4^- calculation. For this case the $3p-3d$ separation is certainly greater than the $3d-4p$ separation in MnO_4^- and $CrO_4^=$.

‖ We are using here a notation similar to that used in atomic structures, designating molecular orbitals of a given symmetry by numbers 1, 2, 3, etc., in order of increasing energy.

TABLE IV. Calculated and observed excitation energies.

	MnO_4^-		$CrO_4^{=}$		ClO_4^-	
	Obs	Calc	Obs	Calc	Obs	Calc
1st transition	2.29 ev	1.68 ev	3.25 ev	2.42 ev	6 ev	5.23 ev
2nd transition	3.96	2.78	4.59	3.15	...	...

bond in the valence bond treatment.¶ There exist further two a_1 orbitals, one bonding ($1a_1$), and one antibonding ($2a_1$). Finally there is one nonbonding oxygen orbital, t_1 (threefold degenerate). Placing the twenty-four electrons in the lowest lying orbitals gives the ground state:

$$(1e)^4(1t_2)^6(1a_1)^2(2t_2)^6(t_1)^6 \text{ with symmetry } A_1.$$

RELATIVE EXCITATION ENERGIES

The first excited orbital has the symmetry T_2, and it is assumed that the first absorption maximum corresponds to the excitation of a nonbonding t_1 electron to the lowest-lying unfilled orbitals $3t_2$. Such an excitation

$$(1e)^4(1t_2)^6(1a_1)^2(2t_2)^6(t_1)^6 \rightarrow (1e)^4(1t_2)^6(1a_1)^2(2t_2)^6(t_1)^5(3t_2)^1$$

gives rise to four states of symmetries A_2, E, T_1, and T_2 (both singlet and triplet states, of which only the singlet states are considered here since no spectral evidence of the triplet states has been found). For the symmetry T_d only the transitions $A_1 \rightarrow T_2$ are allowed for dipole radiation. Transitions to the other levels are forbidden in the absence of perturbing effects but are observable in the spectra of solid solutions of MnO_4^- in $KClO_4$, $NaClO_4$, etc., and will be discussed later in connection with these spectra.

The second excitation according to Fig. 2 should involve the excitation of an electron from the $2t_2$ orbital to the $3t_2$ orbital which is at least partially bonding. The resulting configuration,

$$(1e)^4(1t_2)^6(1a_1)^2(2t_2)^5(t_1)^6(3t_2)^1,$$

gives rise again to four states of symmetries A_1, E, T_1, and T_2. The second maximum characteristic of the MnO_4^- and $CrO_4^{=}$ spectra is accordingly attributed to the transition from the ${}^1A_1$ ground state to the ${}^1T_2$ state of this configuration. The existence of the states to which transitions are forbidden is not so clearly given in this case by the crystal absorption data because of the diffuseness of the spectra. The presence of a T_1 state, however, seems reasonably certain.

If the assumption is made that the energy absorbed in the transition ${}^1A_1 \rightarrow {}^1T_2$ is equal to the difference in energy of the final and initial molecular orbitals for the first transition, $\epsilon_{t_2} - \epsilon_{t_1}$, then the calculations show qualitative agreement with the experimental data. This assumption is incorrect, and in more exact calculations terms of the type neglected here have been shown to be far from negligible. The correct expressions for the energies of the transitions involving the excitation of a t_1 electron to the $3t_2$ orbital are given in the appendix.

Table IV shows the calculated and observed energies of transition in the three ions considered here, and it is seen that the first absorption by ClO_4^- lies well into the ultraviolet and MnO_4^- and $CrO_4^{=}$ show two maxima in the visible and near ultraviolet. The rather surprising quantitative agreement probably arises from the empirical nature of our calculations. This agreement can be improved by changes in the values of the H_{ii}'s and the F_x's, but these changes are limited severely by the requirement that the intensities of absorption be given correctly.

INTENSITIES OF ABSORPTION

Evaluation of the transition moments has been carried out for the allowed transitions neglecting all integrals except those over the same atom. The results can be given in terms of the coefficients of the various atomic orbitals making up the molecular orbitals involved. The f-value (oscillator strength) for the first transition is found (with the above assumptions) to depend on the coefficients of the π-combination in the $3t_2$ orbitals and is

$$f_{t_1 \rightarrow 3t_2} = (8\pi^2 mc\nu/3h)3S^2h^2,$$

where S is the projection of the X—O bond on the x or y or z axis (see Fig. 1) and h is the coefficient of the π-combination in the excited state, ν is the average frequency (observed wave number) in the absorption band. The observed f-values for this first transition are $f(MnO_4^-)=0.032$ and $f(CrO_4^{=})=0.089$; those calculated are 0.076 and 0.22, respectively. Of more significance for our qualitative calculations, perhaps, is the ratio of $f(MnO_4^-)/f(CrO_4^{=})$, which is calculated to be 2.62 and observed 2.9. The ratio of the f-values of the

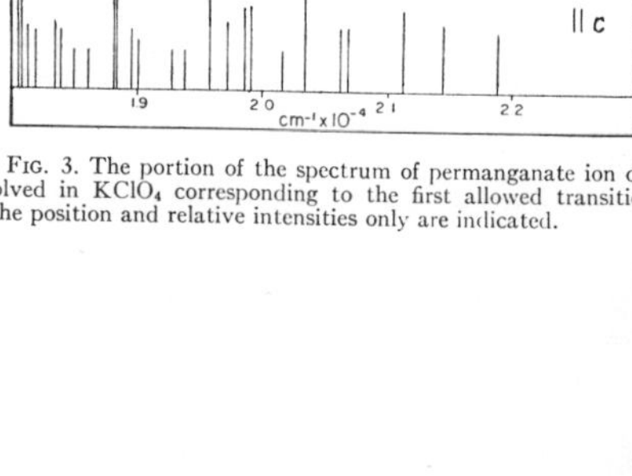

FIG. 3. The portion of the spectrum of permanganate ion dissolved in $KClO_4$ corresponding to the first allowed transition. The position and relative intensities only are indicated.

¶ We have also computed molecular orbital combinations and group overlap integrals for symmetries D_{3h}, O_h, and D_{4h}. It again would appear that bonding between R atom orbitals and π-orbitals on X in compound-RX_n is of no small importance. Thus in predicting configurations of complex molecules one probably has to consider such bonding in addition to the usual single bonding.

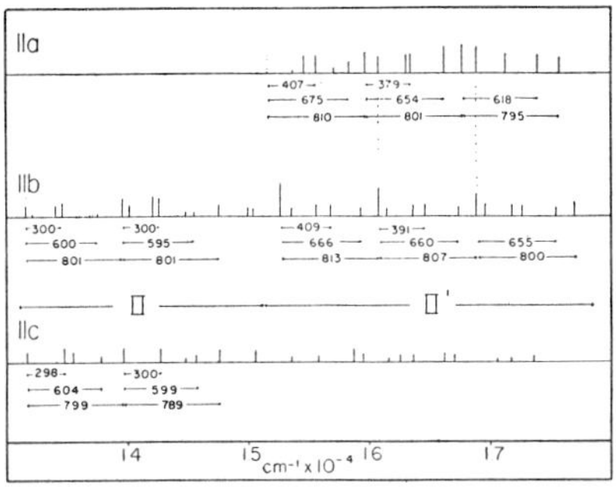

FIG. 4. The spectrum of permanganate ion dissolved in $NaClO_4$ in the region corresponding to forbidden transitions. The dotted line in the spectrum parallel to a is the assumed position of the vibrationless transition A_1-A_1 (of E). The indicated separations were obtained by measuring a similar diagram given by Teltow and are consequently subject to considerable error.

first and second transitions can also readily be calculated with the same assumptions as above and the neglect of the central atom integrals. The ratio is calculated 1.17 for permanganate and observed 1.99. The results for chromate are qualitatively in error giving 0.71 calculated and 1.20 observed.

SPECTRA OF MnO_4^- IN DIFFERENT SOLID SOLUTIONS

Potassium permanganate and potassium perchlorate are isomorphous with very nearly the same unit cell dimensions. It should be possible therefore to dissolve appreciable amounts of MnO_4^- in the perchlorate lattice without great distortion. The site symmetry of the perchlorate (permanganate) ions in this lattice is C_s, that is, of all the symmetry elements of the free MnO_4^- ion the only one remaining in the solid solution is a plane of symmetry which lies perpendicular to the b axis of the crystal. In this environment the T_2 state should be completely split into two A' states and one A'' state of symmetry C_s. The transition $A'\rightarrow A'$ is allowed with the electric vector parallel to the $a-c$ plane and the transition $A'\rightarrow A''$ with the electric vector perpendicular to this plane, that is, parallel to the b axis. The portion of the spectrum of permanganate involving the first allowed transition is shown in Fig. 3. With E parallel to a and c axes one observes doublets with a splitting of about 30 cm^{-1} which one would expect from a slightly perturbed T_2 state in symmetry C_s. With the electric vector parallel to b one finds, again in agreement with prediction, single lines. The relatively large f-value for this absorption indicates that it is probably an allowed transition ${}^1A_1\rightarrow{}^1T_2$. The splitting of the level in the field of symmetry C_s renders this conclusion almost certain.

Of greater interest for our purposes is a discussion of the portion of the spectrum corresponding to the lower energy forbidden transitions of MnO_4^-. A calculation of the relative energies of the states arising from the configuration $(e)^4(1t_1)^6(2t_2)^6(1a_1)^2(t_1)^5(3t_2)^1$ has been carried out (see Appendix), again, as in the calculation of f-values, with the neglect of all integrals except those over the same atom. This calculation gives the energies of the states in the order of increasing energy A_2, E, T_1, and T_2. One should accordingly expect to find evidence of transitions to the A_2, E, and T_1 states on the long wavelength side of the first strong absorption maximum.

This region of the spectrum for MnO_4^- dissolved in anhydrous $NaClO_4$ is shown in Fig. 4. The site symmetry of the tetrahedral ions in this lattice is C_{2v}. The degenerate levels should be completely split in this symmetry, the symmetries of the resulting states (in C_{2v}) and the polarization of the transitions rendered allowed by the perturbation are given in Table V. As indicated in this table a T_1 state in a field of symmetry C_{2v} should show absorption with electric vector parallel to the b and c axes, for an E state absorption with electric vector parallel to the a axis, the direction of the twofold axis at position of the ion. The spectrum in the region II (Teltow's designation) corresponds precisely to what one would expect from a T_1 state.**

There is no way of regarding the region designated by II' as a continuation of the bands II. The strong absorption with E parallel to a would (Table V) then correspond to a transition A_1 of $C_{2v}\rightarrow A_1(C_{2v})$ arising from an E state of T_d. In order to classify the spectra it is necessary to assume that the vibrationless transition $A_1\rightarrow A_1$ (of E) does not appear. This is a phe-

TABLE V. Only the transitions of importance to our calculations are listed in the table. Transitions from the A_1 ground state. The polarizations are given relative to the crystal axes of the indicated structures.

T_d	C_s ($KClO_4$)	Allowed transitions	polarization
A_1	A'		
A_2	A''	$A'-A'$	$\parallel a$, $\parallel c$
E	$A'+A''$		
T_1	$A''+A''+A'$	$A'-A''$	$\parallel b$
T_2	$A''+A'+A'$		
	C_{2v} ($NaClO_4$)		
A_1	A_1	A_1-A_1	$\parallel a$
A_2	A_2	A_1-A_2	forbidden
E	A_1+A_2		
T_1	$A_2+B_1+B_2$	A_1-B_1	$\parallel b$
T_2	$A_1+B_1+B_2$	A_1-B_2	$\parallel c$
	C_{3v} ($LiClO_4$)		
A_1	A_1		
A_2	A_2		
E	E	A_1-A_1	$\parallel c$
T_1	A_2+E	A_1-A_2	forbidden
T_2	A_1+E		
		A_1-E	$\parallel c$

** The a and b axes given in Fig. 4 are the b and a axes given by Teltow. He expressed doubt as to the correct assignment of axes and gave the axes in terms of the optical axes of the crystal. Because of the suggestion contained in our calculations we have correlated x-ray pictures with the optical constants and have found that, indeed, Teltow's a and b axes should be interchanged. For assistance with this work we are indebted to Professor A. F. Fredrickson of the geology department of Washington University.

Table VI.

			Obs		Calc		
	(cm^{-1})	I	E	T_1	E	T_1	Transition
$\parallel b$	14,446	0.5	0		0		$A'-A''(E)$
	14,526	0.3	80		80[a]		$-A''(E)+v_0^1$
	14,592	0.2	146		151[a]		$-A''(E)+v_0^2$
	14,681	0.05		0		0	$-A''(T_1)$
	14,844	0.2	398		...		$-A''(E)+v_2(E)$[b]
	15,172	0.3		491		...	$-A''(T_1)+v_4(T_2)$
	15,246	0.9	800		...		$-A''(E)+v_1(A_1)$
	15,324	1.0	878		880		$-A''(E)+v_1(A_1)+v_0^1$
	15,472	0.2		791		...	$-A''(T_1)+v_1(A_1)$
	15,620	0.2	1174		1198		$-A''(E)+v_1(A_1)+v_2(E)$
	15,958	0.2		1277		1282	$-A''(T_1)+v_1(A_1)+v_4(T_2)$
	16,049	1.0	1603		1600		$-A''(E)+2v_1(A_1)$
	16,117	1.0	1671		1680		$-A''(E)+2v_1(A_1)+v_0^1$
	16,256	0.2		1575		1582	$-A''(T_1)+2v_1(A_1)$
	16,860	0.5	2414		2440		$-\begin{cases}A''(E)+3v_1(A_1)\\A''(E)+3v_1(A_1)+v_0^1\end{cases}$
$\parallel a$	14,460	0.0	0		0		$-A'(E)$
	14,567	0.1	107		80		$-A'(E)+v_0^1$
	14,632	0.2	172		151		$-A'(E)+v_0^2$
	14,670	0.0		0		0	$-A'(T_1)$
	14,793	0.15		123		151	$-A'(T_1)+v_0^2$
	15,002	0.15	542		491		$-A'(E)+v_4(T_2)$
	15,538	0.6		868		867	$-\begin{cases}A'(T_1)+v_1(A_1)\\A'(T_1)+v_1(A_1)+v_0^2\end{cases}$
	15,793	0.1	1333		1291		$-A'(E)+v_1(A_1)+v_4(T_2)$
	16,320	0.6		1650		1658	$-\begin{cases}A'(T_1)+2v_1(A_1)\\A'(T_1)+2v_1(A_1)+v_0^2\end{cases}$
	17,078	0.4		2408		2449	$-\begin{cases}A'(T_1)+3v_1(A_1)\\A'(T_1)+3v_1(A_1)+v_0^2\end{cases}$
$\parallel c$	14,460	0.0	0		0		$A'\rightarrow A'(E)$
	14,605	0.1	145		151		$\rightarrow A'(E)+v_0^2$
	14,670	0.0		0		0	$\rightarrow A'(T_1)$
	14,705	0.1		35		80	$\rightarrow A'(T_1)+v_0^1$
	14,962	0.06	502		491		$\rightarrow A'(E)+v_4(T_2)$
	15,260	0.3	800		800		$-A'(E)+v_1(A_1)$
	15,342	0.2	882		880		$-A'(E)+v_1(A_1)+v_0^1$
	15,460	0.4		790		791	$-A'(T_1)+v_1(A_1)$
	15,640	0.2	1180		1198		$-A'(E)+v_1(A_1)+v_2(E)$
	16,051	0.5	1591		1600		$-A'(E)+2v_1(A_1)$
	16,138	0.3	1678		1680		$-A'(E)+2v_1(A_1)+v_0^1$
	16,252	0.6		1582		1582	$-A'(T_1)+2v_1(A_1)$
	16,402	0.2	1942		1998		$-A'(E)+2v_1(A_1)+v_2(E)$
	16,900	0.5	2440		2440		$-\begin{cases}A'(E)+3v_1(A_1)\\A'(E)+3v_1(A_1)+v_0^1\end{cases}$

[a] v_0^1 and v_0^2 are assumed to be lattice frequencies and are taken from the ${}^1A_1\rightarrow{}^1T_2$ spectrum.

[b] $v_2(X)$ indicates the frequency of the vibration of the tetrahedral ion of symmetry X. The designations are those given by Herzberg, *Infrared and Raman Spectra* (D. Van Nostrand and Company, Inc., New York, 1945), p. 100.

nomenon which is met within all the forbidden transitions in different crystalline fields and one for which we have no explanation. The weakest point in the above argument is the strong absorption parallel to b in the region II'. With the assumption mentioned above the spectra can be ordered, and it is seen that the strong absorptions parallel to b coincide (as indicated by the dotted lines) with lines parallel to a which correspond to transitions $A_1\rightarrow A_1$ (in C_{2v})+lattice vibrations. The lattice perturbations in $NaClO_4$ are certainly large compared to those in $KClO_4$ as may be seen from the splitting of the T_2 state (200 cm^{-1}) and it is perhaps not unreasonable to assume that this large perturbation is responsible for the large intensity of absorption parallel to b in the region II'. The intensity of absorption generally in this region is of the order of five times that in the region II. The difference between the characteristic vibrational intervals in the two regions offers additional evidence that they correspond to transitions to different states.

The ratio of the intensities of the forbidden transitions to those allowed is roughly five times as great in the case of $NaClO_4$ solid solutions as in the case of $KClO_4$ indicating again that the perturbation is greater in the former case causing a greater mixing of the T_2 wave functions with the E and T_1 functions allowed in C_{2v}.

In the $KClO_4$ lattice the site symmetry of the tetrahedral ions is C_s. The symmetries of the resulting states and the directions of polarization of the allowed transitions are shown in Table V. The spectrum shows one very sharp line (half-width 7 cm^{-1}) at the long wavelength extreme of the bands parallel to b. Close to this line there appear two lines, less sharp, which must be attributed to an electronic transition accompanied by the excitation of a lattice vibration since the intervals are too small to correspond to any frequencies of the ion even in an excited state. That this is so is also suggested by the fact that these same intervals occur in the allowed spectrum in which the vibrational frequencies are markedly different.

The sharp line at 14,446 cm^{-1} is attributed to a "vibrationless" transition from the A' ground state to the A'' (of C_s) state arising from E (of T_d) on the basis of the following argument: if this transition were $A'\rightarrow A''$ (of T_1 in T_d), one would expect to observe a doublet since T_1 splits into two A'' states in this symmetry; the splitting might well be small but should be observable because of the sharpness of the line. In the $NaClO_4$ lattice where the crystal perturbations are larger as indicated by the greater splitting of the T_2 level and the greater relative intensity of the forbidden spectrum relative to the allowed spectrum, the transitions to the states arising from E of T_d are five times as strong as those attributed to the A_1 to T_1 transition and in the absence of any evidence to the contrary it seems plausible to attribute a larger f-value to the $A_1\rightarrow E$ transition.

No sharp lines appear when the electric vector is polarized parallel the $a-c$ plane. Internal consistency of all the crystal data demands that there exist at least one A' state in this solid solution; consequently, as mentioned above, it is assumed that the $A'-A'$ vibrationless transition is not observed. With this assumption one may classify the spectra in terms of the set of frequencies observed parallel to the b axis with the A' state arising from E occurring at 14460 cm^{-1}, 14 cm^{-1} from the A'' of E and an A' state arising from T_1 at 14,670 cm^{-1}, 11 cm^{-1} from the unresolved A'' states (from T_1 at 14,681 cm^{-1}). The results of such an analysis are given in Table VI. There is nothing unique about the assignment of frequencies given in Table VI. The table is given merely to indicate that the data may be accounted for in a relatively simple way and that the data which appeared to Teltow to be without regularity can be ordered in terms of the proposed scheme. Particularly is this true of the data in the sodium perchlorate solid solutions.

The spectra obtained in solid solutions $LiCl(Mn)O_4 \cdot 3H_2O$ are also compatible with the proposed assignment, showing sharp absorption perpendicular to the threefold axis of the crystal ($A_1 \rightarrow E$ transitions) and only weak and diffuse absorption parallel to the threefold axis which presumably arises from electronic transitions $A_1 \rightarrow E$ accompanied by vibrations.

The polarization data discussed above seem to show definitely that more than one level exists lower lying than the T_2 level, that one of these is probably T_1 and the other E. There is no evidence for either an A_2 or A_1 state, the data in $LiCl(Mn)O_4$ tending to rule out the A_1 state more definitely than the A_2 state. We feel therefore that these data lend strong evidence in favor of the correctness of the calculations which predict that the first transition involves the excitation of a t_1 electron to a t_2 orbital since this is the only transition which gives rise to the states which are compatible with the observations. A transition $t_2 \rightarrow t_2$ would also give E, T_1, and T_2 states but would in addition give an A_1 state which might be expected to show up in the spectra, particularly in solid solution in which the site symmetry is C_{3v}($LiClO_4 \cdot 3H_2O$, or $Ba(ClO_4)_2 \cdot 3H_2O$).

CONCLUSIONS

The qualitative aspects of the spectra are given correctly by our empirical calculations and we feel that the electronic structure of the ions discussed is essentially correct. The interpretation of the bond distance in terms of "double bonding" is seen to have some justification in view of the importance of π-bonding between the central atom and oxygen atoms. In the case of the transition element compounds all five d orbitals and the $4s$ orbital are apparently employed in bonding whereas in ClO_4^- the $3d$ atomic orbitals would have very small coefficients in the bonding molecular orbitals in an approximate calculation. This would correspond to a more important contribution of "double bonding" in MnO_4^- and $CrO_4^=$ than in ClO_4^-, but the term double bond loses much of its meaning in these cases since the structure can only with difficulty be considered as resulting from the resonance of localized double bonds.

The quantitative agreement gives some hope that a similar treatment applied to more ions might enable us to formulate rules for the empirical estimation of the H_{ii}'s and thus give a very simple method of making molecular orbital calculations for inorganic molecules. The method outlined above has been applied to the ion CrO_3F^- with very satisfactory results both as to relative excitation energies and intensities. Calculations on CrO_2Cl_2 and SO_2Cl_2 are under way at the present time.

We are indebted to Professor S. I. Weissman of Washington University and to Professor R. S. Mulliken of the University of Chicago for generous advice and encouragement in the course of the work.

APPENDIX

The energies of the excited singlet states arising from the configuration $\cdots(t_1)^5(3t_2)^1$ relative to the ground state are given below where

$$(\phi_x{}^{t_1}\phi_y{}^{t_2})(\phi_z{}^{t_1}\phi_w{}^{t_2}) = \int \phi_x{}^{t_1}(1)\phi_y{}^{t_2}(1)e^2/r_{12}\phi_z{}^{t_1}(2)\phi_w{}^{t_2}(2)dr_1dr_2$$

and $\phi_x{}^{t_1}$ is the xth orbital function of the degenerate nonbonding set of symmetry T_1 and $\phi_y{}^{t_2}$ is the yth orbital function of the set $3t_2$.

$$E\text{ (singlet }T_2) - E\text{ (singlet }A_1) = \epsilon\,(3t_2) - \epsilon\,(t_1) - K + 2C + J - 2D$$

$$E\text{ (singlet }E) - E\text{ (singlet }A_1) = \epsilon\,(3t_2) - \epsilon\,(t_1) - H + 2A - 2B + J$$

$$E\text{ (singlet }T_1) - E\text{ (singlet }A_1) = \epsilon\,(3t_2) - \epsilon\,(t_1) - K + 2C - J + 2D$$

$$E\text{ (singlet }A_2) - E\text{ (singlet }A_1) = \epsilon\,(3t_2) - \epsilon\,(t_1) - H + 2A + 4B - 2J$$

$$A = (\phi_1{}^{t_2}\phi_1{}^{t_1})(\phi_1{}^{t_2}\phi_1{}^{t_1}) = (\phi_i{}^{t_2}\phi_i{}^{t_1})(\phi_i{}^{t_2}\phi_i{}^{t_1})$$
$$B = (\phi_1{}^{t_2}\phi_1{}^{t_1})(\phi_2{}^{t_2}\phi_2{}^{t_1}) = (\phi_i{}^{t_2}\phi_i{}^{t_1})(\phi_j{}^{t_2}\phi_j{}^{t_1})$$
$$C = (\phi_1{}^{t_2}\phi_2{}^{t_1})(\phi_1{}^{t_2}\phi_2{}^{t_1}) = (\phi_i{}^{t_2}\phi_j{}^{t_1})(\phi_i{}^{t_2}\phi_j{}^{t_1})$$
$$D = (\phi_1{}^{t_2}\phi_2{}^{t_1})(\phi_2{}^{t_2}\phi_1{}^{t_1}) = (\phi_i{}^{t_2}\phi_j{}^{t_1})(\phi_i{}^{t_2}\phi_j{}^{t_1})$$
$$H = (\phi_1{}^{t_2}\phi_1{}^{t_2})(\phi_1{}^{t_1}\phi_1{}^{t_1}) = (\phi_i{}^{t_2}\phi_i{}^{t_2})(\phi_i{}^{t_1}\phi_i{}^{t_1})$$
$$J = (\phi_1{}^{t_2}\phi_2{}^{t_2})(\phi_1{}^{t_1}\phi_2{}^{t_1}) = (\phi_i{}^{t_2}\phi_j{}^{t_2})(\phi_i{}^{t_1}\phi_j{}^{t_1})$$
$$K = (\phi_1{}^{t_2}\phi_1{}^{t_2})(\phi_2{}^{t_1}\phi_2{}^{t_1}) = (\phi_i{}^{t_2}\phi_i{}^{t_2})(\phi_j{}^{t_1}\phi_j{}^{t_1}).$$

CORRECTIONS

Professor Mulliken has noted the following corrections:

1. Regarding the formulae 14(3) in the paper by Mulliken, Rieke, Orloff, and Orloff, $(-\xi\eta - 1)$ should be $(\xi\eta + 1)$ in 14(3) and $(\xi^2\eta^2 - 1)$ should be $(1 - \xi^2\eta^2)$ in 14(4). The equations derived from 14(3) are correct and those from 14(4) incorrect by a minus sign. All calculated overlap integrals are, however, correct.

2. All formulae and tables that pertain to 5s, $5p\pi$, etc. are really for $n^* = 4$. They are listed as $n = 5$ to correspond to Slater's rules which for $n = 5$ would specify $n^* = 4$.

Dr. M. Wolfsberg has noted the following in the Wolfberg-Helmholz paper:

Typographical Error in Table II: The expression for $G_{T_2}(d, \pi)$ should read:

$$G_{T_2}(d, \pi) = (2\sqrt{2}/3)\, S(d\pi, p\pi)$$

INDEX